# Biomolecular Engineering

# Biomolecular Engineering

Edited by Yen Bishop

SYRAWOOD
PUBLISHING HOUSE
New York

Published by Syrawood Publishing House,
750 Third Avenue, 9th Floor,
New York, NY 10017, USA
www.syrawoodpublishinghouse.com

**Biomolecular Engineering**
Edited by Yen Bishop



International Standard Book Number: 978-1-64740-349-2 (Hardback)



**Cataloging-in-publication Data**

Biomolecular engineering / edited by Yen Bishop.
p. cm.
Includes bibliographical references and index.
ISBN 978-1-64740-349-2
1. Biomolecules. 2. Genetic engineering. 3. Biochemical engineering.
4. Molecular genetics. I. Bishop, Yen.
QH506 .B56 2023
572.8--dc23

# TABLE OF CONTENTS

# PREFACE

Biomolecular engineering is a discipline of engineering that draws from three different fields of study, namely, molecular biology, biophysical chemistry and chemical engineering. It studies the implementation of engineering principles and practices for the intentional manipulation of biomolecules. This field of engineering integrates knowledge of biological processes with the knowledge of chemical engineering to provide molecular level solutions for issues and problems related to environment, agriculture, energy, industry, food production, biotechnology and medicine. Molecules of biological origin include peptides, proteins, nucleic acids, carbohydrates and lipids. The biomolecules are manipulated within this field keeping in mind their structures, functions, properties and their applicability in areas such as environmental remediation, crop and livestock production, biofuel cells, and biomolecular diagnostics. Biomolecular engineering is used to observe and investigate the thermodynamics and kinetics of molecular recognition in enzymes, antibodies, DNA hybridization, bio-conjugation/bio-immobilization, and bioseparations. This book is a valuable compilation of topics, ranging from the basic to the most complex advancements in the field of biomolecular engineering. It will serve as a valuable source of reference for graduate and postgraduate students.

This book is the end result of constructive efforts and intensive research done by experts in this field. The aim of this book is to enlighten the readers with recent information in this area of research. The information provided in this profound book would serve as a valuable reference to students and researchers in this field.

At the end, I would like to thank all the authors for devoting their precious time and providing their valuable contribution to this book. I would also like to express my gratitude to my fellow colleagues who encouraged me throughout the process.

**Editor**

1

# Molecular Tools for Gene Analysis in Fission Yeast

*Irma Pilar Herrera-Camacho, Lourdes Millán-Pérez-Peña, Francisca Sosa-Jurado, Nancy Martínez-Montiel, Rebeca Débora Martínez-Contreras and Nora Hilda Rosas Murrieta*

**Abstract**

*Schizosaccharomyces pombe* or fission yeast has been called micromammal due to the potential application of the knowledge derived from the yeast in the physiology of higher eukaryotes. Fission yeast has been consolidated as an excellent model for the study of highly conserved cellular processes. The possibility of using haploid or diploid strains facilitates the analysis of the dominant or recessive phenotype of an allele as well as its function, making it a model of first choice for the development of any investigation in eukaryotes cells. With a growing community that employs fission yeast as a model system for the study of numerous cellular processes, it has motivated the simultaneous development of molecular tools that facilitate the study of genes and proteins in the yeast. In this review, we present the most used molecular techniques in fission yeast for the analysis of genes, its characterization, as well as the determination of its function.

**Keywords:** fission yeast, gene replacement, mutants, gene expression, CRISPR/Cas, RNAi, yeast two-hybrid, microarrays, NGS, ChiP

## 1. Introduction

*Schizosaccharomyces pombe* (*S. pombe*) is a single-cell, nonpathogenic yeast, described in Germany in 1893 by P. Linder, named "pombe," and was originally isolated from East African millet beer [1]. *S. pombe* is an ascomycete fungus, whose lineage is evolutionarily remote from the yeast *Saccharomyces cerevisiae* [2]. Actually, *S. pombe* is phylogenetically as distant from budding yeasts as it is from humans. In 1950, two homothallic strains, h90 (968) and h40, and two heterothallic strains with opposite mating types, $h^-$ and $h^+$, were isolated [3, 4]. There are several heterothallic strains with different genomic configurations at the mating type locus, but the heterothallic strains commonly used in the laboratory are $h^{+N}$ (975) and $h^{-S}$ (972) [5].

*S. pombe* is also called fission yeast because it is divided by binary fission. However, it has two forms of reproduction: one by binary fission and another by sporulation. Therefore, it is possible to find it in both haploid and diploid states. *S. pombe* cells are cylindrical, 3–4 μm in diameter and 7–15 μm long in haploid state, while in diploid state, they measure from 4 to 5 μm in diameter and 20–25 μm long [6].*S. pombe* was the sixth eukaryote to have its entire genome sequenced [7]. The genome of *S. pombe* has a size of 13.8 Mb and is organized in chromosome I of 5.7 Mb, chromosome II of 4.6 Mb, and chromosome III of 3.5 Mb, along with a mitochondrial genome of 20 Kb [8]. It contains the ribosomal RNA genes 5.8S, 18S, and 25S with a length of approximately 1.1 Mb [9]. Approximately, 4940 genes encoding proteins (including 11 mitochondrial genes) and 33 pseudogenes have been predicted. Almost 50% of fission yeast genes have at least one intron, and in total, there are 5300 introns in 2510 protein-coding genes. The process of splicing also appears to be more similar to splicing in human cells (http://www.pombase.org/status/statistics). The telomeres, centromeres, and origins of replication are more similar to complex eukaryotes than they are with the case of budding yeast. The three centromeres are 35, 65, and 110 Kb in length for chromosomes I, II, and II, respectively, with a total of 0.2 Mb [7]. The complete information of the yeast genome, gene expression data, mutations and proteins, curated literature up to tools for sequence and structure analysis is located in the main databases in the world on the NCBI, EMBL and DDBJ informatics domains as well as the Pombase database [10].

In its haploid state and in favorable conditions, *S. pombe* grows through a mitotic cycle. The optimal growth temperature for *S. pombe* cells is 30°C (25–36°C) with a doubling time of 2–4 hours [11]. In both haploid and diploid cells, *S. pombe* mitotic cell cycle is organized into the G1, S, G2, and M phases. There are two major controls regulating progress through the cell cycle: the G1–S transition and the G2–M transition. Both points are regulated by cyclin-dependent serine/threonine protein kinase Cdc2 [12, 13]. The meiotic cell cycle is a modified mitotic (M) cell cycle [14, 15]. Like other eukaryotes, in the meiotic cell cycle of fission yeast, MI is a reductional division, without intervening S phase before the second meiotic division (MII). The meiosis process concludes with the formation of an ascus including four haploid spores [15].

Under conditions of nutrient restriction especially nitrogen, cells become arrested in G1, and if the two sexual types ($h^+$ and $h^-$) are present, they will conjugate to form a diploid zygote, known as zygotic ascus [16, 17]. In a similar way as in mammals, two cells of opposite sex are recognized by the system of communication of pheromones. A cell experiment a polarized morphogenesis in the direction in the direction of the pheromones source cell by a process called shmooing. Next, two cells will merge by conjugation or “mating” producing a zygote. The zygote is diploid and could be kept in a diploid mitotic cycle if the conditions of the medium improve at this point in the cycle. If the growing conditions are unfavorable, the diploid yeast will enter into meiosis to culminate with the formation of an ascus with four haploid spores. The spores will germinate and enter again into the mitotic cycle when the environmental conditions allow it, thus closing the cycle.

*S. pombe* has become one of the best-studied eukaryotes today. Dr. Forsburg gave it the name of micromammal [18]. In fission yeast, genes and proteins homologous to higher eukaryotes have been described related to recombination, chromosomal organization, chromatin modification, stress response mechanisms, DNA damage response, mitosis, meiosis, cell cycle control, mRNA splicing, cell morphogenesis and polarity, and post-translational modifications of proteins such as glycosylation [19–27].

## 2. Gene replacement

In fission yeast, gene deletion or one-step gene deletion by gene replacement via homologous recombination is probably the most used molecular tool in the functional characterization of the function of the gene and the protein. Gene disruption is a genetic analysis strategy to achieve gene modifications, generation of tagged protein fusions, genetic expression placed under the control of a regulated promoter, specific mutations, insertions, and deletion [28, 29]. Gene replacement by homologous recombination in *S. pombe* has allowed the construction of chromosomal interruptions of genes such as *sts1, gcs1, gsh2, hmt1* [30], and *git2/cyr1* [31].

Gene replacement requires a switch construction that contains 5′ and 3′ homologous regions of the target locus that flank a selection marker gene, and its efficiency in homologous recombination depends largely on the size of these regions [28]. The genetic construct is incorporated into the cells by transformation, then the reporter gene that will be used for gene replacement is inserted into the target gene due to the presence of terminal homologous regions in the construct, thus eliminating a large fragment of the target gene and incorporating instead, the reporter gene. At the beginning of the use of this technique, the protocol was based on obtaining the homologous regions to the target gene by digestion to flank a selection marker gene that was obtained from a plasmid containing the desired selection gene as well as the restriction sites resulting from the digestion of the homologous regions.

For this method, it was essential that DNA fragments share the restriction site for subsequent linkage. With the advance in molecular biology, methods based on PCR were developed [32]. The PCR strategy was improved by Wang et al. [32], and the protocol described the generation of construction switch gene called two-step PCR. Four oligonucleotides are required for the amplification of the homologous regions of the target gene to eliminate or modify. These PCR fragments can be called AB located in the 5′ region and CD located in the 3′ region of the target gene. A novel strategy was a little modification in the 3′ antisense oligonucleotide from AB region and 5′ sense oligonucleotide from CD region, which contains a short complementary sequence and a single restriction site to facilitate the link of the two products generated in the first PCR and then forming a product that serves as template for the second PCR. The final PCR product can be called ABCD and is cloned into a plasmid. At the same time, the gene marker selection used in the gene replacement like leu1 or ura4+ is amplified by PCR with oligonucleotides including the same restriction site in both ends, used in the 3′ antisense oligonucleotide from AB region and 5′ sense oligonucleotide from CD region of target gene to replacement. The PCR product of marker gene is incorporated into a cloning vector, and then it's digested with the unique restriction enzyme selected. Finally, the plasmid containing ABCD fragment is digested with the same restriction enzyme used to prepare the marker gene and linked to produce the AB-selection marker gene-CD gene deletion cassette.

In order to achieve the gene modification, a one-step gene deletion technique by pop-in homologous must be performed [29]. The gene deletion cassette is transformed into a yeast strain with a deletion in the endogenous gene selected like ura+ (ura4-D18) to the gene replacement [33] by the lithium acetate protocol [34]. Then, it is efficiently targeted to its homologous location in the chromosome DNA. Moreover, it is widely known that the efficiency of homologous recombination is greatly stimulated if the incoming DNA sequence has free ends. The DNA flanking to the marker gene, on each side, recombines with the genome, inserting the marker gene into the target gene, therefore disrupting or completely replacing it.

It has been reported that the optimal length of homologous sequences to achieve an efficient elimination of the gene is 80–100 pb. Nevertheless, high efficiency in mutagenesis directed for *S. pombe* has been brought, using long segments of homologia of the gene target (≥250 pb), with efficiencies in the homologous integration of up to 100% [35]. The selection marker genes used in *S. pombe* are based on gene markers of *Saccharomyces cerevisiae*. The most used genes are ura3, leu2, ade6, trp1, and his3 that synthesize enzymes used for the biosynthesis of uracil, leucine, adenine, tryptophan, and histidine, respectively [36]. A high and efficient integration in the strains that have mutations in the locus leu1-32 and Ura4-294 of *S. pombe* with own genes *leu1*$^+$ and *ura4*$^+$ has been showed.

In addition, to make the functional analyses of various genes as well as minimize incidental recombination events between DNA sequences within the marker gene and a chromosomal sequence, gene deletion cassettes consisting entirely of heterologous DNA sequences have been designated. Those gene deletion cassettes even allow multiple gene deletions to be performed. Because the incorporation of *loxP* sites flanking the marker gene allows Cre recombinase-mediated rescue, the marker can be reused for the next gene deletion. Genes can be deleted in sequential order using different gene deletion cassettes carrying different selectable markers. Then, a gene deletion cassette would be removed from the chromosome DNA by mitotic or recombinase-mediated recombination. The strategy allows the use of the recyclable deletion cassettes, useful to disrupt the next gene of interest with the same marker gene [37, 38].

## 3. Heterologous protein expression

Fission yeast is a very popular system for protein expression with potential biotechnological applications. The choice of yeasts for the purification of proteins, their structural analysis, and the generation of mutants aimed at knowing the function of proteins is based on the shared conserved biological processes as cell cycle progression, protein turnover, vesicular trafficking, and signal transduction with cells of higher eukaryotes [39, 40]. In yeasts, the appropriate expression of proteins with the posttranslational modifications required allows to obtain the correct protein structure and function. So, the use of yeast in the industrial production of enzymes employed in food, medicine and health, environment, and other applications has been proposed [41, 42]. To fulfill this purpose, "humanized yeast model systems" have been created as tools to study the molecular mechanisms involved in chronic degenerative diseases such as neurological disorders [43, 44]. Due to the accessibility of the yeast to simple genetic and environmental manipulations, it reduced complexity compared to the mammalian models.

Fission yeast is an excellent system to study the complex intracellular mechanisms underlying neurodegenerative diseases such as Alzheimer's disease (AD). Heterologous expression of Tau and Aβ can provide new insights into the pathobiology of these proteins in vivo as well as the screening of compounds that may be useful in treatment and/or prevention of AD [45]. Recently, it was reported that ginger (dietary condiment) fermented with *S. pombe* had neuroprotective effects on in vivo models of AD. FG improved recognition memory, ameliorated memory impairment in amyloid beta$_{1–42}$ (A$\beta_{1–42}$) plaque-injected mice, reinstated the pre- and postsynaptic protein levels decreased by amyloid plaque toxicity, as well as attenuated memory impairment in Aβ1–42 plaque-induced AD mice [46].

Numerous expression vectors have been used in molecular studies on *S. pombe* [47–49]. A typical plasmid of *S. pombe* contains an origin of bacterial replication, an antibiotic resistance gene to select recombinant cells in bacteria, an autonomous replication sequence (ARS1), and a marker of selection of yeast. More complex plasmids can include a regulated or constitutive promoter, a transcription terminator, or epitope tags [47, 50].

The use of antibiotics to induce genes to antibiotic resistance genes as selection markers into the yeast plasmid is very frequent. The kanamycin/G418, hygromycin B, phleomycin/bleomycin, and nourseothricin/clonNat are excellent markers in fission yeast [51]. Relative to auxotrophy, new markers such as ade7, his1, his2, his3, his5, arg3, arg12, lys1, lys2, and tyr1 are being developed [51–55].

However, ade6, his3+, LEU2, and ura4+ remain the most widely used markers for the selection of multi-copy vectors in common use. The pDUAL series and pJK148 vectors have been used to achieve the conversion of the leucine auxotrophy of leu1.32 to leucine prototrophy to select integration at the leu1 locus by recombination as well as pJK210 has been used to rescue ura4.294 to target integration at the ura4 locus [36].

In regard to the promoters used in the cloning vectors to protein expression, there are many promoters between the most used such as *adh1+*, which is a constitutive promoter. The *fbp1+* is repressed by exogenous cAMP. The *SV40* promoter is of constitutive expression. The CaMV promoter is tetracycline inducible. The *inv1+* is glucose repressible. The *ctr4+* is copper repressible and *nmt1* (strong, intermediate, and weak promoters) is thiamine repressible [47]. The latter is the most used promoter and was the first characterized in the expression of protein heterologous. The *nmt1* (*no message in thiamine*) promoter (Pnmt1) is considered as an inducible/repressible strong promoter that directs the transcription. It can be repressed by the addition of thiamine to a medium or induced in the absence of thiamine [56]. Pnmt1 has excellent dynamic range and a low off-state transcription but takes 14–16 h to induce upon thiamine withdrawal. Pnmt1 responds to the lack of exogenous thiamine and is induced approximately 75-fold when thiamine is removed from the growth media. However, the activity of Pnmt1 is repressed by the yeast extract present in a medium rich in YE and YES. So, some modifications into the TATA box of Pnmt1 have been made. Variants of this promoter were developed to reduce both off-state and on-state transcription [57, 58]. Pnmt4 and Pnmt8 are excellent options to choose the desired level of expression. However, an induction of transcription of 14–16 his also required.

To solve the problem, other promoters were generated to avoid the inactivation of the promoter nmt1 in the YES culture medium. The promoters of the 276-bp eno and 273-bp gpd were modified from eno101 and gpd3 genes in *S. pombe*. Both are stronger and constitutive promoters, which increase 1.5-fold higher expression of lacZ gene than nmt1 promoter. In addition, the 276-bp eno and 273-bp gpd promot-ers were not affected by the components of YES medium like Pnmt1.

As it was mentioned, there are other constitutive promoters widely used in *S. pombe*. The CaMV 35S promoter is a moderate constitutive promoter in *S. pombe* derived from the native 35S promoter of the plant viral cauliflower mosaic virus through deletion of the Tet repressor [59]. The adh1+ promoter of alcohol dehydrogenase is constitutively transcribed at high levels in cells grown in glucose and glycerol. However, the adh1+ promoter is weaker than the nmt1 promoter and may only be useful if a low level of gene expression is desired [58, 60]. Padh1 has two mutant variants, namely Padh41 (a mildly weak version of the *adh1*$^+$ promoter) and Padh81 (a weak version of the *adh1*$^+$ promoter, where the TATA box sequence TATAAATA is changed into TA), and both of these promoters express the downstream gene constitutively. Padh81 has been used in the study of the dynamic of the kinetochores [61, 62].

Therefore, it is necessary to find more efficient promoters for high-expression proteins in *S. pombe*. Other induction systems have rapid response times, but have a short dynamic range or relatively high levels of off-state transcription. The lsd90 promoter that is strongly induced by heat stress was cloned into the pJH5 vector, which contains an ARS element and a truncated URA3m as selectable marker. Following the expression of the luciferase reporter into the vector and making the comparison with other promoters such as Pnmt1, Padh1, and AOX1, it was found that lsd90 promoter promotes a constitutive expression of luciferase, at a level of 19-, 39-, and 10-fold higher than the promoters above mentioned, respectively [63]. The urg1 gene was identified as a rapidly induced transcript, responding to uracil addition in ~30 min and exhibiting low off-state transcription and high dynamic range [64] Other useful constitutive promoters in the protein expression are tif471 (with moderate force) and lys7 (weak promoter) [27, 65].

The pREP series vectors are general-purpose episomal vectors widely used in fission yeast research that contains a replication origin ARS1, ura4+, or LEU2 as the selective marker and kan, nat, hph, and bsd genes as a second type of marker of resistance to the specific antibiotics G418, clonNAT, hygromycin B, and blasticidin S, respectively. The latter are used routinely during chromosomal integration. The pREP vectors have been modified to produce novel and versatile plasmids pREP1 and pREP41. pREP1 contains a promoter derived from the gene nmt1. pREP41 contains a moderate-activity promoter (Pnmt41), whereas pREP81 contains a weaker promoter (Pnmt81). pREP vectors that contain ura4+ along with Pnmt1, Pnmt41, and Pnmt81 are named pREP2, pREP42, and pREP82, respectively [57]. The dominant selection marker genes kan, nat, hph, and bsd, which confer resis-tance against the specific antibiotics G418, clonNAT, hygromycin B, and blasticidin S, respectively, are used routinely during chromosomal integration [66–69].

Other important kinds of vectors of *S. pombe* are those of the pRI series generated from vector pREP, which were produced by deleting the *ars1* origin of replication sequence, and it has been used for the creation and expression of a single copy gene integrated into the chromosome [70].

The pYZ vectors are derivatives from the pREP series, which were designated for general purposes of cloning and large scale random gene cloning, as well as for allowing positive identification of cloning gene insertion and fusion to the GFP gene for analysis of gene expression. The pYZ vectors were constructed by inserting an *E. coli* α-peptide (position 239–684 on the pUC19 plasmid) of the lacZ (β-galactosidase) in opposite orientation to the Pnmt1 on the pREP series, leading to the complementation of the *lacZΔM15* deletion in *E. coli* strains such as DH5α or JM105 [56, 71, 72].

The pREP1, pREP41, pREP81, and pSGA plasmids were generated from the pREP series called pYZ1N, pYZ41N, pYZ81N (N represents an additional *Not* I site), and pYZ3N-GFP, respectively. In those vectors, the distance between the Pnmt1 and the ATG start codon remains the same as in the pREP vectors, and the promoter strength is unchanged [71]. The pYZ vectors have been useful because they were designated to produce a correct positive identification of cloning gene, fusion to the GFP, and large-scale random gene screening. The versatility of the pYZ vectors has allowed their use in numerous researches. HIV-1 vpr is a virionassociated viral protein of about 12.7 kD, whose function is required for efficient viral infection of nondividing mammalian cells such as monocytes and macrophages [73].

The HIV-1 protease (PR) is a viral enzyme encoded by vpr gene that was initially expressed in *S. pombe* from pREP1N. Vpr makes proteolytic processing required to the production of viral enzymes and structural proteins and for maturation of infectious viral particles [74].With the aim to improve the functional studies, HIV-1 *vpr* gene was cloned in the pYZ vectors. The *vpr* gene was fused to GFP in the pYZ3N-GFP vector and expressed in the yeast, where Vpr localizes to the nucleus of fission yeast cells. Expression of the *vpr* gene from the pYZ1N vector allows the analysis of the effects on cell morphology, the cell cycle G2 arrest, and cell killing [75].

In the molecular analysis of the Zika virus (ZIKV) infection, a large-scale molecular cloning and functional characterization of the viral proteins were performed. The Zika virus (ZIKV) is the causal agent of the microcephaly and the Guillain-Barré syndrome after the viral infection. However, there is insufficient knowledge about how ZIKV viral proteins are involved in cell damage. So, *S. pombe* was used to identify ZIKV factors responsible for the ZIKV-mediated cytopathic effects as well as the pathogenic factors associated with the viral infection.

By cloning the 14 coding-genes into the pYZ3N including the N-terminal GFP, it was possible to determine the subcellular localizations (nuclear, ER, Golgi, and cytoplasm) of ZIKV proteins expressed in a wild-type fission yeast strain, SP223 [70]. Importantly, seven ZIKV proteins affect cellular proliferation, which would be related to the microcephaly. So, ZIKV-induced microcephaly was proposed due to the intrauterine growth restriction, reduced cell proliferation, reduced neuronal cell layer volume, or cell death/apoptosis. Also, it was observed that prM, C, M, E, and NS4A proteins cause cell-cycle dysregulation because of cell cycle G2/M phase accumulation. These findings allow to follow the study of ZIKV infection.

Other interesting series of vectors are those that were produced as the pREP-X vectors that lack an ATG start codon [76]. Between them, pREP3X (promoter strength high), pREP41X (promoter strength medium), and pREP81X (promoter strength low), the three vectors lack tags and used Leu2 as marker. The pSLF vectors contain N-terminal or C-terminal triple hemagglutinin (3× HA) epitope tag. Between them, pSLF173 (promoter strength high), pSLF273 (medium), and pSLF373 (low), all of them contain 3xHA as tag and use ura4+ as the selective marker and the inducible promoter nmt1. From the pREP-X series were constructed several vectors with the purpose of being utilized for high-throughput functional analysis of heterologous genes in *S. pombe* such as pDS vectors that add GST tag-gings [50] as well as pSGA vector that includes GFP fusions.

There are many expression vectors constructed containing a destination cassette suitable for high-throughput cloning of target genes via the gateway system. There are vectors with N-terminal tagging such as the pDES173N, 273 N, and 373 N series, which add a 3XHA tag with the ura4+ gene as marker, and the vectors were constructed from the pSLF173, 273, and 373 vectors. The pDES175N, 275 N, and 375 N series add a GFP tag with the LEU2 marker, and those plasmids were built from the pSLF175, 275, and 375 vectors. The pDES177N, 277 N, and 377 N vectors add a GFP tag using ura4+ as marker selection. The pDES5XN, 45XN, and 85XN series add a RFP tag, with the LEU2 marker, which were derived from the pSLF5X, 45X, and 85X vectors. The pDES179, 279, and 379 series add a RFP tag, with the ura4+ marker, which were derived from the pSLF179, 279, and 379 vectors [77].

There are vectors with C-terminal tagging; those in the pDes173C, 273C, and 373C series add a 3XHA tag with ura4+ as marker, and the plasmids were constructed from the pSLF173, 273, and 373 vectors. The pDEs175C, 275C, and 375C series add a GFP tag with the LEU2 as marker, and those were constructed from the pSLF175, 275, and 375 vectors. The pDEs179C, 279C, and 379C series that add an RFP tag with the ura4+ marker were constructed from the pSLF179, 279, and 379 plasmids [77, 78]. These vectors exposed above lead the protein expression with N-terminal or C-terminal tagged, useful for the affinity purification or the functional analysis of target genes [77].

In 2013, an interesting series of vectors was described to PCR-based epitope tagging and gene disruption. The vectors developed were pFA6a-LEU2MX6, pFA6a-his3MX6, and pFA6a-ura4MX6. All of them were designed from the pFA6a-MX6-based plasmid (which contains antibiotic-resistance markers as kan) for amplification of gene- targeting DNA cassettes and integration into specific genetic loci, allowing expression of proteins fused to 12 tandem copies of the Pk (V5) (epitope from the P and V proteins of the paramyxovirus SV5), or 5 tandem copies of the FLAG epitope with a glycine linker. All vectors can use the LEU2, his3+, and ura4 + genes as selection markers. Also, some vectors as pFA6a-G9–5FLAG-kanMX6 and pFA6a-G11–5FLAG-kanMX6 were created, which were generated for studies of proteins when the direct epitope tagging compromises protein conformation and/or function. Other vectors were constructed to add a green fluorescent protein (GFP(S65 T)) or a monomeric red fluorescent protein (mRFP) genomic tagging as FA6A-GFP-bleMX6 [79].

Between the PK-tagging vectors are the pFA6a-6 × GLY-V5-(marker) and C-terminal FLAG-tagging vectors using KanMX6 and hphMX4 as markers. The FLAG-tagging vectors with N-terminal and C-terminal tags included the pFA6a-6 × GLY-FLAG-(maker), with kanMX6, hphMX6, natMX6, bleMX6, and his3MX6 as possible markers. Between the GFP-tagging vectors are pFA6a-GFP(S65 T)-(maker) and N-terminal and C-terminal GFP(S65 T)-tagging, which include kanMX6, hphMX6, natMX6, bleMX6, and ura4MX6. Also, some disruption plasmids as pFA6a-(maker), which has been used for gene deletions using kanMX6, hphMX6, natMX6, bleMX6, ura4MX6, his3MX6, and LEU2MX6, were constructed [79].

A novel system to cloning several DNA fragments, into a plasmid, is the Golden Gate shuffling method. Golden Gate cloning [80–82] is a modular cloning system and was set up for simultaneous overexpression of multiple genes. Some of the applications of the Golden Gate that have been tested in *Pichia pastoris* are the development of strain engineering, pathway expression, and protein production [83].

The use of this methodology for the construction of pREP1-type plasmids that expressed GOI-FPtag was reported *S. pombe*. To apply the Golden Gate cloning, several modules including promoters, tags, marker genes, terminators, and the gene of interest (GOI), which are cloned separately, are produced separately. They are digested with the enzyme BsaI that recognizes a specific sequence GGTCTC and cleaves any four-base sequence after it (such as nNNNN, mMMMM, and kKKKK) at 37°C but generates cohesive ends for various sequences. The Golden Gate method connects all the modules in the order desired in a single reaction. The cleaved fragments are joined by DNA ligase at 16°C. Once complementary four-base overhangs are connected, the site can no longer be cleaved with BsaI. The temperature shift is repeated up to 50 times until circular plasmids are efficiently produced. The system allows the assembly of up to eight expression units on one plasmid with the ability to use different characterized promoters and terminators for each expression unit [84].

In first place, modules were prepared using the pREP1 vector [70]. A segment from pREP1, which includes ars1 and Amp, was amplified by PCR with a pair of oligonucleotides containing BsaI and NotI sites. A typical expression plasmid for *S. pombe* is composed of six modules in total. The modules are a promoter, a terminator, a GOI, an FPtag fused at the N- or C-terminus, a selection marker such as an antibiotic resistance gene, and auxotrophic marker gene required to select colonies that harbor the expression plasmid. With this method, several plasmids were generated. The first plasmid was named pBMod-exv (colEI ori, Amp, ars1, NotI, and KanR sites), and this plasmid was the backbone of all vectors. Plasmids named pRGG (from pRGG-1 to pRGG-5) are expression vectors designed to express GFP-Atb2 from pREP-type multicopy plasmids. For the construction of pRGG-1, LEU2 was chosen as a marker module, whereas for pRGG-2, kan was chosen. To further demonstrate the convenience of the Golden Gate method, a series of plas-mids of variable promoter strength were designed to express GFP-Atb2. The genetic elements included were the promoter (nmt1–41-81 and adh1–41-81 y urg1), an FPTag-N (GFP+ linker, mCherry+ linker, and CFP+ linker), an FPTag-C (linker+ GFP, linker+ mCherry, and linker+ CFP), GOI, and Terminator + marker (Tadh + Kan, Tadh + hpd, Tadh+nat, and Tadh+bsd) [84].

Recently, pheromone-inducible expression vectors for were developed *S. pombe*. By replacing the native Pnmt1+, the promoter regions of the sxa2+ and rep1+ genes were utilized to couple pheromone signaling to the expression of reporter genes for quantitative assessment of the cellular response to mating pheromones. The rep1+ and sxa2+ genes were chosen considering that sxa2+

mRNA increases more than 1600-fold upon pheromone perception in M-type cells [85, 86]. The EGFP open reading frame was placed downstream of the pheromone-inducible promoters, yielding pJR1-rep1-EGFP and pJR1-sxa2-EGFP, respectively [87].

In some cases of the heterologous protein expression, the better way to obtain the right protein production host is through its ability to secrete high titers of properly folded post-translationally processed and active recombinant proteins into the culture media. Proteins secreted in their native hosts will also be secreted in the culture medium. Some signal sequences used to secrete the protein into the extracellular space include α-MF and SUC2 invertase. Both are derived from *S. cerevisiae α*. α-MF is composed of a pre- and proregion and has proven to be most effective in directing protein through the secretory pathway. Other signal peptides to sorting are PHO1 P.p. acid phosphatase, SUC2 S.c. invertase, PHA-E phytohemagglutinin, KILM1 Kl toxin, pGKLpGKL killer protein, CLY and CLY-L8 C-lysozyme and syn., leucin-rich peptide, and K28 pre-pro-toxin K28 virus toxin, to produce molecules such as human interferon, α-amylase, α-1-antitrypsin, and human lysozyme [88].

One of the major problems to the correct production and purification of heterol-ogous proteins from fission yeast is the proteolytic degradation of the recombinant gene product by host-specific proteases. To avoid that problem, a proteasedeficient disruptant was constructed set by disruption of 52 *S. pombe* protease genes using the PCR-mediated single gene-targeted gene disruption method. This technique was used to delete the full open reading frame (ORF) sequence of each target protease gene, using ura4+ as the selection marker [89].

In the first place, the protease-deficient disruptant was obtained, which was amplified from genomic DNA of the *S. pombe* ARC010 strain, using appropriate adapter designed to fuse with the 5′ and 3′ termini of *ura4* (1762 bp), respectively. Then, by fusion extension PCR, *ura4* was sandwiched with the resultant PCR products to obtain the gene disruption fragment (2.2–2.3 bp). The resultant DNA fragments were then introduced into competent cells of the ARC010 strain, using the lithium acetate-based transformation method. Then, the efficient protecting activity of protease of the mutant strains was analyzed. A chromosome-integrative hGH expression vector using the pXL4 plasmid was constructed [89].

To analysed the levels of the secretory production of human growth hormone (hGH), that its known to be a proteolytically sensitive model protein. The results indicated that some of the resultant disruptants were effective in reducing hGH degradation. Although in some cases, added inhibitors of proteasas like Antipain, bestatin, Chymostatin, E-64, Leupeptin, pepstatin, Phosphoramidon, EDTA, apro-tininto avoid protein degradation were necessary. Eight protease coding genes useful for reducing degradation of recombinant proteins [isp6 (subtylase type 9 proteinase), pgp1 (endopeptidase), psp3 (subtylase type peptidase), sxa2 (serine carboxy-peptidase), ppp51 (aminopeptidase), ppp53 were identified (zinc metallopeptidase), ppp60 (metalloprotease) and ppp80 (peptidase)], the use of a strain lacking the aforementioned enzymes allowed a high level of recombinant hGH production. This publication raised the need to evaluate different proteases to identify those that are the best candidates for the production of recombinant proteins, as well as for functional screening, specification, and modification of proteases in *S. pombe* [89].

In relation to the methods for the transformation of *S. pombe*, the lithium acetate and polyethylene glycol-based transformation of plasmid DNA are the most popular and temperature stresses. With these methods, it is possible to achieve transformation efficiencies between $1.0 \times 10^3$ and $1.0 \times 10^4$ transformants per microgram of the plasmid with $10^8$ *S. pombe* cells [90, 91].

## 4. Mutants to analyze the function of genes

The use of mutants to analyze the function of genes has been a tool widely used in *S. pombe*. In this yeast, several types of mutants have been produced such as the temperature-sensitive mutants with conditional defects in the ability to participate in some cellular process in the cell cycle, cytokinesis, lipid metabolism, or DMSO-sensitive [92]. The use of temperature changes to impose a restrictive condition is a strategy widely employed. But, there are methods such as altered sensitivity to drugs, pheromones, and changes in ionic strength, among others. For mutational analysis, the haploid state offers the advantage to observe the effect of specific mutations [93].

In the case of the essential genes, a lethal phenotype is frequently observed. To achieve the study of essential genes, there are two strategies. First, the mutations or gene deletions are created in the diploid state and then the synthetic lethality is studied in the haploid state. Sometimes, it's possible to observe a slowgrowth phenotype, in which haploid cells can partially survive without function of the inactivated gene. Second, the creation of the conditional lethal mutations allows to study a relatively normal gene function under permissive conditions, and then the loss of function is observed under nonpermissive conditions. The most used conditional mutants are the temperature sensitivity, sensitivity to DNA-damaging agents, sensitivity to drugs and inhibitors, and dependence on amino acids or certain carbon sources for viability. Three methods highly used to produce mutants are gene knockouts, random mutagenesis, and site-directed mutagenesis [94].

## 5. CRISPR/Cas9

The CRISPR/Cas system is a bacterial defense mechanism, and its main function is to identify and degrade exogenous nucleic acid sequences [95]. CRISPR-CAs is organized in an operon, which codes the CAS proteins, and a series of identical repeated sequences separated by other sequences known as spacers, which are recognized by intruding DNA molecules [96]. A part of the nucleic acid stranger is incorporated into the spacer's zone of the operon using the Cas proteins, which degrade the strange DNA. Next, the transcription of CRISPR-Cas generates a precursor CRISPR-RNA or pre-crRNA, which is then processed to generate crRNAs of small size, which are complementary to the sequence of the foreign DNA. In the last known phase of interference, Cas proteins, using as a guide to crRNAs, detect intruding sequences and degrade them [96].

The CRISPR/Cas technology allows to identify a specific segment of DNA, remove, or replace it using always the same tools: a duplex RNA with the copy of the DNA to be identified (sgARN) and a short sequence adjacent to the protospacer (PAM) that will bind to DNA and stabilize the protein Cas9, protein with endonuclease activity, and helicase guided by the sgARN that separates and cuts the two strands of DNA. A Cas9-gRNA plasmid expressing the active Cas9 enzyme and sgRNA, as well as another plasmid with donor DNA for each deletion are required. The CRISPR-Cas technology allows targeting of multiple genetic manipulations to the same strain, it avoids indirect physiological effects, and it limits the perturbation of the local chromatin and transcriptional environment to the gene manipulation of interest. In fission yeast, this technique has allowed to produce genetic modifications as point mutation knock-in, endogenous N-terminal tagging, and genomic sequence deletion [97].

Recently, a web-tool called CRISPR4P CRISPR for *Pombe* or CRISPR *Pombe* PCR Primer Program was developed as freely available from the website (bahlerlab.info/crispr4p) [98]. This tool was created to support the design of all kinds of primers required for the deletion of any genomic region: PCR-based sgRNA cloning, PCR-based synthesis of DNA template for the deletion by homologous recombina-tion, and checking primers to confirm the deletion. Through CRISPR/Cas9-based approach in *S. pombe*, the success in the deletion of over 80 different noncoding RNA genes that were lowly expressed was reported. Using the web tool, the preparation of G1-synchronized and cryopreserved *S. pombe* cells was achieved, whose major property was the efficiency and speed for transformations. The steps to achieve the deletions reported by Rodríguez-López et al., 2016, are: (1) identify better sgRNAs to target region of modification using CRISPR4P tool. (2) Design primers required for whole process using CRISPR4P including sgRNA cloning; synthesis DNA template for homologous recombination (HR template) for gene deletion; and check primers to confirm gene deletion. (3) Clone sgRNAs into nourseothricin-selectable plasmid pMZ379 that contains Cas9 enzyme gene, the *natMX6* selection marker, and the *rrk1* promoter/leader. (4) Produce the HR template by PCR using primers with sequences flanking the region of modification (deletion) and over-lapping at their 3′ ends. (5) Delete region of interest by cotransforming sgRNA/Cas9-plasmid and HR template into *S. pombe* cells, previously synchronized and cryopreserved to increase transformation efficiency [99].

A gap-repair-based CRISPR/Cas9 procedure allows to efficiently knockin a point mutation in fission yeast. The rpl42-P56Q mutation confers cycloheximide resistance (CYHR) [100]. Employing this technique, a CCC codon for proline was changed, and with the use of a pair of 90-nt complementary oligos as donor DNA, the gap repair procedure achieved a high editing efficiency (84%).

Using the CRISPR-Cas9, yeast strains, functional and successfully complemented with the markers ura4-D18, leu1-Δ0, his3- Δ0, and lys9-Δ0, were created. To achieve the goal, all the components were assembled with the “BsaI-pad,” a single 42 bp region containing two BsaI cutting sites to produce the plasmids pYZ182, pYZ183, and pYZ184 with nmt1, nmt41, and nmt81 cassettes, respectively. Using that design, the marker genes ura4, leu1, his3, and lys9 were integrated separately. Later, the plasmids were transformed into yeast [101].

Recently, the type VI CRISPR system, Cas13a from Leptotrichia shahii (LshCas13a), was employed to introduce genetic changes on the DNA, disrupting or editing to target and knockdown endogenous gene transcripts with different efficiencies in *S. pombe* [102].

## 6. RNAi

RNA interference (RNAi) is a highly conserved eukaryotic gene regulatory mechanism, which uses small noncoding RNAs to mediate posttranscriptional gene silencing as a host defense mechanism. It was described that *S. pombe* has the entire RNAi machinery (Dcr1, DICER ribonuclease; the Rdp1, RNA-dependent RNA polymerase 1; and the Ago1, Argonaute family member). In *S. pombe*, the role of the RNAi pathway on the heterochromatin assembly has been widely studied [103]. RNAi plays a role in regulating expression of Tf2 retrotransposons, and it is also involved in the RNAi-dependent heterochromatin assembly by the Hsps, Hsp90 and Mas5 (a nucleocytoplasmic type-I Hsp40 protein).

siRNA is generated by the Dicer family endoribonuclease Dcr1, from double-stranded noncoding RNA that is complementary to heterochromatin. The siRNA duplex is loaded onto a non–chromatin-associated complex called Argonaute, small interfering RNA chaperone (ARC), which contains the Ago1 endoribonuclease. The loading of the siRNA duplex onto the Ago1 subunit requires the two ARC-specific subunits, Arb1 and Arb2, which also inhibit the release of the passenger strand [104]. Thus, this complex changes its subunits' composition to form a chromatin-associated effector complex called RNA-induced transcriptional silencing (RITS) [105]. The RITS complex is composed of Ago1, now binding single-stranded siRNA as a guide for target recognition, and the two RITS-specific subunits: Chp1 and Tas3. Chp1 uses a chromodomain to recognize H3K9me, whereas Tas3 bridges Ago1 and Chp1 [106].

To analyze the role of the RNAi in fission yeast, the lacZ fission yeast system was employed. With this system, it was possible to know that the gene inhibition is dependent on the dose of the antisense RNA, the size of the antisense transcript, as well as the targeted region. Any of them can affect the efficacy of target gene inhibition. The generation of dsRNA through either intermolecular or intramolecular hybridization is central to make the antisense RNA-mediated gene silencing in *S. pombe* [107]. As a genetic tool to analyze the function of genes, the ura4-based RNAi-based selective assay was developed using a repressible thiamine promoter [108]. The RNAi must be optimized in order to know the minimum requirements to achieve the knockdown of a specific gene. U-HP construct was produced as a hairpin complementary to 200 bp of ura+ gene expressed from the nmt1 promoter and integrated at ars1 on chromosome 1. U-HP silences ura4+ inserted nearby to centromere 1, but not the endogenous ura4+ gene. Interestingly, in *S. pombe*, exogenous siRNAs can only silence efficiently in *trans*, when the target locus is near endogenous sites of heterochromatin.

An interesting proposal to analyze the role of the siRNAs in *S. pombe* was achieved with the development of a GFP-HP construct. This system was generated under control of the Pnmt1, and it contains two GFP open reading frames arranged in an inverted orientation, around the first intron from the rad9 gene. When it was probed, it was demonstrated that GFP-HP induces trans-silencing of target genes. GFP siRNAs generated by the expression of a GFP-HP can act in *trans* to establish heterochromatin on target genes bearing homology to GFP siRNAs and silencing their expression. This silencing does not require other manipulations, such as deletion of eri1+ or increased expression of Swi6HP1, a heterochromatin component, to promote RNAi-mediated silencing in *trans* [109].

## 7. Yeast two-hybrid system

The yeast two-hybrid system (Y2H) is a method widely employed to study the physical interaction of proteins by the downstream activation of a reporter gene. Considering that many eukaryotic transcription factors are organized in a modular way with at least two domains, it is possible to separate them into their domains [110].

In this assay, two plasmids are created; the first is named the bait plasmid including the DNA-binding domain of a transcription factor joined to one of the proteins to analyze and it is named Bait. In this vector, a selection marker is included such as HIS3, ADE2 (Gal4 system), or LEU2 (LexA system with binding sites for the DNA-binding domain). The second vector is named prey including the activation domain of the transcription factor joined to the second protein to study in the interaction, named Prey. As in the other vector, a different selection marker is included. When the Bait and Prey proteins are put together by protein interaction, they restored the organization of the transcription factor, and then they can activate the transcription of the reporter gene as the *E. coli lacZ* gene. The transcription factors more frequently used are *Escherichia coli* LexA protein and the yeast Gal4 protein, as well as herpes simplex virus VP16 protein and the B42 acid blob from *E. coli* [111].

Gal4 is a transcriptional activator in yeast that binds to UAS (upstream activation domain), a specific DNA sequence, and activates transcription in the presence of galactose. The separation of Gal4 in two fragments produces N-terminal DNA-binding domain (DBD) and C-terminal transcriptional activation domain (AD), but did not activate transcription in the presence of galactose until both domains are associated to reconstitute a fully functional Gal4. Some disadvantages of the assay consider that in some cases, it's necessary to modify the bait proteins because a protein with both DNA-binding and transcriptional activating properties is possible to be found. Some fused proteins may not be able to enter or be expressed in the yeast nucleus. The GAL4 BD has its own nuclear localization signal (NLS). If the GAL4-based Y2H system fails, the interaction could be analyzed and detected successfully using a LexA-based Y2H system [110, 111].

The Y2H system has been widely used. In *S. pombe*, its use in the searching of the new determinants of aging was reported. Chen et al. described a method to select long-lived mutants from *S. pombe* bar code-tagged insertion mutant library (each insertion had a unique sequence tag called a bar code produced by random barcode). With this strategy, it was possible to identify an insertion mutation or deletion in the cyclin gene *clg1+* that extended the chronological aging of the yeast. At the same time, it was determined that depletion of Clg1p also decreases the cyclin-dependent kinase Pef1p and an extended longevity was observed. To analyze if the phenotype was produced by direct or indirect contact, a yeast two-hybrid analysis and immunoprecipitation assay were performed [112].

To the assay, the entire *pef1+* ORF was fused to the Gal4p DNA-binding domain and the entire *clg1+* ORF was fused to the Gal4p activation domain. A physical interaction was observed between Clg1 and Pef1. To perform this assay, the pGBT9-Pef1 and pGAD424-Clg1(full length) or pGAD424-Clg1(1–590) plasmids were constructed and transformed into the *Saccharomyces cerevisiae* two hybrid indicator strain Y187 (*MATα*, *ura3–52*, *his3–200*, *ade2–101*, *trp1–901*, *leu2–3112*, *gal4Δ*, *met-*, *gal80Δ*, *MEL1*, and *URA3::GAL1UAS-GAL1TATAlacZ*, Clontech). Positive transfor-mants were selected on complete medium plates without leucine and tryptophan at 30°C for 3 days. The reporter gene *lacZ* expression was probed from five individual colonies from each transformation and was patched on plates that require both plasmids for growth and incubated at 30°C for 2 days. Then, the coimmunopre-cipitation was performed with FLAG-tagged Clg1p, which was expressed in cells that also expressed triple HA (3HA)-tagged Pef1p [113]. Using Western blotting of FLAG-Clg1p immunoprecipitates revealed the presence Pef1p-3HA. Chen et al. concluded that Clg1p interacts with the cyclin-dependent kinase Pef1p in *S. pombe* cells. In addition, a third Pef1p cyclin named Psl1p was identified. Genetic and coimmunoprecipitation assays indicated Pef1p controls lifespan by downstreaming the protein kinase Cek1p [114].

## 8. DNA microarray

DNA microarray is an orderly set of segments of genes that are immobilized on a surface called chip. The DNA arrangements allow the massive study of the gene expression of an organism, and it allows to know the differences of gene expression between two samples of RNA in a given cellular condition. In cells that present some mutation or elimination in some genes or cells derived from individuals with some infectious disease or not, the microarrays allow the identification of sets of genes related to the gene or genes under study or the condition of disease. Comparing RNA prepared from diseased cells and normal cells can lead to the identification of sets of genes that play key roles in diseases. Genes that are overexpressed or underexpressed in the diseased cells often present excellent targets for therapeutic drugs.

The application of DNA microarray technology requires a genomic library conformed by a set of DNA segment derived from each of the genes of the model of interest, which is generated from PCR products or synthetic oligonucleotides, as well as the design and construction of the arrangement, to determine the physical location and accurate identification for the analysis and interpretation of gene expression data. Microarray analysis requires total RNA extraction from control and the problem obtained by any strategy optimized for certain cell type [115]. Total RNA control and the problem should be submitted to retrotranscription incorporating uracil marked with a fluorescent molecule as dUTP-Cy3, dUTP-Cy5, dUTP-Alexa 555, dUTP-Alexa 647, and biotin, among others. The labeling of the cDNA must be differentiable between the two tissues to be analyzed [116]. The hybridization of the microarray containing probe sets that represent a finite number of transcripts is carried out. Fluorescence reading is obtained with a microarray reader. The quantification of the signal produced by the fluorescence of the spots allows to calculate for each point the mean density value of the nucleotides marked cDNA (g. e. of Alexa555, Alexa647) and the average value of the background. To identify the genes expressed differentially in the experiment, it is necessary to perform a statistical analysis, from the normalization of the data. The goal is to analyze those genes that move away from normalization through the value of Z [117]. The genes with the value of $Z > 2$ present a statistically significant change between the experimental condition and the control (genes with greater or lesser expression). [116]. Easy and useful software for data analysis of microarrays is GenArise (computer unit of the Institute of Cellular Physiology of UNAM.

From the data that record a significant change, it is necessary to determine its association to some biological processes by clustering analysis for gene expression [118A].

With this molecular tool, it was possible to analyze in fission yeast the effect of Spc1, a mitogen-activated protein kinase in the stress responses. Spc1 is an activator of transcription factors that control gene expression in response to extracellular stimuli and is also known to interact with the translation machinery. Using microarrays of Affymetrix GeneChip Yeast Genome 2.0 Array, it was possible to know the set of genes that is regulated by SPC1, and this analysis was carried out without and with a stress condition to evaluate the effect of the wild-type SPC1 kinase and Spc1K49R, a mutant of this enzyme. Spc1 and Spc1K49R were separately overexpressed in *S. pombe* cells, and gene expression was compared with the control cells (which are transformed with the empty with the Pnmt1). Interestingly, only 42 genes were found with differential expression after Spc1 overexpression, while 132 genes were found to be differentially expressed after Spc1K49R overexpression. Some of the genes up-regulated after Spc1 overexpression were Mitogen-activated protein kinase sty1 and M cell-type agglutination protein mam3. The downregulated genes were NAD-dependent malic enzyme, meiotic cohesin complex subunit Rec8, and aph1 bis(5'-nucleosidyl)-tetraphosphatase. Between genes differentially expressed after Spc1K49R overexpression, those upregulated included pheromone p-factor receptor, RNA-binding protein involved in meiosis Mei2, MAP kinase Spk1, cell agglutination protein Mam3, M-factor precursor Mfm1, and M-factor precursor Mfm3. And some downregulated were serine/threonine protein kinase Gsk3, RNA-binding protein Sap49, and Argininosuccinate lyase [119].

In 2016, the role of the putative NO dioxygenase SPAC869.02c (Yhb1) and the S-nitrosoglutathione reductase Fmd2 was analyzed. Both proteins are NO-detoxification enzymes. In the study, it was found that exogenous NO protects *S. pombe* cells against $H_2O_2$-induced oxidative stress by inhibition of $Fe^{(3+)}$ to $Fe^{(2+)}$ conversion, upregulation of the $H_2O_2$-detoxifying enzymes, as well as downregulation of the MRC genes. Transcriptomic analysis was carried out with an Affymetrix Gene Chip Yeast Genome 2.0 Array [120].

The fission yeast *S. pombe* generally reproduces by mitosis. To know the role of the fhl1 protein in meiosis, a microarray analysis of the fhl1Δ strain was performed. Interestingly, it was found that nitrogen starvation-response genes are controlled by fhl1. Some of them are genes of mating and sporulation such as isp4, mfm1, mfm2, Mat-Mc, ste4, ste11, map1, map3, mei2, and mcp7 [121].

## 9. Next-generation sequencing

Next-generation sequencing (NGS) involves the parallel mass sequencing of thousands of DNA fragments. Sample processing for NGS can be summarized as follows: First, nucleic acid extraction (DNA or RNA). Second, selection of the type of NGS sequencing (targeted sequencing, whole exome sequencing, and whole genome sequencing). Third, library generation by DNA fragmentation, ligation of adaptors, and amplification and sample enrichment. Fourth, template generation or cluster generation according to the platform of sequencing. Fifth, sequencing (using a specific platform as Illumina, PacBio). Sixth, data analysis. Data analysis includes the quality evaluation of the sequence, alignment to reference sequence to identify some possible variations such as single nucleotide polymorphism (SNP) or insertion-deletion (indel) identification, phylogenetic or metagenomic analysis, as well as the identification, interpretation, and classification of pathogenic variants [122, 123].

Splicing is an essential step in eukaryotic gene expression. Introns are excised by the spliceosome, composed of five uridine-rich small nuclear RNAs (U1, U2, U4, U5, and U6 snRNAs) and several polypeptides. To characterize the U2·U5·U6 complex of *S. pombe*, cell lysates were obtained. A large-scale isolation of the U2·U5·U6 complex was performed using double-affinity purification using a split TAP-tag approach [124], with protein A attached to U2 snRNP protein Lea1 (U2 A′ in humans) and calmodulin-binding peptide (CBP) attached to U5 snRNP protein Snu114 (U5 116K in humans). After the purification of the complexes, the content of protein and RNA associated to the U2·U5·U6 complexes was analyzed. By denaturing PAGE and high-throughput sequencing (RNAseq), the presence of U4, U1, and heterogeneous higher molecular weight species was shown. In addition, the U2·U5·U6 snRNA complex contains excised introns, indicating that it is primarily the ILS (intron lariat spliceosome) complexes. The protein content of the ILS complex of *S. pombe* was similar to the spliced product of humans and the ILS complexes assembled on single premRNAs in vitro from *Saccharomyces cerevisiae* [112].

There are some other techniques to study several aspects of the physiology of *S. pombe*. Chromosome conformation capture (Hi-C) is a technique widely used to identify long-range chromatin interactions. The spatial organization of mitotic chromosomes with the greatest compaction during mitosis is an interesting aspect of the cell cycle. In *S. pombe*, it is known that condensin, a structural maintenance of chromosomes (SMC) family member, has a role on the chromatin architecture. Biochemical studies have been applied to discover the more relevant points of the mechanism. By chromosome conformation capture (Hi-C), it was demonstrated that condensin is able to replace short-range local contacts in the interphase with longer-range interactions in the mitosis. Condensin achieves this by setting up longer-range, intrachromosomal DNA interactions, which compact and individualize chromosomes. Even local chromatin contacts are constrained by condensin during mitosis [125].

Finally, it is necessary to mention that Rallis & Bähler offered to the world pombe community an excellent review showing the relevance of *S. pombe* in the eukaryotic studies employing a wide genome screen and phenomic assays, ranging from growing conditions to metabolomics [126, 127].

## 10. Conclusion

*Schizosaccharomyces pombe* is an excellent model to study highly conserved processes between eukaryotes, its versatility, ease of manipulation, its accessibility to genetic manipulations, making it a great model system increasingly used by a growing scientific community interested in fission yeast. At the same time, this interest has promoted the technological development, the implementation, and the continuous improvement of new molecular tools that when applied to *S. pombe* will allow to elucidate new mechanisms of cellular processes with potential application to the Eukaryotic kingdom including the human being.

## Author details

Irma Pilar Herrera-Camacho[1], Lourdes Millán-Pérez-Peña[1], Francisca Sosa-Jurado[2], Nancy Martínez-Montiel[3], Rebeca Débora Martínez-Contreras[3] and Nora Hilda Rosas Murrieta[1*]

1 Biochemistry and Molecular Biology Laboratory, Chemistry Center, Science Institute, Autonomous University of Puebla, Puebla, México

2 Virology and Molecular Biology Laboratory, Eastern Biomedical Research Center, Mexican Social Security Institute, Puebla, México

3 Microbial Molecular Ecology Laboratory, Center for Research in Microbiological Sciences (CICM) - Science Institute, Autonomous University of Puebla, Puebla, Mexico

*Address all correspondence to: nora.rosas@correo.buap.mx

## References


[1] Lindner P. Schizosaccharomyces *pombe sp.* nov., a new ferment. Wochenschrift für Brauerei. 1893;**10**:1298-1300

[2] Heckman DS, Geiser DM, Eidell BR, Stauffer RL, Kardos NL, Hedges SB. Molecular evidence for the early colonization of land by fungi and plants. Science. 2001;**293**(5532):1129-1133

[3] Leupold U. Die Vererbung von Homothallie und Heterothallie bei *Schizosaccharomyces pombe*. Comptes-Rendus des Travaux du Laboratoire Carlsberg. 1949;**24**:381-480

[4] Hayles J, Nurse P. Introduction to fission yeast as a model system. Cold Spring Harbor Protocols. 2018;**2018**(5):pdb.top079749

[5] Egel R. Mating-type genes, meiosis and sporulation. In: Nasim A, Young P, Johnson BF, editors. Molecular Biology of the Fission Yeast. San Diego: Academic Press; 1989. pp. 31-73

[6] Wixon J. Featured organism: *Schizosaccharomyces pombe*, the fission yeast. Comparative and Functional Genomics. 2002;**3**(2):194-204

[7] Wood V, Gwilliam R, Rajandream MA, Lyne M, Lyne R, Stewart A, et al. The genome sequence of *Schizosaccharomyces pombe*. Nature. 2002;**415**(6874):871-880

[8] Lang BF, Cedergren R, Gray MW. The mitochondrial genome of the fission yeast, *Schizosaccharomyces pombe*. Sequence of the large-subunit ribosomal RNA gene, comparison of potential secondary structure in fungal mitochondrial large-subunit rRNAs and evolutionary considerations. European Journal of Biochemistry. 1987;**169**:527-537

[9] Schaak J, Mao J, Soll D. The 5.8S RNA gene sequence and the ribosomal repeat of *Schizosaccharomyces pombe*. Nucleic Acids Research. 1982;**10**:2851-2864

[10] Lock A, Rutherford K, Harris MA, Wood V. PomBase: The scientific resource for fission yeast. Methods in Molecular Biology. 2018;**1757**:49-68

[11] Zhao RY. Yeast for virus research. Microbial Cell. 2017;**4**(10):311-330

[12] Hagan IM, Grallert A, Simanis V. Analysis of the *Schizosaccharomyces pombe* cell cycle. Cold Spring Harbor Protocols. 2016;**2016**(9). DOI: 10.1101/pdb.top082800

[13] Gómez EB, Forsburg SL. Analysis of the fission yeast *Schizosaccharomyces pombe* cell cycle. Methods in Molecular Biology. 2004;**241**:93-111

[14] Gutiérrez-Escribano P, Nurse P. A single cyclin-CDK complex is sufficient for both mitotic and meiotic progression in fission yeast. Nature Communications. 2015;**6**:6871

[15] Yamashita A, Sakuno T, Watanabe Y, Yamamoto M. Analysis of *Schizosaccharomyces pombe* meiosis. Cold Spring Harbor Protocols. 2017;**2017**(9):pdb.top079855

[16] Egel R. Fission yeast on the brink of meiosis. BioEssays. 2000;**22**(9):854-860

[17] Merlini L, Dudin O, Martin SG. Mate and fuse: How yeast cells do it. Open Biology. 2013;**3**(3):130008

[18] Forsburg SL, Rhind N. Basic methods for fission yeast. Yeast. 2006;**23**(3):173-183

[19] Cam HP, Whitehall S. Analysis of heterochromatin in *Schizosaccharomyces pombe*. Cold Spring Harbor Protocols. 2016. DOI: 10.1101/pdb.top079889

[20] Nandakumar J, Cech TR. Finding the end: Recruitment of telomerase to

telomeres. Nature Reviews. Molecular Cell Biology. 2013;**14**(2):69-82

[21] French B, Straight AF. Swapping CENP-A at the centromere. Nature Cell Biology. 2013;**15**:1028-1030

[22] Davis L, Smith G. Meiotic recombination and chromosome segregation in *Schizosaccharomyces pombe*. PNAS. 2001;**98**(15):8395-8402

[23] Burke DJ. Interpreting spatial information and regulating mitosis in response to spindle orientation. Genes and Development. 2009;**23**(14):1613-1618

[24] Goyal A, Takaine M, Simanis V, Nakano K. Dividing the spoils of growth and the cell cycle: The fission yeast as a model for the study of cytokinesis. Cytoskeleton (Hoboken). 2011;**68**(2):69-88

[25] Yamamoto M. The selective elimination of messenger RNA underlies the mitosis-meiosis switch in fission yeast. Proceedings of the Japan Academy. Series B, Physical and Biological Sciences. 2010;**86**(8):788-797

[26] Brunner D, Nurse P. New concepts in fission yeast morphogenesis. Philosophical Transactions of the Royal Society of London. Series B, Biological Sciences. 2000;**355**(1399):873-877

[27] Hoffman CS, Wood V, Fantes PA. An ancient yeast for young geneticists: A primer on the Schizosaccharomycespombe model system. Genetics. 2015;**201**(2):403-423

[28] Wendland J. PCR-based methods facilitate targeted gene manipulations and cloning procedures. Current Genetics. 2003;**44**(3):115-123

[29] Gao J, Kan F, Wagnon JL, Storey AJ, Protacio RU, Davidson MK, et al. Rapid, efficient and precise allele replacement in the fission yeast *Schizosaccharomyces pombe*. Current Genetics. 2014;**60**(2):109-119

[30] Kaur R, Ingavale SS, Bachhawat AK. PCR-mediated direct gene disruption in *Schizosaccharomyces pombe*. Nucleic Acids Research. 1997;**25**(5):1080-1081

[31] Wang L, Kao R, Ivey FD, Hoffman CS. Strategies for gene disruptions and plasmid constructions in fission yeast. Methods. 2004;**33**(3): 199-205

[32] Wach A. PCR-synthesis of marker cassettes with long flanking homology regions for gene disruptions in *S. cerevisiae*. Yeast. 1996;**12**(3):259-265

[33] Grimm C, Kohli J, Murray J, Maundrell K. Genetic engineering of *Schizosaccharomyces pombe*: A system for gene disruption and replacement using the ura4 gene as a selectable marker. Molecular and General Genetics. 1988;**215**(1):81-86

[34] Sabatinos SA, Forsburg SL. Molecular genetics of *Schizosaccharomyces pombe*. Methods in Enzymology. 2010;**470**:759-795

[35] Krawchuk MD, Wahls WP. High-efficiency gene targeting in *Schizosaccharomyces pombe* using a modular, PCR-based approach with long tracts of flanking homology. Yeast. 1999;**15**(13):1419-1427

[36] Keeney JB, Boeke JD. Efficient targeted integration at leu1-32 and ura4-294 in *Schizosaccharomyces pombe*. Genetics. 1994;**136**(3):849-856

[37] Rothstein R. Targeting, disruption, replacement, and allele rescue: Integrative DNA transformation in yeast. Methods in Enzymology. 1991;**194**:281-301

[38] Hegemann JH, Heick SB, Pöhlmann J, Langen MM, Fleig U. Targeted gene deletion in *Saccharomyces cerevisiae* and

*Schizosaccharomyces pombe*. Methods in Molecular Biology. 2014;**1163**:45-73

[39] Zhao Y, Lieberman HB. *Schizosaccharomyces pombe*: A model for molecular studies of eukaryotic genes. DNA and Cell Biology. 1995;**14**(5):359-371

[40] Petranovic D, Tyo K, Vemuri GN, Nielsen J. Prospects of yeast systems biology for human health: Integrating lipid, protein and energy metabolism. FEMS Yeast Research. 2010;**10**(8):1046-1059

[41] Giga-Hama Y, Tohda H, Takegawa K, Kumagai H. *Schizosaccharomyces pombe* minimum genome factory. Biotechnology and Applied Biochemistry. 2007;**46**(Pt 3):147-155

[42] Kashiwagi K, Shigeta T, Imataka H, Ito T, Yokoyama S. Expression, purification, and crystallization of *Schizosaccharomyces pombe* eIF2B. Journal of Structural and Functional Genomics. 2016;**17**(1):33-38

[43] Winderickx J, Delay C, De Vos A, Klinger H, Pellens K, Vanhelmont T, Van Leuven F, Zabrocki P. Protein folding diseases and neurodegeneration: lessons learned from yeast. Biochimica et Biophysica Acta. 2008; 1783(7): 1381-1395.

[44] Tenreiro S, Munder MC, Alberti S, Outeiro TF. Harnessing the power of yeast to unravel the molecular basis of neurodegeneration. Journal of Neurochemistry. 2013;**127**(4):438-452

[45] Seynnaeve D, Vecchio MD, Fruhmann G, Verelst J, Cools M, Beckers J, et al. Recent insights on alzheimer's disease originating from yeast models. The International Journal of Molecular Sciences. 2018;**19**(7):E1947

[46] Huh E, Lim S, Kim HG, Ha SK, Park HY, Huh Y, et al. Ginger fermented with *Schizosaccharomyces pombe* alleviates memory impairment via protecting hippocampal neuronal cells in amyloid beta1-42 plaque injected mice. Food and Function. 2018;**9**(1):171-178

[47] Siam R, Dolan WP, Forsburg SL. Choosing and using *Schizosaccharomyces pombe* plasmids. Methods. 2004;**33**(3):189-198

[48] Adams C, Haldar D, Kamakaka RT. Construction and characterization of a series of vectors for *Schizosaccharomyces pombe*. Yeast. 2005;**22**(16):1307-1314

[49] Van Driessche B, Tafforeau L, Hentges P, Carr AM, Vandenhaute J. Additional vectors for PCR-based gene tagging in *Saccharomyces cerevisiae* and *Schizosaccharomyces pombe* using nourseothricin resistance. Yeast. 2005;**22**(13):1061-1068

[50] Forsburg SL, Sherman DA. General purpose tagging vectors for fission yeast. Gene. 1997;**191**(2):191-195

[51] Fennessy D, Grallert A, Krapp A, Cokoja A, Bridge AJ, Petersen J, et al. Extending the *Schizosaccharomyces pombe* molecular genetic toolbox. PLoS One. 2014;**9**(5):e97683

[52] Kikuchi Y, Kitazawa Y, Shimatake H, Yamamoto M. The primary structure of the leu1+ gene of *Schizosaccharomyces pombe*. Current Genetics. 1988;**14**(4):375-379

[53] Apolinario E, Nocero M, Jin M, Hoffman CS. Cloning and manipulation of the *Schizosaccharomyces pombe* his7+ gene as a new selectable marker for molecular genetic studies. Current Genetics. 1993;**24**(6):491-495

[54] Fujita Y, Giga-Hama Y, Takegawa K. Development of a genetic transformation system using new selectable markers for fission yeast *Schizosaccharomyces pombe*. Yeast. 2005;**22**(3):193-202

[55] Ma Y, Sugiura R, Saito M, Koike A, Sio SO, Fujita Y, et al. Six new amino acid-auxotrophic markers for targeted gene integration and disruption in fission yeast. Current Genetics. 2007;**52**(2):97-105

[56] Maundrell K. nmt1 of fission yeast. A highly transcribed gene completely repressed by thiamine. The Journal of Biological Chemistry. 1990;**265**:10857-10864

[57] Basi G, Schmid E, Maundrell K. TATA box mutations in the *Schizosaccharomyces pombe* nmt1 promoter affect transcription efficiency but not the transcription start point or thiamine repressibility. Gene. 1993;**123**(1):131-136

[58] Forsburg SL. Comparison of *Schizosaccharomyces pombe* expression systems. Nucleic Acids Research. 1993;**21**(12):2955-2956

[59] Faryar K, Gatz C. Construction of a tetracycline-inducible promoter in *Schizosaccharomyces pombe*. Current Genetics. 1992;**21**(4-5):345-349

[60] McLeod M, Stein M, Beach D. The product of the mei3+ gene, expressed under control of the mating-type locus, induces meiosis and sporulation in fission yeast. The EMBO Journal. 1987;**6**(3):729-736

[61] Kawashima SA, Tsukahara T, Langegger M, Hauf S, Kitajima TS, Watanabe Y. Shugoshin enables tension-generating attachment of kinetochores by loading Aurora to centromeres. Genes and Development. 2007;**21**(4):420-435

[62] Yokobayashi S, Watanabe Y. The kinetochore protein Moa1 enables cohesion-mediated monopolar attachment at meiosis I. Cell. 2005;**123**(5):803-817

[63] Verma HK, Shukla P, Alfatah M, Khare AK, Upadhyay U, Ganesan K, et al. High level constitutive expression of luciferase reporter by lsd90 promoter in fission yeast. PLoS One. 2014;**9**(7):e101201

[64] Watson AT, Werler P, Carr AM. Regulation of gene expression at the fission yeast *Schizosaccharomyces pombe* urg1 locus. Gene. 2011;**484**(1-2):75-85

[65] de Medeiros AS, Kwak G, Vanderhooft J, Rivera S, Gottlieb R, Hoffman CS. Fission yeast-based high-throughput screens for PKA pathway inhibitors and activators. Methods in Molecular Biology. 2015;**1263**:77-91

[66] Bähler J, Wu JQ , Longtine MS, Shah NG, McKenzie A 3rd, Steever AB, et al. Heterologous modules for efficient and versatile PCR-based gene targeting in *Schizosaccharomyces pombe*. Yeast. 1998;**14**(10):943-951

[67] Hentges P, Van Driessche B, Tafforeau L, Vandenhaute J, Carr AM. Three novel antibiotic marker cassettes for gene disruption and marker switching in *Schizosaccharomyces pombe*. Yeast. 2005;**22**(13):1013-1019

[68] Kimura M, Kamakura T, Tao QZ, Kaneko I, Yamaguchi I. Cloning of the blasticidin S deaminase gene (BSD) from *Aspergillus terreus* and its use as a selectable marker for *Schizosaccharomyces pombe* and *Pyricularia oryzae*. Molecular and General Genetics. 1994;**242**(2):121-129

[69] Sato M, Dhut S, Toda T. New drug-resistant cassettes for gene disruption and epitope tagging in *Schizosaccharomyces pombe*. Yeast. 2005;**22**(7):583-591

[70] Maundrell K. Thiamine-repressible expression vectors pREP and pRIP for fission yeast. Gene. 1993;**123**(1):127-130

[71] Zhao Y, Elder RT, Chen M, Cao J. Fission yeast expression vectors adapted for positive identification of gene

insertion and green fluorescent protein fusion. BioTechniques. 1998;**25**(3):438-440, 442, 444

[72] Maundrell K. nmt1 of fission yeast. A highly transcribed gene completely repressed by thiamine. Journal of Biological Chemistry. 1990;**265**(19):10857-10864

[73] He J, Choe S, Walker R, Di Marzio P, Morgan DO, Landau NR. Human immunodeficiency virus type 1 viral protein R (Vpr) arrests cells in the G2 phase of the cell cycle by inhibiting p34cdc2 activity. Journal of Virology. 1995;**69**(11):6705-6711

[74] Benko Z, Elder RT, Li G, Liang D, Zhao RY. HIV-1 protease in the fission yeast *Schizosaccharomyces pombe*. PLoS One. 2016;**11**(3):e0151286

[75] Zhao Y, Cao J, O'Gorman MR, Yu M, Yogev R. Effect of human immunodeficiency virus type 1 protein R (vpr) gene expression on basic cellular function of fission yeast *Schizosaccharomyces pombe*. Journal of Virology. 1996;**70**(9):5821-5826

[76] Forsburg SL. Codon usage table for *Schizosaccharomyces pombe*. Yeast. 1994;**10**(8):1045-1047

[77] Ahn J, Choi CH, Kang CM, Kim CH, Park HM, Song KB, et al. Generation of expression vectors for high-throughput functional analysis of target genes in *Schizosaccharomyces pombe*. Journal of Microbiology. 2009;**47**(6):789-795

[78] Ahn J, Won M, Kyun ML, Kim YS, Jung CR, Im DS, et al. Development of episomal vectors carrying a nourseothricin-resistance marker for use in minimal media for *Schizosaccharomyces pombe*. Yeast. 2013;**30**(6):219-227

[79] Gadaleta MC, Iwasaki O, Noguchi C, Noma K, Noguchi E. New vectors for epitope tagging and gene disruption in *Schizosaccharomyces pombe*. BioTechniques. 2013;**55**(5):257-263

[80] Engler C, Kandzia R, Marillonnet S. A one pot, one step, precision cloning method with high throughput capability. PLoS One. 2008;**3**(11):e3647

[81] Engler C, Gruetzner R, Kandzia R, Marillonnet S. Golden Gate shuffling: A one-pot DNA shuffling method based on type IIs restriction enzymes. PLoS One. 2009;**4**(5):e5553

[82] Werner S, Engler C, Weber E, Gruetzner R, Marillonnet S. Fast track assembly of multigene constructs using Golden Gate cloning and the MoClo system. Bioeng Bugs. 2012;**3**(1):38-43

[83] Prielhofer R, Barrero JJ, Steuer S, Gassler T, Zahrl R, Baumann K, et al. GoldenPiCS: A Golden Gate-derived modular cloning system for applied synthetic biology in the yeast Pichia pastoris. BMC Systems Biology. 2017;**11**(1):123

[84] Kiriya K, Tsuyuzaki H, Sato M. Module-based systematic construction of plasmids for episomal gene expression in fission yeast. Gene. 2017;**637**:14-24

[85] Mata J, Bähler J. Global roles of Ste11p, cell type, and pheromone in the control of gene expression during early sexual differentiation in fission yeast. Proceedings of the National Academy of Sciences of the United States of America. 2006;**103**(42):15517-15522

[86] Xue-Franzén Y, Kjaerulff S, Holmberg C, Wright A, Nielsen O. Genomewide identification of pheromone-targeted transcription in fission yeast. BMC Genomics. 2006;**7**:303

[87] Hennig S, Hornauer N, Rödel G, Ostermann K. Pheromone-inducible expression vectors for fission yeast *Schizosaccharomyces pombe*. Plasmid. 2018;**95**:1-6

[88] Ahmad M, Hirz M, Pichler H, Schwab H. Protein expression in *Pichia pastoris*: Recent achievements and perspectives for heterologous protein production. Applied Microbiology and Biotechnology. 2014;**98**(12):5301-5317

[89] Idiris A, Bi K, Tohda H, Kumagai H, Giga-Hama Y. Construction of a protease-deficient strain set for the fission yeast *Schizosaccharomyces pombe*, useful for effective production of protease-sensitive heterologous proteins. Yeast. 2006;**23**(2):83-99

[90] Rai SK, Atwood-Moore A, Levin HL. High-frequency lithium acetate transformation of *Schizosaccharomyces pombe*. Methods in Molecular Biology. 2018;**1721**:167-177

[91] Morita T, Takegawa K. A simple and efficient procedure for transformation of *Schizosaccharomyces pombe*. Yeast. 2004;**21**(8):613-617

[92] Poloni D, Simanis VA. DMSO-sensitive conditional mutant of the fission yeast orthologue of the *Saccharomyces cerevisiae* SEC13 gene is defective in septation. FEBS Letters. 2002;**511**(1-3):85-89

[93] Rajagopalan S, Liling Z, Liu J, Balasubramanian M. The N-degron approach to create temperature-sensitive mutants in *Schizosaccharomyces pombe*. Methods. 2004;**33**(3):206-212

[94] Zhang L, Radziwon A, Reha-Krantz LJ. Targeted mutagenesis of a specific gene in yeast. Methods in Molecular Biology. 2014;**1163**:109-129

[95] Horvath P, Barrangou R. CRISPR/Cas, the immune system of bacteria and archaea. Science. 2010;**327**(5962):167-170

[96] Chylinski K, Le Rhun A, Charpentier E. The tracrRNA and Cas9 families of type II CRISPR-Cas immunity systems. RNA Biology. 2013;**10**(5):726-737

[97] Zhang XR, He JB, Wang YZ, Du LL. A cloning-free method for CRISPR/Cas9-mediated genome editing in fission yeast. G3 (Bethesda). 2018;**8**(6):2067-2077

[98] Rodríguez-López M, Cotobal C, Fernández-Sánchez O, Borbarán Bravo N, Oktriani R, Abendroth H, et al. A CRISPR/Cas9-based method and primer design tool for seamless genome editing in fission yeast. Wellcome Open Research. 2017;**1**:19

[99] Jacobs JZ, Ciccaglione KM, Tournier V, Zaratiegui M. Implementation of the CRISPR-Cas9 system in fission yeast. Nature Communications. 2014;**5**:5344

[100] Shirai A, Sadaie M, Shinmyozu K, Nakayama J. Methylation of ribosomal protein L42 regulates ribosomal function and stress-adapted cell growth. The Journal of Biological Chemistry. 2010;**285**(29):22448-22460

[101] Zhao Y, Boeke JD. Construction of designer selectable marker deletions with a CRISPR-Cas9 toolbox in *Schizosaccharomyces pombe* and new design of common entry vectors. G3 (Bethesda). 2018;**8**(3):789-796

[102] Jing X, Xie B, Chen L, Zhang N, Jiang Y, Qin H, et al. Implementation of the CRISPR-Cas13a system in fission yeast and its repurposing for precise RNA editing. Nucleic Acids Research. 2018;**46**(15):e90

[103] Allshire RC, Ekwall K. Epigenetic regulation of chromatin states in *Schizosaccharomyces pombe*. Cold Spring Harbor Perspectives in Biology. 2015;**7**:a018770

[104] Buker SM, Iida T, Buhler M, Villen J, Gygi SP, Nakayama J, et al. Two different Argonaute complexes

are required for siRNA generation and heterochromatin assembly in fission yeast. Nature Structural and Molecular Biology. 2007;**14**:200-207

[105] Verdel A, Jia S, Gerber S, Sugiyama T, Gygi S, Grewal SI, et al. RNAi mediated targeting of heterochromatin by the RITS complex. Science. 2004;**303**:672-676

[106] Petrie VJ, Wuitschick JD, Givens CD, Kosinski AM, Partridge JF. RNA interference (RNAi)-dependent and RNAi-independent association of the Chp1 chromodomain protein with distinct heterochromatic loci in fission yeast. Molecular and Cellular Biology. 2005;**25**:2331-2346

[107] Holoch D, Moazed D. RNAi in fission yeast finds new targets and new ways of targeting at the nuclear periphery. Genes and Development. 2012;**26**(8):741-745

[108] Simmer F, Buscaino A, Kos- Braun IC, Kagansky A, Boukaba A, Urano T, et al. Hairpin RNA Induces secondary small interfering RNA synthesis and silencing in trans in fission yeast. EMBO Reports. 2010;**11**(2):112-118

[109] Iida T, Nakayama J, Moazed D. siRNA-mediated heterochromatin establishment requires HP1 and is associated with antisense transcription. Molecular Cell. 2008;**31**(2):178-189

[110] Lin JS, Lai EM. Protein-protein interactions: Yeast two-hybrid system. Methods in Molecular Biology. 1615;**2017**:177-187

[111] Rodríguez-Negrete E, Bejarano ER, Castillo AG. Using the yeast two-hybrid system to identify protein-protein interactions. Methods in Molecular Biology. 2014;**1072**:241-258

[112] Chen W, Shulha HP, Ashar-Patel A, Yan J, Green KM, Query CC, et al. Endogenous U2·U5·U6 snRNA complexes in *S. pombe* are intron lariat spliceosomes. RNA. 2014;**20**(3):308-320

[113] Tanaka K, Okayama H. A pcl-like cyclin activates the Res2p-Cdc10p cell cycle "start" transcriptional factor complex in fission yeast. Molecular Biology of the Cell. 2000;**11**(9):2845-2862

[114] Chen BR, Li Y, Eisenstatt JR, Runge KW. Identification of a lifespan extending mutation in the Schizosaccharomyces pombe cyclin gene clg1+ by direct selection of long-lived mutants. PLoS One. 2013;**8**(7):e69084

[115] Chomczynski P, Sacchi N. The single-step method of RNA isolation by acid guanidinium thiocyanate-phenol-chloroform extraction: Twenty-something years on. Nature Protocols. 2006;**1**(2):581-585

[116] Ramírez J, Chávez L, Santillán JL, Guzmán S. Microarreglos de DNA. Mensaje Bioquímico, Vol XXVII. Depto Bioquímica, Fac Medicina, Universidad Nacional Autónoma de México. Cd Universitaria, México, DF, MÉXICO. 2003. Available from: http://bq.unam.mx/mensajebioquimico. ISSN-0188-137X

[117] Cheadle C, Vawter MP, Freed WJ, Becker KG. Analysis of microarray data using Z score transformation. The Journal of Molecular Diagnostics. 2003;**5**(2):73-81

[118] Rodríguez-Cruz M, Coral-Vázquez RM, Hernández-Stengele G, Sánchez R, Salazar E, Sanchez-Muñoz F, et al. Identification of putative ortholog gene blocks involved in gestant and lactating mammary gland development: A rodent cross-species microarray transcriptomics approach. International Journal of Genomics. 2013;**2013**:624681

[119] Paul M, Sanyal S, Sundaram G. Genome wide transcription profiling

of the effects of overexpression of Spc1 and its kinase dead mutant in *Schizosaccharomyces pombe*. Genomics Data. 2015;**6**:241-244

[120] Astuti RI, Watanabe D, Takagi H. Nitric oxide signaling and its role in oxidative stress response in *Schizosaccharomyces pombe*. Nitric Oxide. 2016;**52**:29-40

[121] Pataki E, Weisman R, Sipiczki M, Miklos I. fhl1 gene of the fission yeast regulates transcription of meiotic genes and nitrogen starvation response, downstream of the TORC1 pathway. Current Genetics. 2017;**63**(1):91-101

[122] Alekseyev YO, Fazeli R, Yang S, Basran R, Maher T, Miller NS, et al. A next-generation sequencing primer-how does it work and what can it do? Academic Pathology. 2018;**5**:2374289518766521

[123] Reddy RRS, Ramanujam MV. High throughput sequencing-based approaches for gene expression analysis. Methods in Molecular Biology. 1783;**2018**:299-323

[124] Ren L, McLean JR, Hazbun TR, Fields S, Vander Kooi C, Ohi MD, et al. Systematic two-hybrid and comparative proteomic analyses reveal novel yeast pre-mRNA splicing factors connected to Prp19. PLoS One. 2011;**6**(2):e16719

[125] Matsuda A, Asakawa H, Haraguchi T, Hiraoka Y. Spatial organization of the *Schizosaccharomyces pombe* genome within the nucleus. Yeast. 2017;**34**(2):55-66

[126] Rallis C, Bähler J. Cell-based screens and phenomics with fission yeast. Critical Reviews in Biochemistry and Molecular Biology. 2016;**51**(2):86-95

[127] Migeot V, Hermand D. Chromatin immunoprecipitation-polymerase chain reaction (ChIP-PCR) detects methylation, acetylation, and ubiquitylation in *S. pombe*. Methods in Molecular Biology. 2018;**1721**:25-34

# 2

# The Chemistry Behind Plant DNA Isolation Protocols

*Jina Heikrujam, Rajkumar Kishor and Pranab Behari Mazumder*

**Abstract**

Various plant species are biochemically heterogeneous in nature, a single deoxyribose nucleic acid (DNA) isolation protocol may not be suitable. There have been continuous modification and standardization in DNA isolation protocols. Most of the plant DNA isolation protocols used today are modified versions of hexadecyltri-methyl-ammonium bromide (CTAB) extraction procedure. Modification is usually performed in the concentration of chemicals used during the extraction procedure according to the plant species and plant part used. Thus, understanding the role of each chemical (*viz.* CTAB, NaCl, PVP, ethanol, and isopropanol) used during the DNA extraction procedure will benefit to set or modify protocols for more precisions. A review of the chemicals used in the CTAB method of DNA extraction and their probable functions on the highly evolved yet complex to students and researchers has been summarized.

**Keywords:** DNA extraction, CTAB buffer, polysaccharide, organic phase, RNase A

## 1. Introduction

The isolation of good-quality DNA is the prerequisite for molecular research. Maintaining yield and quality of DNA during plant DNA extraction is one of the difficult tasks compared to that of animals, because of its rigid cell wall, which is made up of cellulose along with other variable levels of chemical components such as polysaccharides, polyphenols, proteins, and lipids that act as a contaminant during DNA extraction. The amount of these components varies according to plant species, plant part used, environmental condition, and growth stage and it is very problematic when isolating DNA. For example, cereals are rich in carbohydrates whereas medicinal plants are rich in the polyphenols wherein stressed plants have higher polyphenols. These contaminants can be removed during extraction by standardizing basic DNA extraction protocol [1–3].

Generally fresh leaves aged 15–20 days are preferred for plant tissues (fresh, freeze-dried, or frozen in liquid nitrogen) and usually ruptured by mechanical force in pestle and motor or TissueLyser. If liquid nitrogen is unavailable, CTAB buffer can be used directly or prewarmed for grinding. The main objective of various DNA isolation methods is development of relatively quick, inexpensive, and consistent protocol to extract high-quality DNA with better yield. Generally, leaf samples contain large quantities of polyphenols, tannins, and polysaccharides. The basic principle of DNA isolation is disruption of the cell wall, cell membrane, and nuclear membrane to release the highly intact DNA into solution followed by precipitation of DNA and removal of the contaminating biomolecules such as the proteins, polysaccharides, lipids, phenols, and other secondary metabolites by enzymatic or chemical methods [4].

The plant DNA is extracted by either CTAB-based [5, 6] or sodium dodecyl sulfate (SDS)-based methods [7]. The majority of the protocols developed for DNA extraction are modified versions of hexadecyltrimethylammonium bromide (CTAB) extraction [8]. The role of various chemicals involved in CTAB extraction method has been described in the present communication.

## 2. CTAB buffer

The CTAB buffer mainly includes CTAB, sodium chloride (NaCl), and ethylene-diaminetetraacetic acid (EDTA) Tris2-amino-2-hydroxymethyl-1,3-propanediol (TRIS), polyvinylpyrrolidone (PVP), and β mercaptoethanol.

### 2.1 CTAB

The plant cells enclose themselves in complex polysaccharide cell wall, of which cellulose is a major constituent [9], which is crystalline in nature, due to chain-like structure and intermolecular hydrogen bonding. This can be weakened to open the cell wall, by applying mechanical force exerted during grinding along with CTAB buffer or liquid nitrogen.

Cell membrane lies next to the cell wall and cellulose and is composed of a diverse set of phospholipid molecules and proteins. It dissolves in surfactant, detergents, which are amphipathic (hydrophobic tail and hydrophilic head) in nature, very much similar to phospholipid membranes. Surfactants are characterized based on their hydrophilic group, that is, ionic, nonionic, and zwitterionic. Ionic surfac-tant has been always better in denaturing protein molecules, and thus in dissolving the membranes [10].

CTAB, a cationic detergent, constitutes a long hydrophobic hydrocarbon chain and a hydrophilic head. It forms micelle in water because of the amphipathic nature. During DNA extraction, under aqueous condition, CTAB comes in contact with the biological membrane, captures the lipids (**Figure 1**), and results in the release of nucleus, which is devoid of membrane [11]. Plant tissue, which is rich in complex polysaccharides and secondary metabolites, interfere and co-precipitate with DNA; CTAB along with some other chemicals like PVP is used to minimize the effect of these metabolites.

CTAB works differently based on the ionic strength of the solution. At a low ionic strength, it precipitates nucleic acid and acidic polysaccharides (pectin, xylan, and carrageenan), while protein and neutral polysaccharides (dextran, gum locust bean, starch, and inulin) remain in the solution [12]. However, at high ionic concentration, it gets bound to the polysaccharides and forms complexes that are removed during subsequent chloroform extraction. It also denatures or inhibits the activity of proteins and/or enzymes [13].

### 2.2 NaCl

NaCl helps to remove proteins that are bound to the DNA. It also helps to keep the proteins dissolved in the aqueous layer so they do not precipitate in the alcohol along with the DNA by neutralizing the negative charges on the DNA so that the molecules can come together.

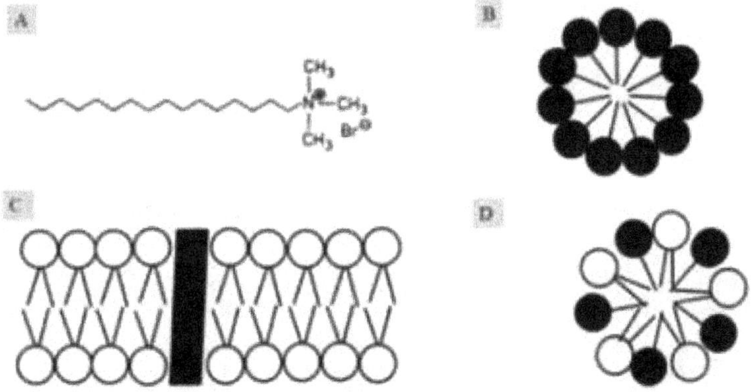


**Figure 1.**
*CTAB's role in removing membrane [25]. (A) CTAB is amphipathic in nature. It has long hydrocarbon chain (hydrophobic tail) and positively charged trimethylammonium group (hydrophilic head); (B) CTAB forms the micelle into the aqueous solution. Polar heads (hydrophilic) face outside and nonpolar (hydrophobic) hydrocarbon tail hides inside making micelle; (C) biological membrane made up of amphipathic lipids with integral protein; and (D) CTAB captures the membrane lipids and forms the hybrid micelle.*

Osmosis occurs when cell is subjected to hypo or hypertonic solution. If the cells are kept in hypotonic solution, water enters inside the cell that leads to swelling, rising internal pressure and eventually bursting. On the other hand, in a hypertonic solution, water tends to ooze out from the cell and eventually plant cell shrinks and crumples, which leads to plasmolysis. Therefore, salt concentration plays a significant role in cell lysis.

The salt concentration of more than 0.5 M provides the ionic strength needed for CTAB to precipitate polysaccharides [8, 14]. In several protocols, 1.4 M concentration of NaCl has been suggested; however, in the protocols developed for getting rid of polysaccharides, higher concentration of the NaCl and/or CTAB has been recommended.

### 2.3 Tris

Tris is a (hydroxymethyl) aminomethane with the molecular formula $(HOCH_2)_3CNH_2$, which has three primary alcohols and an amine group with a pKa of 8.1, is an effective buffer between pH 7 and 9. When the pH is adjusted to 8, with HCl, it contains a mixture of weak base and its conjugate weak acid (**Figure 2**), which can act as a buffer and further increases the permeability of the cell wall. When the cell wall and membranes are broken during tissue grinding, compartmentalization ends, cytoplasmic material is released, because of which the pH gets altered, and consequently the stability of biomolecules like nucleic acid is disturbed. The buffer plays a major role under such situations, and the Tris buffer maintains the pH of the solution.

### 2.4 EDTA

EDTA ($C_{10}H_{16}N_2O_8$) chelates divalent cations, such as $Mg^{2+}$ and $Ca^{2+}$ (**Figure 3**), which is present in the enzymes and reduces the enzyme activity of DNase and RNase. Divalent cations are the cofactors for many enzymes that increase the activity of the enzyme. For example, DNase enzyme requires $Mg^{2+}$ ions as a cofactor for its activity. Chelating $Mg^{2+}$ ions with EDTA makes enzyme DNase nonfunctional, and thereby protects the DNA. The $Mg^{2+}$ ions are also required for aggregation of nucleic acid with protein; whereas $Ca^{2+}$ions are required for cementing of cell wall's

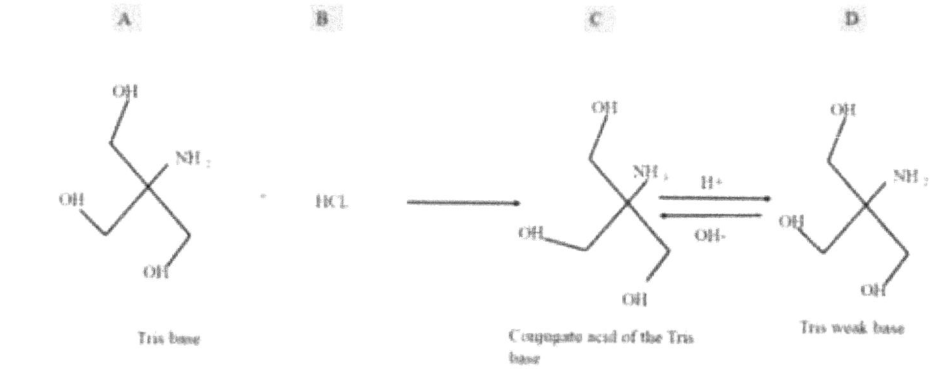


**Figure 2.**
*Tris buffer after titration of Tris base solution [25]: (A) with HCL; (B) around pH 8, it contains Tris weak base; (C) its conjugate acid; and (D) in equilibrium it acts as buffer near physiological pH range.*

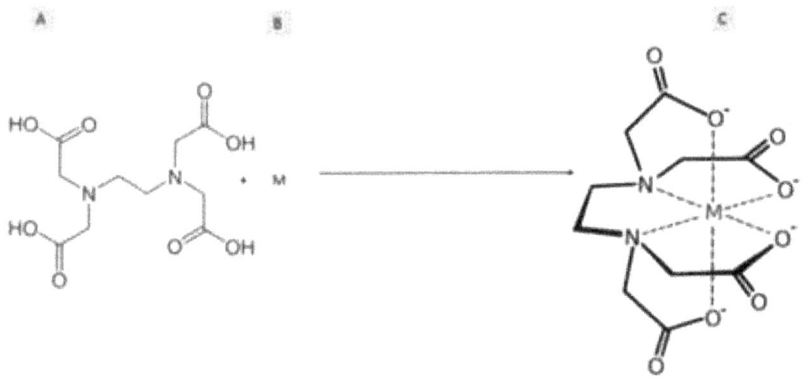


**Figure 3.**
*EDTA chelates divalent cations like magnesium and calcium [25]. (A) Structure of EDTA; (B) "M" depicts the free divalent cations like magnesium and calcium; and (C) EDTA chelates the divalent cations, thereby making unavailable to the DNase and some other activity like cell wall binding and histone-DNA complex formation.*

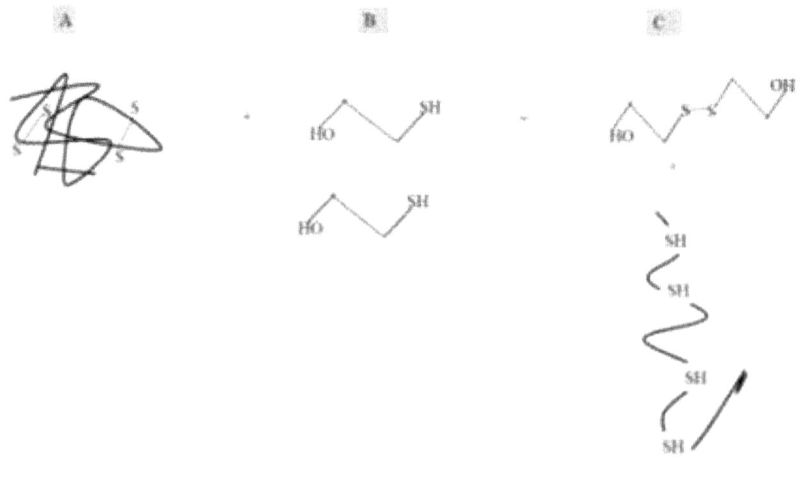

**Figure 4.**
*β-Mercaptoethanol reduces disulfide linkage of protein, thus denaturing it [25]. (A) Protein tertiary structure with disulfide bonds; (B) β-mercaptoethanol; and (C) oxidized β-mercaptoethanol and protein denatured by β-mercaptoethanol via its ability to cleave disulfide bonds.*

middle layer and membrane stability. Thus, harnessing them by EDTA results in destabilization of the enzyme's integrity.

### 2.5 β-Mercaptoethanol

Plants are rich in phenolics compounds and to get a quality DNA these should be removed. β-Mercaptoethanol ($HOCH_2CH_2SH$) is added most of the time in extraction buffers and is a strong reducing agent to clean tannins and other polyphenols present in the crude plant extract.

Globular proteins get dissolved in water. To make them insoluble, their denaturation is one of the alternatives that can be done at tertiary and quaternary structure level of protein by reducing intermolecular disulfide linkages. β-Mercaptoethanol reduces disulfide bonds of the protein (**Figure 4**) and thus the proteins are denatured.

### 2.6 PVP

PVP is added to remove phenolic compounds from plant DNA extracts. Polyphenol is a major component in medicinal plants, woody plants, and mature plant parts. It is present in the vacuole, while its oxidizing enzyme, polyphenol oxidase (PPO) is located in plastid [15]. During grinding of the tissue, compartmentalization breaks and PPO convert polyphenols into quinone, which gives brown coloration. Polyphenols bind DNA and make downstream processing difficult as they get co-precipitated with the nucleic acid. PVP removes polyphenolic contamination by binding it through hydrogen bond [16, 17]. Thus, it prevents polyphenol oxidation, and thereby browning of DNA samples [18]. When the extract is centrifuged with chloroform, PVP complexes get accumulated at the interphase.

Cell lysate mixture with CTAB buffer should be kept in the water bath at 65° C, which irreversibly inhibits enzyme DNase. After removing the sample from water bath, it should be allowed to cool at room temperature, then chloroform:isoamyl alcohol (24:1) or phenol:chloroform:isoamyl alcohol (25:24:1) shall be added. Chloroform:octanol (24:1) can also be used instead of chloroform:isoamyl alcohol (24:1).

## 3. Phenol

Phenol is an organic solvent, so it is not miscible with water and is used along with chloroform and isoamyl alcohol for purification of the DNA to remove proteins and polysaccharide contaminants. When phenol is shaken with cell extract, the nonpolar components of the cell will be fractionated in phenol, leaving polar ones in water. DNA is insoluble in phenol because phenol is a nonpolar solution. On the other side, protein has both polar and nonpolar groups present in it because of the long chain of different amino acids. Different amino acids have different groups present on their side chain. Also, the folding of the protein into the secondary, tertiary, and quaternary structure depends on the polarity of the amino acids. The bonds between amino acids are broken by the addition of phenol and protein gets denatured and ultimately the protein becomes unfolded.

Centrifugation after phenol:chloroform:isoamyl alcohol in 25:24:1 ratio steps gives three layers, that is aqueous, interphase, and at bottom organic phase. At neutral to alkaline pH, the nucleic acids are negatively charged and polar. Therefore, it is hydrophilic and remains in an aqueous phase. In aqueous solution, hydrophobic amino acid forms a protective core. However, after denaturation, nonpolar cores (hydrophobic) get exposed, causing precipitation of protein as well as some poly-saccharides at interphase.

The phenol-chloroform combination reduces the partitioning of poly (A) and mRNA into the organic phase and reduces the formation of insoluble RNA protein complexes at the interphase. Phenol retains about 10–15% of the aqueous phase, which results in a similar loss of RNA; chloroform prevents this retention of water and thus improves yields.

Only neutral phenol should be used, as acidic phenol dissolves DNA within, or phenol turns into quinones by oxidation and it forms free radical, degrading nucleic acid. Simple observation of phenol's pink color will state its acidic nature. The centrifugation after chloroform:isoamyl alcohol step should be done under room temperature, because below 15°C, CTAB/nucleic acid forms irreversible aggregates and may precipitate. During this step, the DNA shall be in aqueous phase [19].

## 4. Chloroform

Chloroform ($CHCl_3$) or trichloromethane is a nonpolar (hydrophobic) solvent, in which nonpolar proteins and lipids get dissolved to promote the partitioning of lipids and cellular debris into the organic phase, leaving isolated DNA protected in the aqueous phase. Chloroform ensures phase separation of the two liquids because it has a higher density (1.47 $g/cm^3$) and forces a sharper separation of the organic and aqueous phases, thereby assisting in the removal of the aqueous phase with minimal cross contamination from the organic phase. As chloroform is volatile in nature, it does not hinder the downstream process.

## 5. Isoamyl alcohol

Chloroform comes in contact with the air and forms gas phosgene ($COCl_2$, carbonyl chloride), which is harmful. If we simply use chloroform only, the gas entrapment causes foaming or frothing, it foams up between interphase during extraction process and makes it difficult to properly purify the DNA, which is prevented when chloroform is used along with isoamyl alcohol or isopentanol $\{(CH_3)_2CHCH_2CH_2OH\}$ or octanol $\{CH_3(CH_2)_7OH\}$ by preventing the emulsification of a solution. Isoamyl alcohol or isopentanol is not miscible in the aqueous solution because it is a long-chain aliphatic compound, containing five carbon atoms and stabilizes the interphase between organic and aqueous layer. The aqueous phase contains DNA and the organic phase contains lipid, proteins, and other impurities. Isoamyl alcohol helps to inhibit RNase activity and to help prevent the solubilization in the phenol phase of long RNA molecules with long poly (A) portions. This will increase the purity of DNA.

## 6. Ribonuclease A

Genomic DNA should be treated with Ribonuclease A (RNase A) to remove the contamination of RNA for DNA purification. RNase A is an endoribonuclease that catalyzes the hydrolysis of the 3′,5′-phosphodiester linkage of RNA at the 5′-ester bond in a two-step reaction. The first step is a transphosphorylation to give an oligonucleotide terminating in a pyrimidine 2′,3′-cyclic phosphate. The second is the hydrolysis of the cyclic phosphate to give a terminal 3′-phosphate. Numerous chemical studies have suggested that histidine 12, histidine 119, and lysine 41 are involved in the active site of the enzyme and the DNA is devoid of 2′OH group (deoxy), it remains secure (**Figures 5** and **6**) [20].

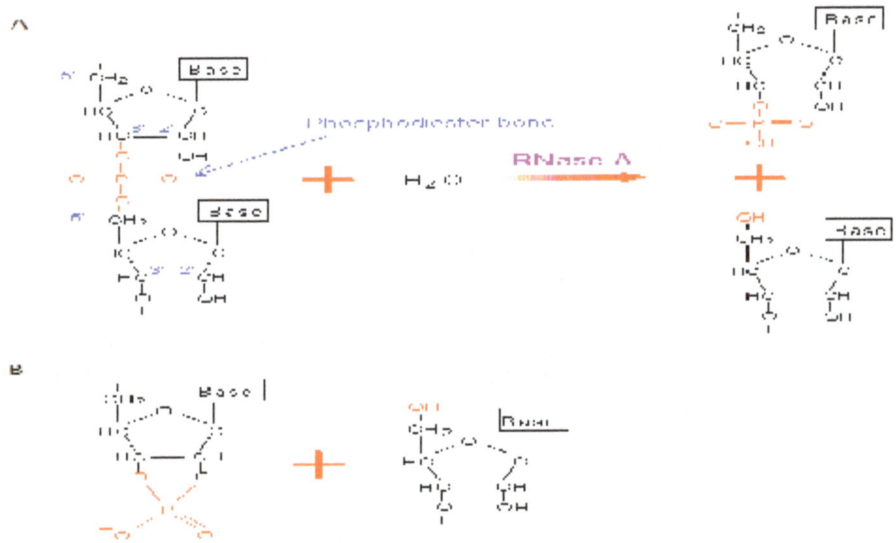

**Figure 5.**
*(A) The hydrolysis reaction catalyzed by RNase A. An RNA molecule is a chain of nucleotides linked by the phosphodiester bond, which may be cleaved by RNase [27]. (A) This figure shows only two nucleotides adjacent to the cleavage site and (B) the intermediate product (transition state) of this reaction.*

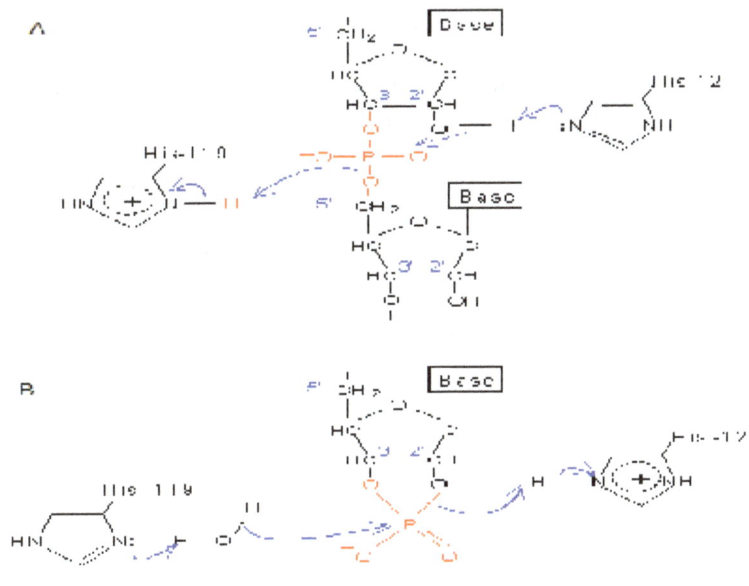

**Figure 6.**
*The catalytic mechanism of RNase A, which contains two critical residues: His-12 and His-119 [27]. (A) The transition state is formed by electron transfer from His-12 to His-119, passing through 2'-OH and (B) after the transition state is formed, the electron can move from His-119 to His-12, generating the final product. DNA lacks the critical 2'-OH and thus cannot be catalyzed by RNase A.*

## 7. Isopropanol/ethanol

Alcohol is used to precipitate the DNA out of the extraction solution, so we can wash all those salts and chemicals away and then dissolve it in our final solvent—usually water or some variant of Tris-EDTA solution. DNA remains dissolved in aqueous solution because DNA has phosphodiester backbone, which is hydrophilic in nature. Water molecule forms hydration shell around DNA by forming hydrogen bonds. Isopropanol/ethanol is used in precipitation of DNA, which breaks the hydration shell. Isopropanol is a good choice for precipitation of DNA. The amount of isopropanol requirement is less (0.6–0.7 volume of supernatant), as isopropanol has a higher capacity to reduce the dielectric constant of water than the ethanol (2–3 volume) and also requires a fair amount of salt to work. RNA which has extra 2′OH remains hydrogen bounded with water more strongly than DNA tends to stay soluble in it, thus selective precipitation of DNA can be done. Isopropanol also dissolves nonpolar solvents such as chloroform, thus the impurities form previous step can also be removed.

Using ice-cold isopropanol is generally practiced, but many researchers say that it should be used at room temperature, otherwise it will precipitate polysaccharides also [21]. Though the yield of DNA will be increased at low temperature, it may increase impurities [22].

## 8. Sodium acetate/ammonium acetate/potassium acetate/sodium chloride/lithium chloride/potassium chloride

The role of the salt in the extraction protocol is to neutralize the charges on the sugar phosphate backbone of the DNA. Sodium acetate with pH 5.2 is commonly used for precipitation of nucleic acid along with ethanol [23]. In solution, sodium acetate dissociates into $Na^+$ and $[CH_3COO]^-$. The positively charged sodium ions neutralize the negative charge on the $PO^{3-}$ groups on the sugar phosphate backbone of nucleic acids reducing repulsion between DNA molecules, making the DNA molecule far less hydrophilic, and therefore much less soluble in water. The electrostatic attraction between the $Na^+$ ions in solution and the $PO^{3-}$ ions on the nucleic acid are dictated by Coulomb's Law, which is affected by the dielectric constant of the solution. Water has a high dielectric constant, which makes it fairly difficult for the $Na^+$ and $PO^{3-}$ to come together. This is useful in aggregation and formation of tangled mass. It is also called as salting out. Nevertheless, it is not seen when salt alone is used. It requires the solution with low dielectric constant, which allows this interaction. This is affected by either ethanol or isopropanol, which has a much

**Figure 7.**
*Role of salt in DNA precipitation [25]. (A) DNA molecules in aqueous solution have the negative charge and repel each other; (B) sodium acetate dissociates into the water into sodium and acetate ion; and (C) sodium ion shields the negative charge on the DNA molecules by neutralizing it and helps in aggregation and precipitation.*

lower dielectric constant, making it much easier for $Na^+$ to interact with the $PO^{3-}$, shield its charge, and make the nucleic acid less hydrophilic, causing the DNA to drop out of solution (**Figure** 7).

## 9. Ethanol

DNA precipitate is washed again with 70% ethanol to rinse excess salt that might come along with the extraction buffers from the pellet [24], centrifuged, and ethanol is discarded, leaving DNA in the precipitate. Precipitate is air-dried or vacuumdried. Over drying should be avoided as DNA converts B form to D form, which is difficult to dissolve later [25].

## 10. Tris-EDTA (TE) buffer/sterile water

In older times in DNA isolation methods, DNA used to be stored dry and diluted when required. Nowadays, for long-term storage, it is prudent to store DNA in a buffer that maintains its pH and keeps it from getting degraded. TE buffer contains Tris (10 mM) and EDTA (1 mM), where Tris is the buffering component and EDTA the chelating component. For DNA isolation, the pH is usually set to 7.5–8.5, the slight alkalinity of TE buffer also prevents chances of acid hydrolysis that may further disrupt the stability of DNA stored in water. Tris amino constituent of TE buffer has the ability to protect DNA strands from radiation damage, in both solid state and fluid solution. As radiation produces free radicals, it may break DNA strands. Thus, in the fluid solution at ambient temperature Tris acts by scavenging hydroxyl radicals [26]. The purpose of EDTA is to chelate $Mg^{2+}$ ions in solution necessary for DNase or RNase action, thus protecting the DNA from DNases or RNase.

Sterile water can be utilized for short-duration storage of DNA. If TE buffer is used for storage of DNA, it should be diluted further with sterile water to dilute EDTA concentration for making magnesium ions available for polymerase activity during PCR because if DNA has to be sent for sequencing afterward, the buffer components in TE hinders the process. The same EDTA that chelates ions to degrade magnesium also hinders the action of DNA polymerases during PCR, which can be overcome by adding more magnesium to the master mix, or perhaps diluting the DNA sample so that the already low concentrations of EDTA do not actually disrupt PCR. In fact, in a large number of cases, they do not.

## Author details

Jina Heikrujam[1,2]*, Rajkumar Kishor[2] and Pranab Behari Mazumder[1]

1 Plant Biotechnology Laboratory, Department of Biotechnology, Assam University, Silchar, Assam, India

2 Kwaklei and Khonggunmelei Orchids Pvt. Ltd., Imphal, Manipur, India

*Address all correspondence to: jina.heikrujam@gmail.com

## References

[1] Sushma T, Tomar RS, Tripathi MK, Ashok A. Modified protocol for plant genomic DNA isolation. Indian Research Journal of Genetics and Biotechnology. 2017;**9**(4):478-485

[2] Nasir A, Rita de Cássia PR, DTC A, Marco AK. Current nucleic acid extraction methods and their implications to point-of-care diagnostics. BioMed Research International. 2017:1-13

[3] Siun CT, Beow CY. DNA, RNA, and protein extraction: The past and the present. Journal of Biomedicine and Biotechnology. 2009:1-10

[4] Kamirou CS, Timnit K, Hubert AS, Leonard A, Aliou S, Lamine BM, et al. A simple and efficient genomic DNA extraction protocol for large scale genetic analyses of plant biological systems. Plant Gene. 2015;**1**:43-45

[5] Saghai-Maroof M, Soliman K, Jorgensen RA, Allard R. Ribosomal DNA spacer-length polymorphisms in barley: Mendelian inheritance, chromosomal location, and population dynamics. Proceedings of the National Academy of Sciences. 1984;**81**(24):8014-8018

[6] Doyle JJ. Isolation of plant DNA from fresh tissue. Focus. 1990;**12**:13-15

[7] Dellaporta SL, Wood J, Hicks JB. A plant DNA minipreparation: Version II. Plant Molecular Biology Reporter. 1983;**1**:19-21

[8] Murray M, Thompson WF. Rapid isolation of high molecular weight plant DNA. Nucleic Acids Research. 1980;**8**(19):4321-4325

[9] Cosgrove DJ. Growth of the plant cell wall. Nature Reviews Molecular Cell Biology. 2005;**6**:850-861

[10] Nick C, Mary T, Malcolm C. Molecular biology of the plant cell wall: Searching for the genes that define structure, architecture and dynamics. Plant Molecular Biology. 2001;**47**:1-5

[11] Vinod K. Total genomic DNA extraction, quality check and quantitation. In: Training Programme on Classical and Modern Plant Breeding Techniques—A Hands on Training. Coimbatore, India: Tamil Nadu Agricultural University; 2004. p. 109

[12] Sambrook J, Russell DW, Russell DW. Molecular Cloning: A Laboratory Manual. Cold Spring Harbor, New York: Cold Spring Harbor Laboratory Press; 2001

[13] Available from: https://geneticeducation. co. in/ctab-dna-extraction-buffer -for- plant-dna-extraction

[14] Paterson AH, Brubaker CL, Wendel JF. A rapid method for extraction of cotton (*Gossypium spp.*) genomic DNA suitable for RFLP or PCR analysis. Plant Molecular Biology Reporter. 1993;**11**(2):122-127

[15] Holderbaum DF, Kon T, Kudo T, Guerra MP. Enzymatic browning, polyphenol oxidase activity, and polyphenols in four apple cultivars: Dynamics during fruit development. Hort Science. 2010;**45**(8):1150-1154

[16] Varma A, Padh H, Shrivastava N. Plant genomic DNA isolation: An art or a science. Biotechnology Journal. 2007;**2**(3):386-392

[17] Henry RJ. Plant Genotyping II: SNP Technology. Cambridge, MA: CABI North American Office; 2008. ISBN: 978-1-84593-382-1

[18] Loomis W. Overcoming problems of phenolics and quinones in the

isolation of plant enzymes and organelles. Methods in Enzymology. 1974;**31**:528-544

[19] de León DG. Laboratory Protocols. CIMMYT: CIMMYT Applied Molecular Genetics Laboratory; 1994

[20] Gordon CKR, Edward AD, Donella HMJS, Cohen OJM. The mechanism of action of ribonuclease. PNAS. 1969;**62**:1151-1158

[21] Shepherd LD, McLay TG. Two micro-scale protocols for the isolation of DNA from polysaccharide-rich plant tissue. Journal of Plant Research. 2011;**124**(2):311-314

[22] Michiels A, Van den Ende W, Tucker M, Van Riet L, Van Laere A. Extraction of high-quality genomic DNA from latex-containing plants. Analytical Biochemistry. 2003;**315**(1):85-89

[23] Maniatis T, Fritsch EF, Sambrook J. Molecular Cloning: A Laboratory Manual. Cold Spring Harbor, NY: Cold Spring Harbor Laboratory; 1982

[24] Tan SC, Yiap BC. DNA, RNA, and protein extraction: The past and the present. BioMed Research International. 2009:1-10

[25] Jadhav KP, Ranjani RV, Senthil N. Chemistry of plant genomic DNA extraction protocol. Bioinfolet. 2015;**12**(3A):543-548

[26] Cullis P, Elsy D, Fan S, Symons M. Marked effect of buffers on yield of single-and double-strand breaks in DNA irradiated at room temperature and at 77 K. International Journal of Radiation Biology. 1993;**63**(2):161-165

[27] Available from: http://www.web-books. c om/MoBio/Free/Ch2E3. htm

# 3

# Latest Implications of Next-Gen Sequencing in Diagnosis of Acute and Chronic Myeloid Leukemia

*Oana Maria Boldura, Cristina Petrine, Alin Mihu and Cornel Balta*

**Abstract**

The spectacular progress which was present in the past few years in the field of genome sequencing, together with the appearance on the market of some high performance devices in this field, the reduction of the costs regarding the analysis of the samples and the standardization of some protocols, has led to the establishment and introduction of the new generation of sequencing techniques in clinical diagnostic labs. An important role is played by the implementation of this technique in the oncology clinics. In this context, we found it appropriate to discuss in this chapter about the role of next-gen sequencing in determining the genetic probabilities of occurrence of oncological pathologies in the healthy population, the screening of these diseases at the population level, the diagnosis and classification of this pathology, the establishment of the therapeutic conduct using the technique, as well as the progression of the disease. In this chapter, we intend to discuss in particular the involvement of this technology in hemato-oncological diseases.

**Keywords:** next-generation sequencing, acute myeloid leukemia, chronic myeloid leukemia, molecular diagnosis

## 1. Introduction

The rapid development of the new sequencing techniques, the development of databases for the analysis and comparison of the various pathologies, as well as the reduction of the costs related to their exploitation and the total value of the cost related to the analysis, will lead to the implementation of this technique in the clinical diagnostic laboratories. Also the implementation of this technique leads to the need to develop and comply with standards, which will result in the development of valid and useful results to the clinician. This technique is used especially for the diagnosis and monitorization of hereditary diseases, being able to evaluate the genetic changes that appeared in both germline and somatic cell lines, focusing on evaluating mutations occurring at the level of a single gene, evaluating gene panels involved in the molecular pathobiology of various disorders, as well as the evaluation of the various protein-coding genes involved in this process [1]. In view of the above, we can conclude that this technique brings important information to the clinician, both about the presence of a possible mutation which will cause an affection, as well as about the metabolic interactions and the play of gene expression involved in the pathobiological mechanism of diseases, thus offering not only an early and accurate diagnosis but also the possibility of highlighting molecular targets for therapy, as well as a precise assessment of the progression of the disorders, being necessary correlated with standardized clinical and paraclinical examinations.

The Food and Drug Administration (FDA) has prepared and finalized a document guiding the use of next-generation sequencing in germline disease assess-ment, which was published on April 12, 2018, under the name of "Considerations for Design, Development and Analytical Validation of Next Generation Sequencing (NGS) - In Vitro Based Diagnostics (IVDs) Intended to Aid in the Diagnosis of Suspected Germline Diseases," thereby trying to take a step forward to standardize and introduce this technique into the current practice of diagnostic laboratories, but at the same time making sure that the patient's safety is the number one priority in front of technological innovations and possible analytical errors [2].

Lately, the progress in this field has had as a direct consequence on the drastic decrease of the cost with this analysis, at the same time developing over 55,000 genetic tests for more than 11,000 pathological conditions.

Remarkable results could be noticed after the implementation of this technology in the diagnosis and follow-up of the progression of oncological pathologies, among them being noted Hodgkin's lymphoma, breast cancer, and chronic myelogenous leukemia. Also this technique brings real benefits in the diagnosis, understanding, and study of the progression of cardiovascular diseases in direct correlation with the therapy administered to digestive, respiratory, and nervous disorders. Particular importance must also be given to the power of this new technology to aid microbiological diagnosis, as well as its usefulness in establishing the resistance of pathogenic microorganisms to various anti-infectious agents.

## 2. Acute myeloid leukemia: patho-molecular mechanism and diagnosis

The introduction of the new generation sequencing techniques has led to the development of knowledge of the mechanisms that govern the gene mechanisms that trigger and lead to the progression of malignant oncological diseases of the myeloid line. Besides the chromosomal mutations revealed by classical cytogenetic methods, observed on a larger scale, the next-generation sequencing revealed numerous other genetic alterations, which could not be revealed using classical methods. These studies have revealed some genetic similarities in various morphologically distinct conditions, suggesting that they have a similar molecular mechanism, these mechanisms being represented by cell signaling, transcription, regulation of the cell cycle, regulation of DNA methylation, changes occurred in histone regulation, RNA splicing, and alterations of the components of the sister chromatid cohesion complex [3]. All these genetic alterations can represent a starting point in the development of molecular biomarkers, which can be easily monitored by the new sequencing techniques; in **Figure 1**, the complex genetic substrate, involved in the induction and evolution of the malignant pathologies of the myeloid line is presented, hence being instrumental in establishing the diagnosis, prognosis, and therapeutic option, some of them being already validated and used in current practice.

Acute myeloid leukemia is part of the myeloid hemato-oncological disorders with high aggression, which affects the blood cells, being the leukemia with the highest weight in the adult population, having an unfavorable prognosis, despite the spectacular progression of the new therapies applied to this pathology, which leads to the in-depth study and a better understanding of the molecular processes that occur during the evolution of the disease.

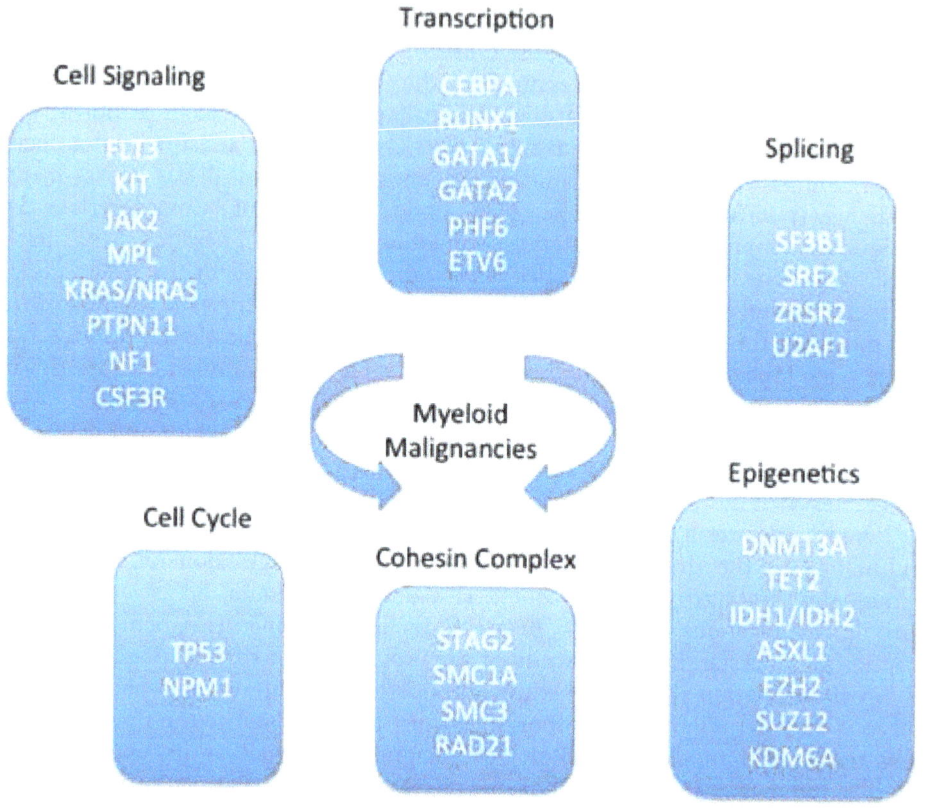


**Figure 1.**
*Metabolic pathways affected in the oncological malignant pathology of the myeloid lineage that could be used as potential biomarkers in the diagnosis of these disorders [3].*

An important first step in understanding the molecular and genetic mechanisms that governed the occurrence and progression of AML was the introduction of chromosome analysis by banding; this analysis provides relationships on the chromosome level changes, with direct resonance in the molecular pattern modification and clinically being specific for the disease. The first genetic alteration discovered and correlated with the evolution of promyelocytic leukemia was the translocation of t (15;17), but in the progression of this pathology, other chromosomal alterations appear, such as the translocation of t (8;21), inversion of 16, all of which are associ-ated with a favorable prognosis, while the association of these alterations together with the existence of structural alterations will lead to the establishment of an unfavorable prognosis.

Mutations affecting the cell lines involved in AML can be classified into two main categories: Class I mutations leading to the promotion of monoclonal cell proliferation and Class II mutations leading to the inhibition of myeloid differentiation into mature, immunocompetent cell stages; this classification is illustrated in **Figure 2** [4].

Highlighting these mutations will lead to the diagnosis of most acute myeloid leukemias with normal cytogenetic profile. In this context, the new generation sequencing is a useful element in the discovery of leukemias with normal cytogenetic profile, being able to discover even new mutations involved in the progression of this pathology. Also in the pathogenesis of these diseases, not only the DNA substrate modifications are involved, represented by the gene mutations or chromosomal translocations, but also epigenetic mechanisms that dictate the expression of these genes, such as the changes produced at the histone level and DNA methylation, may also be implicated. There are also miRNAs that can act as oncogenes or as tumor suppressor genes [5, 6]. Thus, the combination of new generation sequencing techniques, functional genomics and proteomics, will contribute to a better understanding, highlighting new therapeutic targets and new treatment modalities for AML in the future.

The alterations produced in the genetic material of the myeloid cell lines involved in the appearance of AML, lead directly to both functional and numerical alterations of these cells, as well as to structural and morphological alterations, alterations that are important in the primary hematological diagnosis, highlighted by smears made from peripheral blood samples or samples from the hematogenous spinal cord (**Figure 3**).

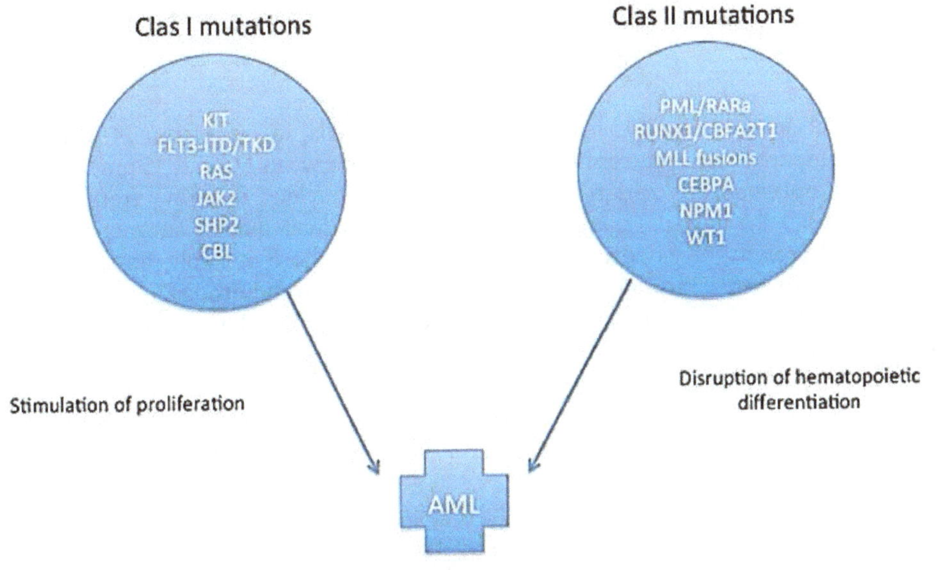


**Figure 2.**
*Molecular models of mechanism involved in acute myeloid leukemia [4].*

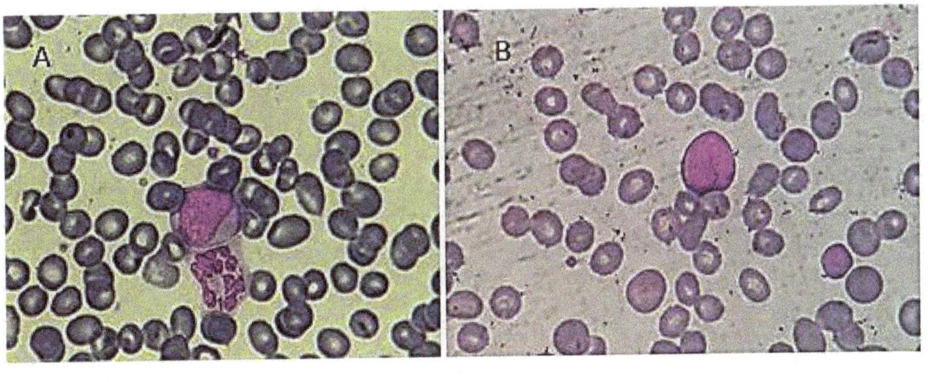


**Figure 3.**
*Peripheral blood samples from a patient with AML, May Grunwald Giemsa stain (MGG), magnification stage 1000X. Picture (A) contains a myeloblast next to a dysplastic agranular hypersegmented neutrophil and picture (B) shows a myeloblast surrounding RBC that shows discrete anisochromia, slight anisocytosis, and a polychromatophilic RBC on the bottom right.*

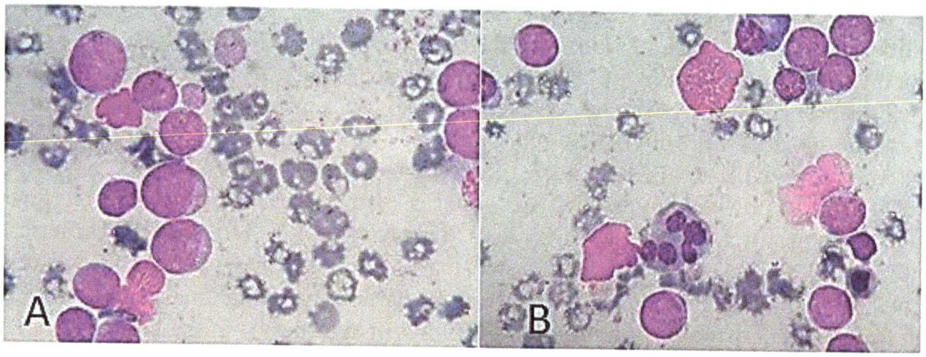


**Figure 4.**
*Bone marrow aspirate from a patient with AML: in picture (A), on the right middle part, a myeloblast displaying a Auer rod surrounded by multiple myeloblasts is shown, and in picture (B), a dysplastic oxyphilic erythroblast showing nuclear abnormalities(multiple nuclei, some being incompletely divided) along with myeloblasts (MGG X 1000) is shown.*

Myeloblasts are described as intermediate sized cells, ranging from 14 to 18 μm (when compared to a neutrophil that has the size between 10 and 15 μm) with a nucleo-cytoplasmatic ratio largely in favor of the nucleus (ranging from 4/1 to 5/1 in favor of the nucleus) containing a large, mostly oval-shaped nucleus containing very fine nonagreggated chromatin in which 2 or more nucleoli can usually be seen; the cytoplasm is strongly basophilic and may contain Auer rods [7, 8].

Auer rods are rod-shaped crystalline structures that are derived from primary granules of the myeloid cells. They are mainly reported in AML. They were first reported by John Auer in 1906 and, interestingly, were considered to be inclusions inside lymphoblasts. In the current day, they are considered of diagnostic importance to indicate both the linage and the neoplastic nature of the condition observed [7, 8] (**Figure 4**).

The European Leukemia Network recommends genetic testing of people diagnosed with acute myeloid leukemia, in order to have a complete picture of the risk initiation for each patient, thus being able to use the most appropriate strategy in the fight with this disease. The main genetic markers that can be used are: t (8;21) (q22; q22.1)/RUNX1-RUNX1T1, t (15;17)/PML-RARA, t (9;11) (p21.3; q23.3)/MLLT3-KMT2A, other types of translocation may occur that affect KMT2A genes, t (6;9) (p23; q34.1)/DEK-NUP214, inv (3) (q21.3; q26.2) or t (3;3) (q21.3; q26.2)/GATA2, inv (16) (p13.1q22) and/or t (16;16) (p13.1; q22)/CBFB-MYH11, MECOM, chromosome loss 5/5q, 7, or 17/17p, mutations in CEPBA (biallelic), NPM1, RUNX1, ASXL1 and TP53, and internal tandem duplications (ITD) in the FLT3 gene [9–11].

The most common Next-Gen Sequencing platform currently on the market are offered by Illumina (San Diego, CA, USA), being represented by iSeq100, miniSeq, miSeq, nextSeq System, HiSeq2500, HiSeqX Ten, and NovaSeq, and Thermo Fisher Scientific (Waltham, MA, USA) offers the Ion Proton System, Ion PGM System, Ion S5 System, Ion S5 XL System, Ion GeneStudio S5 System, and the HID GeneStudio S5 System.

For myeloid disease, various NGS Gene panels were designed and validated. Those panels are represented by the proposal of:

a. SureSeq myPanel™ NGS Custom AML (Oxford Gene Technology, Begbroke, Oxfordshire, UK);

b. Leuko-Vantage Myeloid Neoplasm Mutation Panel (Quest Diagnostics, Madison, NJ, USA);

c. AmpliSeq® Myeloid Sequencing Panel (Illumina); and

d. Human Myeloid Neoplasms Panel (Qiagen, Venlo, the Netherlands).

A comparison between these panels is described succinctly by Matynia et al. [12] in **Table 1**.

From this panel, it is recommended to choose the genes to be analyzed according to the choice of the diagnostician. This panel also includes RNA markers for fusion driver genes and expression genes that are not completely listed. The genes listed here come from three other panels and represent combinations between the genes listed. The Web source of this panel presents no other information about the hotspot or the complete gene [13, 14].

The use of NGS techniques can detect mutations in the pretreatment phase, thus having utility in assessing the risk of these patients, establishing the prognosis, making an appropriate and personalized therapeutic decision for each patient, and may even lead to changes in the classification of these types of diseases by WHO.

## 3. Chronic myeloid leukemia: patho-molecular mechanism and diagnosis

Chronic myeloid leukemia is a hemato-oncologic disease, in which a monoclonal line proliferates. The cells are derived from the hematopoietic stem cell, being characterized by aberrant expression of the BCR/ABL oncogene, arising from the chromosomal translocation t (9;12) (q34; q11). This mutation leads to disruption of the fusion protein, increasing the activity of tyrosine kinase, which will lead to proliferation out of control of the myeloid line [15].

This pathology represents about 15–20% of the cases of leukemia in adults, the main clinical features being represented by leukocytosis, the deviation to the left of the leukocyte formula, with splenomegaly, having an progression in three phases: the initial chronic phase, which can last several years, manifested by increasing the number of myeloid cells, but will retain their differentiation capacity and functions, most patients being asymptomatic. The second phase is an intermediate step of acceleration that can last from several months to several years, difficult to diagnose, being most often discovered following routine blood checks, which highlight the increase in the number of immature and frequent blood cells associated symptoms. In the final blastic phase, immature blood cells predominate, and the hope of survival is several months. In this phase, the genetic instability increases, accumulating these defects, and together with them will increase the resistance to drug therapy (**Figure 5**) [16].

The first line in the diagnosis of CML, right after the cell blood count (CBC) done by an automated analyzer, is the blood smear.

Morphology of the peripheral blood smear plays a crucial role in CML due to the differential diagnosis. A well-done blood smear could exclude a leukemoid reaction (in CML, basophilia is found) and can fastly assess the severity of the disease (high basophil count could be a clue that the disease is heading toward an accelerated phase that, eventually, turns into acute leukemia) as well as a blast crisis (blasts more than 20% that shows the change into acute leukemia) [17].

| | | | | |
|---|---|---|---|---|
| | | **Qiagen Human Myeloid Neoplasms Panel** | | |
| ASXL1 (full) | NPM1 (full) | CBL (full) | SETBP1 (full) | GATA2 (full) |
| CEBPA (full) | NRAS (full) | CSF3R (full) | SF3B1 (full) | HRAS (full) |
| DNMT3A (full) | RUNX1 (full) | ETV6 (full) | SRSF2 (full) | IKZF1 (full) |
| FLT3 (full) | TET2 (full) | EZH2 (full) | ZRSR2 (full) | KMD6A (full) |
| IDH1 (full) | TP53 (full) | GATA1 (full) | ABL1 (full) | MYC (full) |
| IDH2 (full) | U2AF1 (full) | JAK2 (full) | BRAF (full) | MYD88 (full) |
| KIT (full) | WT1 (full) | MPL (full) | CREBBP (full) | NF1 (full) |
| KMT2A (full) | BCOR (full) | PHF6 (full) | DDX41 (full) | NTRK3 (full) |
| KRAS (full) | CALR (full) | PTPN11 (full) | EGFR (full) | PDGFRA (full) |
| PRPF8 (full) | RB1 (full) | SH2B3 (full) | SMC1A (full) | STAG2 (full) |
| Total | | 50 genes available | | |
| | | **Illumina AmpliSeq Myeloid Panel** | | |
| ASXL1 (full) | NRAS (hotspot) | — | ZRSR2 (full) | MYC (expression) |
| CEBPA (full) | RUNX1 (full) | EZH2 (full) | ABL1 (hotspot) | MYD88 (hotspot) |
| DNMT3A (hotspot) | TET2 (full) | — | BRAF (hotspot) | NF1 (full) |
| FLT3 (hotspot) | TP53 (full) | JAK2 (hotspot) | CREBBP (fusion) | NTRK3 (fusion) |
| IDH1 (hotspot) | U2AF1 (hotspot) | MPL (hotspot) | — | PDGFRA (fusion) |
| IDH2 (hotspot) | WT1 (hotspot) | PHF6 (full) | EGFR (fusion) | PRPF8 (full) |
| KIT (hotspot) | BCOR (full) | PTPN11 (hotspot) | GATA2 (hotspot) | RB1 (full) |
| KMT2A (fusion) | CALR (full) | SETBP1 (hotspot) | HRAS (hotspot) | SH2B3 (full) |
| KRAS (hotspot) | CBL (hotspot) | SF3B1 (hotspot) | IKZF1(full) | SMC1A (expression) |
| NPM1 (hotspot) | CSF3R (hotspot) | SRSF2 (hotspot) | — | STAG2 (full) |
| Total | | 46 genes available | | |
| | | **Quest Diagnostics LeukoVantage Panel** | | |
| ASXL1 | NRAS | — | ZRSR2 | — |
| CEBPA | RUNX1 | EZH2 | — | — |
| DNMT3A | TET2 | GATA1 | — | — |
| FLT3 | TP53 | JAK2 | — | — |
| IDH1 | U2AF1 | MPL | DDX41 | — |
| IDH2 | WT1 | — | — | — |
| KIT | — | PTPN11 | — | — |
| KMT2A | CALR | SETBP1 | — | — |
| KRAS | CBL | SF3B1 | — | — |
| NPM1 | CSF3R | SRSF2 | KMD6A | — |
| Total | | 30 genes available | | |

| Oxford Gene Technology SureSeq myPanel NGS Custom AML | | | | |
|---|---|---|---|---|
| ASXL1 (full) | NRAS (full) | ETV6 (full) | — | — |
| CEBPA (full) | RUNX1 (full) | — | — | — |
| DNMT3A (full) | TET2 (full) | GATA1 (full) | — | — |
| FLT3 (full) | TP53 (full) | — | — | — |
| IDH1 (full) | U2AF1 (full) | — | — | — |
| IDH2 (full) | WT1 (full) | PHF6 (full) | — | — |
| KIT (full) | BCOR (full) | — | — | — |
| KMT2A (full) | — | — | — | — |
| KRAS (full) | — | — | — | — |
| NPM1 (full) | — | — | — | — |
| Total | 20 genes available | | | |

*The term "full" indicates all exons, and the term "hotspot" indicates hotspot exons (unmentioned here).*
*The term "fusion" indicates the RNA fusion partner; this genes has not been analyzed as DNA sequence, and the panel does not include all the RNA fusion partners.*
*The term "expression" indicates the analyses of quantification of gene expression at mRNA level, those genes not being analyzed at the DNA level.*

**Table 1.**
*Overview of commercially available NGS panels for AML with a list of included genes [12].*

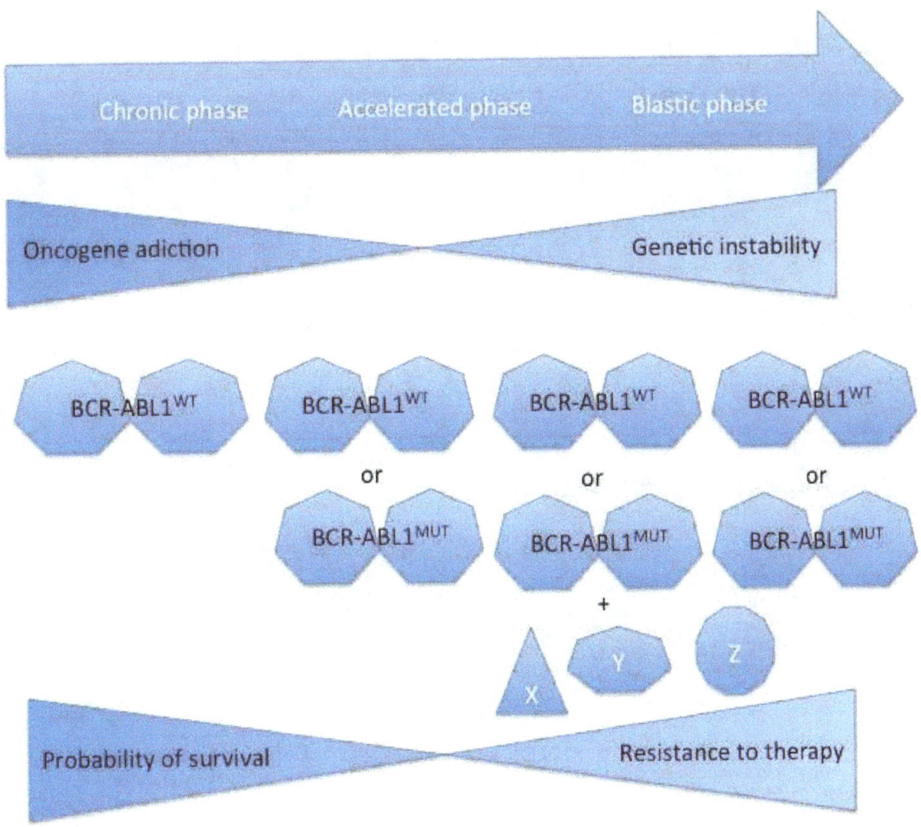


**Figure 5.**
*Evolution of chronic myeloid leukemia [16].*

The bone marrow aspirates in CML are hypercellular with an expansion of the granulocytes (e.g., neutrophils, eosinophils, and basophils) and their progenitor cells. In most cases, megakaryocytes are prominent and most often their size is increased (**Figure 6**) [18].

Most cases of chronic myeloid leukemia show the Philadelphia chromosome, which appeared after alteration of chromosome 22, produced by a reciprocal translocation t (9;22) q (34; 11), thus forming the BCR-ABL1 fusion gene, which became the main one diagnostic marker in chronic myeloid leukemia [19].

The phenotype types associated with this condition are closely correlated with the size of the proteins encoded by the different transcripts of the BCR-ABL1 fusion gene, thus noting that the most frequent rearrangements are represented by b2a2, followed by b3a2, following the rare alterations considered to occur in less than 2% of all cases of chronic myeloid leukemia. Depending on the rearrangements that have been undergone, theories have been issued that claim that transcript b2a2 is responsible for lowering the optimal response rate, and transcript b3a2 is associated with better therapeutic response and longer post treatment remission [20].

NGS is now commonly used to detect mutations in the ABL1 kinase. This represents a mechanism of CML resistance to TKIs, being about half percent of the acquired resistance for CML cases where treatment failed. It has been shown that many unique and different kinase domain mutations are associated not only with imatinib resistance but also with resistance to nilotinib (Y253H, E255K, E255V, F359V, and F359C), dasatinib (V299L, T315A, F317L, F317I, F317V, F317V, F317V, and F317C), or bosutib (Y253H, V299L, and F317V) [21].

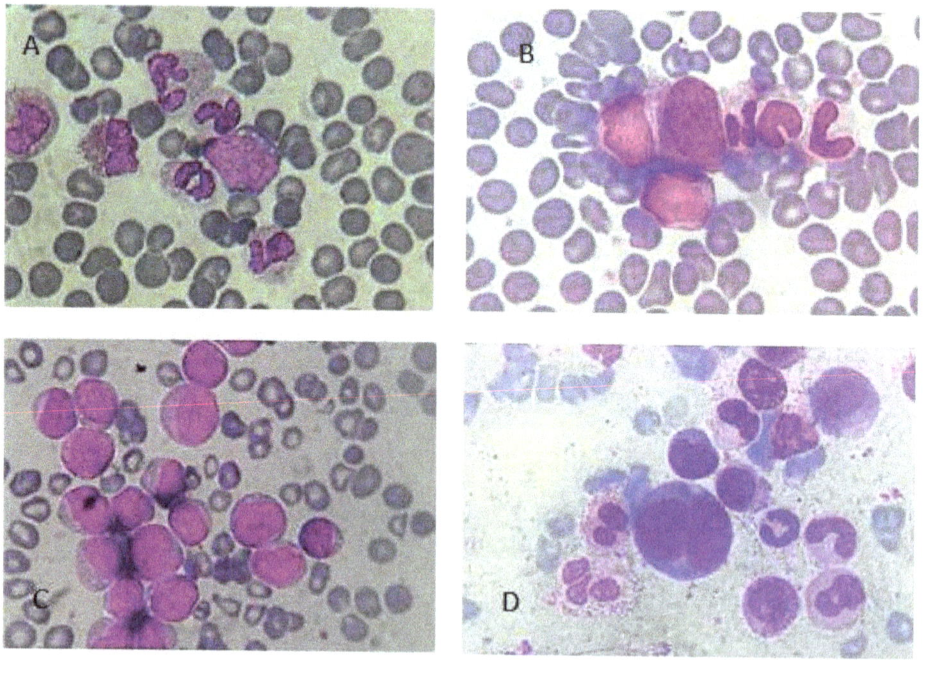


**Figure 6.**
*Chronic myeloid leukemia staining MGG, magnification stage 1000X: (A) Patient with chronic myeloid leukemia, a blast accompanied by dysplastic granulocyte precursors (above the blast, there is unsegmented hypogranular neutrophil, and above, we see an abnormal segmented neutrophil). (B) Patient with chronic myeloid leukemia that displays most of granulocyte precursors (band cell, metamyelocyte, promyelocyte, a myelocyte with nucleo-cytoplasmatic asynchronism and below a blast cell). (C) Blast crisis in chronic myeloid leukemia. (D) Bone marrow aspirate of a patient with CML showing erythroid dysplasia (basophilic giant binucleated erythroblast) accompanied by several hypogranular granulocyte precursors.*

Even if these specific TKI mutations are present in less than 10% of the cases in which the treatment fails, the identification followed by the characterization of these mutations is especially important in choosing the optimal type of TKI that could be used even when the resistance was acquired.

In the hybridization-based NGS technique, artificial oligonucleotides specially designed for BCR and ABL1 marker sequences are used. This specific amplification is followed by sequencing. By in silico analyzes, with the help of software, the fusion junctions are identified on these sequences, whether they are determined by different types of chromosomal structural rearrangements such as chromosomal translocations, inversions, or deletions.

## 4. Conclusions

In the last decades, the correct evaluation of MRD has been of major importance for the superior management of the treatment with TKIs for patients suffering from CML. This process became easier and more accurate than in the case of other hematological malignancies, precisely because the fundamental pathogenetic mechanism of this disease was studied and deciphered, which led to the use of the BCR-ABL1 transcript as the main target of all MRD tests.

An important point in the identification and treatment of CML was the development and adoption of NGS instruments in the clinical field to evaluate mutations undergone in the ABL1 kinase domain, as these mutations are responsible for resistance to the treatment of TKIs in any phase of the disease, either it is chronic or at an advanced stage. Although the Sanger sequencing method has been the most commonly used, the development of the NGS technique, which has a much higher sensitivity, enabling the detection of mutations at the subclonal level, and compound mutations that are responsible for resistance to ponatinib, has led to notable advances in diagnosis and treatment of this disease.

In the near future, it is expected that the use of SNG will be increasingly adopted for patients whose first line of treatment fails but also for those who do not respond optimally to the additional line of treatment. However, there is still much to be done in this area; for example, for allogeneic transplantation, there is no NGS-generated data available, here Sanger is still the technique used.

The existing scientific data so far indicate that the successful therapeutic management of patients with CML is, without doubt, the close collaboration between biologists, technicians, and doctors, which involves primarily the use of scientific evidence data and innovative techniques such as those based on DNA analysis. Existing and ongoing networks, online databases, and ongoing development of methods and equipment will help to achieve these goals.

## Appendices and nomenclature

| | |
|---|---|
| FDA | Food and Drug Administration |
| NGS | next-generation sequencing |
| IVDs | in vitro-based diagnostics |
| AML | acute myeloid leukemia |
| MGG | May-Grunwald-Giemsa stain |
| RBC | red blood cells |
| WHO | World Health Organization |
| CML | chronic myeloid leukemia |
| TKIs | tyrosine kinase inhibitors |

## Author details

Oana Maria Boldura[1], Cristina Petrine[2], Alin Mihu[2] and Cornel Balta[2,3]*

1 Faculty of Veterinary Medicine, Department of Chemistry, Biochemistry and Molecular Biology, Banat University of Agricultural Sciences and Veterinary Medicine "King Mihai I of Romania", Timisoara, Romania

2 Emergency County Clinical Hospital, Arad, Romania

3 Institute of Life Sciences, Vasile Goldis Western University of Arad, Romania

*Address all correspondence to: baltacornel@gmail.com

## References

[1] Hartman P, Beckman K, Silverstein K, Yohe S, Schomaker M, et al. Next generation sequencing for clinical diagnostics: Five year experience of an academic laboratory. Molecular Genetics and Metabolism Reports. 2019;**19**:100464. DOI: 10.1016/j. ymgmr.2019.100464

[2] Luh F, Yen Y. FDA guidance for next generation sequencing-based testing: Balancing regulation and innovation in precision medicine. Genomic Medicine. 2018;**3**:28. DOI: 10.1038/s41525-018-0067-2

[3] Tefferi A. Classification, diagnosis and management of myeloproliferative disorders in the JAK2V617F era. Hematology American Society of Hematology Education Program. 2006; (1):240-245. DOI: 10.1182/ asheducation-2006.1.240

[4] Thol F, Ganser A. Molecular pathogenesis of acute myeloid leukemia: A diverse disease with new perspectives. Frontiers of Medicine in China. 2010;**4**(4):356-362. DOI: 10.1007/s11684-010-0220-5

[5] Medeiros BC, Othus M, Fang M, Roulston D, Appelbaum FR. Prognostic impact of monosomal karyotype in young adult and elderly acute myeloid leukemia: The southwest oncology group (SWOG) experience. Blood. 2010;**116**(13):2224-2228

[6] Dash A, Gilliland DG. Molecular genetics of acute myeloid leukaemia. Best Practice & Research. Clinical Haematology. 2001;**14**(1):49-64

[7] Bain BJ. Auer rods or McCrae rods? American Journal of Hematology. 2011;**86**(8):689. DOI: 10.1002/ ajh.21978

[8] Barbara J. Bain, Blood Cells: A Practical Guide. 5th ed. Wiley-Blackwell. 2015. ISBN: 978-1-118-81733-9

[9] Vardiman JW, Harris NL, Brunning RD. The World Health Organization (WHO) classification of the myeloid neoplasms.Blood.2002;**100**(7):2292-2302 .DOI: 10.1182/blood-2002-04-1199

[10] Karbasian Esfahani M, Morris EL, Dutcher JP, Wiernik PH. Blastic phase of chronic myelogenous leukemia. Current Treatment Options in Oncology. 2006;**7**(3):189-199. DOI: 10.1007/s11864-006-0012-y

[11] Kon A, Shih L, Minamino M, et al. Recurrent mutations in multiple components of the cohesin complex in myeloid neoplasms. Nature Genetics. 2013;**45**:1232-1237. DOI: 10.1038/ng.2731

[12] Matynia AP, Szankasi P, Shen W, Kelley TW. Molecular genetic biomarkers in myeloid malignancies. Archives of Pathology & Laboratory Medicine. 2015;**139**(5):594-601

[13] Leisch M, Jansko B, Zaborsky N, Greil R, Pleyer L. Next generation sequencing in AML—On the way to becoming a new standard for treatment initiation and/or modulation? Cancers (Basel). 2019;**11**(2):252. DOI: 10.3390/ cancers11020252

[14] Dohner H, Estey E, Grimwade D, et al. Diagnosis and management of AML in adults: ELN recommendations from an international expert panel. Blood. 2017;**129**(4):424-447

[15] Huang X, Li Y, Shou L, Li L, Chen Z, Ye X, et al. The molecular mechanisms underlying BCR/ABL degradation in chronic myeloid leukemia cells promoted by Beclin1-mediated autophagy. Cancer Management and Research. 2019;**11**:5197-5208. DOI: 10.2147/CMAR.S202442

[16] Soverini S, Mancini M, Bavaro L, et al. Chronic myeloid leukemia: The paradigm of targeting oncogenic

tyrosine kinase signaling and counteracting resistance for successful cancer therapy. Molecular Cancer. 2018;**17**:49. DOI: 10.1186/s12943-018-0780-6

[17] Ivan Damjanov et al. Pathology Secrets. 3rd ed. 2009. pp. 161-202. Available from: https://morfopatologie.usmf.md/wp-content/blogs.dir/78/files/sites/78/2016/09/Pathology-Secrets-3rd-Edition.pdf

[18] Besa EC, Buehler B, Markman M, Sacher RA, Krishnan K. Chronic Myelogenous Leukemia. Medscape Reference. WebMD; 2014. Available from: https://emedicine.medscape.com/article/199425-treatment

[19] Izzo B, Gottardi EM, Errichiello S, Daraio F, Baratè C, Galimberti S. Monitoring chronic myeloid Leukemia: How molecular tools may drive therapeutic approaches. Frontiers in Oncology. 2019;**9**:833. DOI: 10.3389/fonc.2019.00833

[20] Shanmuganathan N, Branford S, Yong ASM, Hiwase DK, Yeung DT, Ross DM, et al. The e13a2 BCR-ABL1 transcript is associated with higher rates of molecular recurrence after treatment-free remission attempts: Retrospective analysis of the Adelaide cohort. Blood. 2018;**132**:1731. DOI: 10.1182/blood-2018-99-111083

[21] Soverini S, De Benedittis C, Machova Polakova K, Brouckova A, Horner D, Iacono M, et al. Unraveling the complexity of tyrosine kinase inhibitor-resistant populations by ultra-deep sequencing of the BCR-ABL kinase domain. Blood. 2013;**122**:1634-1648. DOI: 10.1182/blood-2013-03-487728

# 4

# Molecular Markers and Their Optimization: Addressing the Problems of Nonhomology Using Decapod COI Gene

*Deepak Jose and Mahadevan Harikrishnan*

**Abstract**

Advancements in DNA sequencing and computational technologies influenced almost all areas of biological sciences. DNA barcoding technology employed for generating nucleotide sequences (DNA barcodes) from standard gene region(s) is capable of resolving the complexities caused due to morphological characters. Thus, they complement taxonomy, population analysis, and phylogenetic and evolutionary studies. DNA barcodes are also utilized for species identification from eggs, larvae, and commercial products. Sequence similarity search using Basic Local Alignment Search Tool (BLAST) is the most reliable and widely used strategy for characterizing newly generated sequences. Similarity searches identify "homologous" gene sequence(s) for query sequence(s) by statistical calculations and provide identity scores. However, DNA barcoding relies on diverse DNA regions which differ considerably among taxa. Even, region-specific variations within barcode sequences from a single gene leading to "nonhomology" have been reported. This causes complications in specimen identification, population analysis, phylogeny, evolution, and allied studies. Hence, the selection of appropriate barcode region(s) homologous to organism of interest is inevitable. Such complications could be avoided using standardized barcode regions sequenced using optimized primers. This chapter discusses about the potential problems encountered due to the unknown/unintentional/intentional use of nonhomologous barcode regions and the need for primer optimization.

**Keywords:** DNA barcodes, BLAST, homologous gene sequences, nonhomology, standardized barcode regions, primer optimization

## 1. Introduction

Deoxyribonucleic acid (DNA) is considered as the prime genetic material of the living world as it stores complete set of information for dictating the structure of every gene product. The order of nucleotide bases (viz. adenine, guanine, cytosine, and thymine) contains these instructions for genetic inheritance along DNA [1]. "DNA sequencing" refers to a technique for understanding the language of DNA by determining the order of nucleotide bases present within the genome of organism(s) of interest [2, 3]. During the 1970s, researchers utilized two-dimensional chromatography for obtaining the first DNA sequence in laboratories. Later, dye-based sequencing methods with automated analysis were developed for easier and faster DNA sequencing. With the continued improvement in sequencing approaches, DNA sequence data derived from genes and genomes of organisms have become indispensable in basic research and allied fields.

Advancements in DNA sequencing and computational technologies influenced almost all areas of biological sciences. Taxonomy and systematics, the science for identifying organisms up to "species" level followed by classifying them based on their relationships, are also well complimented by DNA sequence database. Traditionally, "species," the basic unit of taxonomy, is distinguished on the basis of certain unified external characters within a sufficient number of specimens termed as "morphological characters" [4]. Later, morphological-type specimens were complimented with molecular data from molecular markers (allozymes, nuclear DNA, mitochondrial DNA) specifically in morphologically problematic groups [5, 6]. As molecular markers, gene type sequences (referred to as DNA barcodes) are developed using a technology called "DNA barcoding." Thus, fundamental information from conventional taxonomy is complimented with genetic information from molecular taxonomy for scientific inferences. More than a decade, this technique has been subjected for prime consideration in molecular research due of its capability to distinguish closely related species. It is also applicable to a broad spectrum of taxa for extensive biodiversity assessment studies. DNA barcoding remains as a standard method for specimen identification and allied studies, and DNA barcodes serve as an inevitable tool in understanding genetic relationships of organisms [7–11].

An ideal DNA barcode should possess certain qualities like higher universality and resolution. Since DNA barcoding relies on different DNA regions that vary between organisms (like bacteria, plants, animals, birds, etc.) [12–14], selection of barcode region is dependent on the selected sample type. DNA barcode sequences are normally compared with a DNA reference library of morphologically pre-identified vouchers to assess the rate of similarities/dissimilarities, followed by assignment of taxonomic names to unknown specimens according to the percentage of identity [15, 16]. Since homology relations are proportional to the origin and relations of taxa, focusing on molecular characters to examine homology relations is more direct than on morphology due to the discrete and "simple" nature of the characters in the latter. Thus, comparative sequence analyses are apparent to analyze the biological relationships of DNA sequences. Two major disciplines that work at both interspecific and intraspecific level are molecular phylogenetics and population genetics. Molecular phylogenetics deals typically with evolutionary relationships of different species, while population genetics is applied to characterize variations within and among populations of a single species [17, 18]. In short, DNA barcodes from standard gene region(s) compliment taxonomy, population analysis, and phylogenetic and evolutionary studies at genetic level. They are also utilized for identification of species, particularly for eggs, larvae, and commercial products [19, 20].

Among DNA barcodes, mitochondrial genes gained preference due to their higher stability, mutation rate, copy number per cell, and absence of introns that provide higher genetic information [21]. Among mitochondrial genes, cytochrome c oxidase I (COI) is considered as the primary barcode sequence for animal kingdom [10, 15]. Mitochondrial DNA (mtDNA) has been used for carrying out phylogenetic studies in a large number of animals including crustaceans, in a short span of time. Hitherto, numerous reports in support of broad benefits of DNA barcoding are available [15, 16, 22–25]. Even though DNA barcoding has completed a decade as one of the versatile techniques in addressing numerous concerns in the field of life science, authors like [26–28] have also pointed out many drawbacks with respect to this technique. A recent study [7] reported issues regarding the usage of nonhomologous barcode sequences for molecular studies. This chapter discusses on the nonhomologous barcode regions of COI gene region, available in public database (like NCBI) and issues arousing due to their unknown/intentional/unintentional use in molecular analyses. Molecular results inferred from mitochondrial COI gene sequences (amplified using "Folmer" and "Palumbi" primers) of *Macrobrachium rosenbergii* are used to demonstrate the combined effect of "nonhomologous" sequences over specimen identification, population analysis, and molecular phylogeny.

## 2. An overview of mitochondrial cytochrome oxidase subunit 1 (mtCOI) gene of *Macrobrachium rosenbergii*

DNA was first detected in mitochondria in the year 1963. It was found in association with proteins and lipids, localized to the mitochondrial matrix [29]. Almost all eukaryotic cells possess mitochondrial genome that contains genetic information utilized in systematic and population genetics for the past two decades [30]. Complete mitochondrial DNA (mtDNA) sequence having approximately 17,000 base pairs (bp) has been developed in many species, including humans [31]. Maternal inheritance, relatively rapid mutation rate, and lack of intermolecular recombination are considered as major characteristic features for their extensive use in population structure and phylogenetic studies at different taxonomic levels [21]. Hitherto, more than 1100 complete mitochondrial genome sequences or similar derivatives have been published [32]. However, crustaceans, one of the most morphologically diverse animal life forms, are represented only by limited number of complete mitochondrial sequences. Within crustaceans, decapods represent an extremely diverse group with many commercially important taxa including prawns, shrimps, lobsters, and crabs [7, 30]. Two major COI barcode regions are amplified for them using two sets of primers, namely Folmer (aka 5′ COI; LCO-HCO) [33] and Palumbi (aka 3′ COI; Jerry-Pat) [34], which are nonhomologous with limited

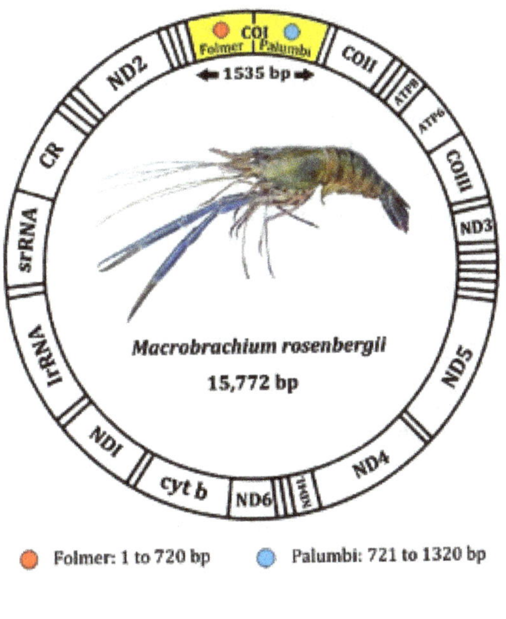


**Figure 1.**
*Gene map of Macrobrachium rosenbergii mitochondrial genome. Yellow color indicates cytochrome c oxidase subunit 1 (COI) gene having 1535 bp length with two nonhomologous regions (viz., Folmer and Palumbi).*

overlaps [7, 35]. These two regions are widely used in decapod molecular taxonomy and associated research. In public database (e.g., NCBI), several decapod species possess COI sequences derived from both these regions. Among them, the giant freshwater prawn *Macrobrachium rosenbergii* (*Crustacea*: *Decapoda*: *Palaemonidae*) is having sufficient mtDNA data including its whole genome (**Figure 1**) and other marker gene sequences [7].

## 3. Impact of nonhomologous barcode regions in molecular taxonomy and allied studies

### 3.1 Specimen identification

DNA-based taxon identification for recognition of known species and discovery of new species is reported in many studies [7, 15, 23]. Mitochondrial cytochrome oxidase I (COI) gene is recommended as an efficient DNA barcode for identifying all kinds of animals [15, 16, 23], including cryptic species [18, 24]. Pairwise comparison of COI sequences of congeneric species generates a divergence rate of >2% [23], reaching up to 3.6% in species complexes, and exceeds 5% in rare cases [15, 16, 23, 24]. The region of the 5′ end of COI ("Folmer" portion) is considered as the "DNA barcode" sequence which might be no better than that of the 3′ end of COI sequences, i.e., Palumbi sequence [7, 35, 36]. Even though these two regions are considered as related fragments, even within crustaceans [36, 37], the region-specific conservation for "Folmer" and "Palumbi" sequences creates nonhomology. This creates diversity within the same gene region, causing misinterpretations if it is used unknowingly.

Here, the results inferred from nonhomologous COI gene regions of *M. rosenbergii* are given for demonstrating the issues related to specimen identification. **Figure 2** depicts a phylogenetic tree constructed based on neighborhood joining (NJ) analysis from sequences of "Folmer" and "Palumbi" regions of *M. rosenbergii*. Tree topology could be expected to array these sequences as barcode regions of a single species (*M. rosenbergii*) within a major clade with sufficient bootstrap value corresponding to their monophyly and the selected out-group as another entity.

Results inferred from the NJ tree exhibited reciprocal monophyly in its array, differentiating "Folmer" and "Palumbi" regions as two different entities. Out-group species that was expected to have higher divergence than the rest showed affinity toward the "Palumbi" sequences of *M. rosenbergii* in the first tree. In the second case, a relationship was established between the "Folmer" sequences of *M. rosenbergii* and the out-group. This indicated a gene specific relationship between the barcode regions of the test and out-group organisms based on their homology. These results focus over the conservative nature of barcode region(s) of COI gene and its dominance over the species-level conservation within individual (genus or species).

Inferences from phylogenetic tree will also be reflected in genetic distance data since the substitution accounted for calculating intraspecific divergence within *M. rosenbergii* is higher than at interspecific level. Substitutions will be more among the nonhomologous sequences since they represent different regions within the same gene, accounting for higher distance. Out-group with homologous gene sequence could provide considerable genetic distance with the homologous sequences of species of interest (here it is *M. rosenbergii*). Further, the genetic distance provided by the nonhomologous sequences of species of interest will be more than that of the genetic distance provided by the homologous sequences of species

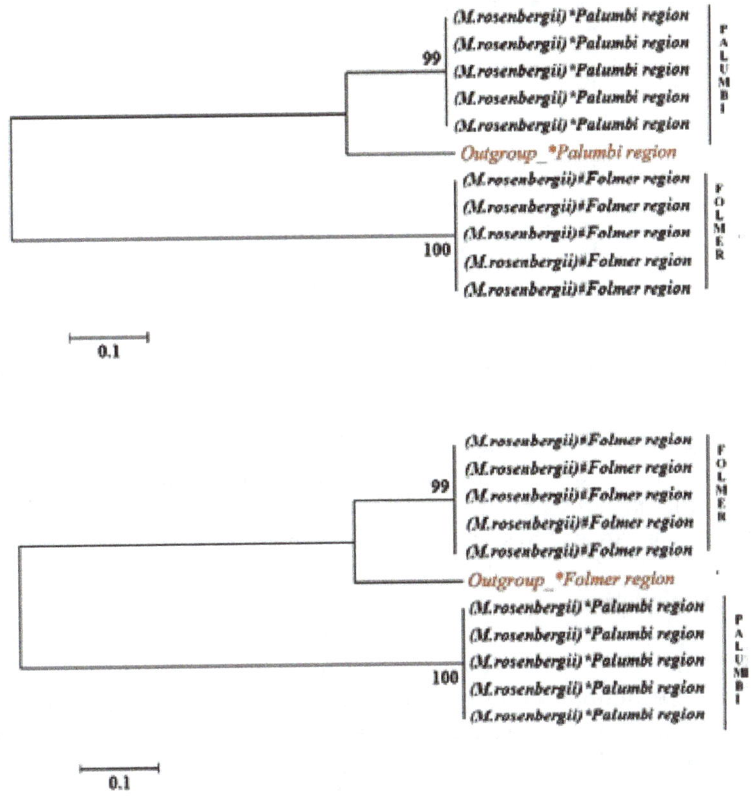


**Figure 2.**
*NJ tree showing different clades for with respect to the "nonhomologous" regions present in COI region of M. rosenbergii.*

of interest and out-group. It could be concluded that the existence of region-specific conservation within the COI barcode gene of decapod crustaceans could dominate species-level conservation causing serious errors in molecular results. Hence, the use of precise mitochondrial gene fragment(s) with respect to the homology of available nucleotide sequences is recommended for specimen identification and species confirmation for avoiding potential errors and erroneous results [7].

## 3.2 Population analysis

COI gene sequences are well considered for population analysis of many species including decapods. Here, the impact of nonhomologous regions in population studies is discussed using "Folmer" and "Palumbi" sequences of the genus *Macrobrachium*. Two populations were selected: both "Folmer" and "Palumbi" sequences were selected for Population 1, while for Population 2, only "Folmer" regions were considered. "Palumbi" sequence of an out-group organism was also considered.

The tree topology was expected to reveal only two highly diverged popula-tions of *M. rosenbergii*, viz., Populations 1 and 2. However, the exhibited pattern showed three populations, differentiating Population 1 into two populations with regard to the nonhomologous barcode regions. The Folmer regions of Populations 1 and 2 were arrayed according to their population diversity, while the "Palumbi" region of Population 1 arrayed along with the "Palumbi" region of the out-group, indicating the presence of a third population, which is virtual (**Figure 3**) and was due to region-specific conservation (for "Folmer" and "Palumbi") in COI gene.

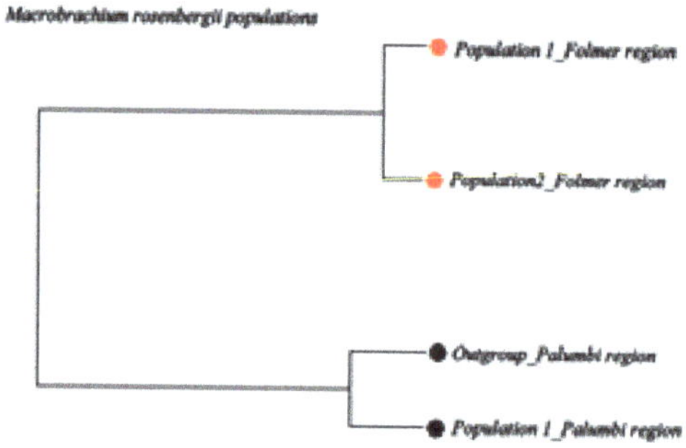


**Figure 3.**
*NJ tree generated for M. rosenbergii populations using nonhomologous barcode sequences and out-group.*

These findings were confirmed using AMOVA analysis using sequences of "Folmer" and "Palumbi" as two different populations which produced significant differences in support of the existence of two populations. This clarified that the nonhomology of barcode regions can lead to serious erroneous inferences.

### 3.3 Molecular phylogeny

The influence of nonhomology in phylogenetic studies was examined using "Folmer" and "Palumbi" sequences of *M. rosenbergii* and other selected congeneric species. Three types of sequence selections were done: (i) incorporation of "Palumbi" region of *M. rosenbergii* along with the "Palumbi" regions of selected congeneric species and out-group (**Figure 4a**), (ii) incorporation of both "Folmer" and "Palumbi" sequences of *M. rosenbergii* along with the "Palumbi" sequences of all other individuals (**Figure 4b**), and (iii) incorporation of "Folmer" sequences of *M. rosenbergii* along with "Palumbi" sequences of other species (excluding "Palumbi" region of *M. rosenbergii*) (**Figure 4c**).

Tree topology exhibited cladistic array of selected organisms in accordance with the previous findings of specimen identification and population analysis, i.e., with respect to the region-specific conservation persisting within "nonhomologous" barcode regions of COI gene. Monophyly of *Macrobrachium* species was exhibited by the first NJ tree (**Figure 4a**) in which only homologous sequences of "Palumbi" region were used. The rest of the phylogenetic trees exhibited absence of mono-phyly and erroneous cladistic array due to the impact of nonhomologous barcode regions, i.e., "Folmer" and "Palumbi" regions (**Figure-4b** and **c**). These incongruences within the phylogenetic trees will be well reflected in pairwise distance data because of the impact of nonhomologous sequences. Due to the higher rate of substitution among "Folmer" and "Palumbi" regions (as they belong to different regions of same gene), the genetic distance was higher among them even though considerable distance was accounted among the congeners. These findings demon-strated problems in molecular phylogeny by incorporating nonhomologous barcode regions of COI.

## 4. Discussion

After the first discovery of mitochondrial DNA in 1963, more than 5300 complete mtDNA sequences of different taxa were submitted in NCBI till date [29, 31, 32]. These sequences are well utilized for addressing different fields of molecular taxonomy [38]. Among the preferred gene regions of mitochondrial

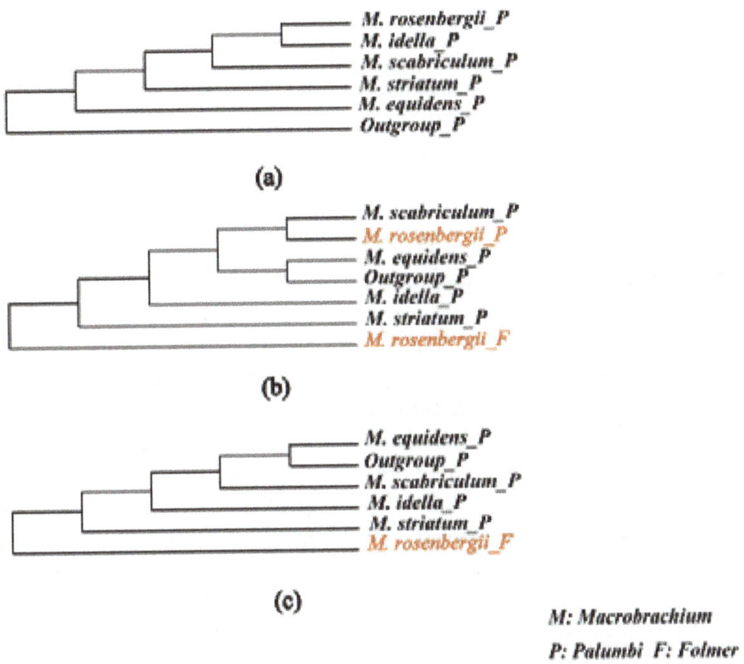


**Figure 4.**
*NJ tree based on Kimura two-parameter model (1000 bootstraps) generated using (a) "Palumbi" regions of M. rosenbergii and other selected congeners and out-group, (b) "Palumbi" and "Folmer" regions of M. rosenbergii and "Palumbi" region of other selected congeners and out-group, and (c) "Folmer" regions of M. rosenbergii and "Palumbi" region of other selected congeners and out-group.*

DNA, cytochrome c oxidase subunit 1 (COI) remains as one of the most recommended molecular markers because of its ability to generate sequence data within a reasonable time in a cost-effective way. These data could be well utilized for sorting collections into identified species, biodiversity assessments, delineation of cryptic species, detection of population structure, gene flow pattern identification, phylo-geographic studies, molecular phylogeny, evolution, etc. [7, 17, 39–41]. Altogether, this protein-coding gene mitochondrial gene has acquired great acceptance in large-scale projects of diverse taxa [42–44].

Usually, an ideal COI barcode region is reported to possess about 648–700 nucleotides that are used for similarity searches in nucleotide database for identification of known/unknown samples. Barcode of Life Data System (BOLD) refers to a freely available database which acquires analyses and releases DNA barcode data. Researchers interested in DNA barcoding and allied studies can submit sequence(s) to the public database (NCBI/DDBJ/EMBL) or the consortium for the Barcoding Life website. Similarity search using nucleotide BLAST (BLASTN) or BOLD search (www.barcodinglife.org) is usually used for identifying the status of DNA sequence of interest. This will lead to the corresponding homologous nucleotide sequence(s) for your DNA sequence that has been sequenced previously or will give homologous sequences of its close relative(s). But there exists some hiding factors that could confuse a researcher to identify a species from the available database, i.e., nonhomologous sequences with region-specific conservation (e.g., COI gene) which may alter results to a great extent. Hence, these results are to be scrutinized carefully since there may be nonhomologous sequences for a taxa of interest that will not appear in the BLAST search because of their nonhomology. Altogether, similarity searches with the available database results in top species matches, where the name of the species having reference sequence accessioned in the database or the name of the closest related taxa in the absence of reference sequence for that particular species, are enlisted [45].

This chapter has clearly discussed about the presence of "Folmer" and "Palumbi" regions of *M. rosenbergii* in its COI gene (1535 base pairs) within the 15,772 base paired complete mitochondrial DNA (NCBI accession no.'s AY659990 and NC_006880) [30]. Within, the COI gene, first approximate 720 base pairs are amplified by "Folmer" primers and the rest (approximately 721–1535) by "Palumbi" primers. Being a recommended barcode region, "Folmer" region is recognized as a universal barcode fragment, and at the same time, the other COI fragment sequenced using "Palumbi" primers dated from the early 1990s is also well-known and utilized for DNA barcoding [34, 35]. Hence, these two fragments are sequenced for studying molecular aspects of crustaceans [36, 38].

However, the presence of these two barcode regions within a single target gene (COI) with "Folmer" region as the first and "Palumbi" as second barcode region for a broad class of organisms (particularly crustaceans) could result in severe problems with respect to molecular studies. It could be helpful if full COI sequences of taxa are included as a scaffold for containing the two fragments prior to molecular analy-sis. But many organisms are still lacking the whole genome mitochondrial sequence data, and instead they are having either "Folmer" or "Palumbi" sequence(s). In such a scenario, if a "Palumbi" region for a specimen is sequenced and the public data-base is having only the "Folmer" region for the same organism, BLAST search will fail to identify that particular sequence. It will indicate only the closest organism on the basis of sequence homology. For "Folmer" region also the case will be the same if database is having "Palumbi" sequences. This mainly affects specimen identification as it could be hard for a researcher to identify the sample, particularly for those specimens lacking major morphological characters. Even in specimens with little morphological variations from its type descriptions, there could be failures in identifying the existing species causing misinterpretation of the same as a novel species. Regarding the impact of nonhomology in population analysis, the chance of misinterpretation of "nonhomologous" fragments as a different population exists.

Molecular phylogeny could also be affected from severe errors due to dual barcode regions. Even if both barcode sequences are contained within the nucleo-tide database, BLAST search will enlist sequences according to the homology of our sequence(s) only. In such cases, the chance for errors could be minimized even though the full dataset is not explored. However, there could be possibilities of missing dataset of taxa supplied to the database, due to nonhomology. Another case is that, even among congeneric species, monophyly could not be established due to the impact of "nonhomologs" (refer to **Figure 4b** and **c**). Tree topology could be altered due to the impact of dual barcode regions. As a result, relationship between morphologically similar species and species groups could be altered.

## 5. Conclusions and recommendations

GenBank accounts for an enormous amount of molecular data within which more than 90% of mtDNAs belongs to metazoans and the remaining sequences represent fungi and terrestrial plants. About 3% of available mitochondrial genomes represents protists. Despite the usually discussed issues like misidentifications and pseudogenes, lack of primer pair(s) data, particularly for certain unpublished data-set, remains as a major drawback. Moreover, designing and using of multiple primer pairs for various objective of molecular taxonomy have generated multiple DNA fragment from the same gene. Hence, under a single species name, there could be multiple DNA sequences from a single gene, which are "nonhomologous" in nature.

It is very basic that selection of nucleotide sequences could be done on the basis of their homology but still, in the present scenario, it is hard to identify homologs by BLAST search. Most trace samples, used in forensic studies, remain undetected due to the lack of standardization in barcode regions. Even though “Folmer” region is considered as a universal barcode region, there are numerous reports regarding the use of “Palumbi” sequences as better barcodes based on species specificity. This issue could be well resolved with the use of complete genome sequences which, however, are developed only for limited taxa. A better way to resolve this problem of “nonho-mology” is to provide data regarding primer pair(s) used along with the nucleotide sequence data. Even if someone is concerned about the privacy of research, they could opt the embargo period provided for releasing the nucleotide sequence data to public database. It is also true that researchers develop diverse primers for amplifying specific genes. One more recommendation is to update the nucleotide submission data after publication of the corresponding manuscript so that the entire research community could get the proper information regarding the background of the nucleotide without interfering one’s privacy. DNA barcoding has crossed the bound-aries of academics and has made use of in food authentication, medical applications, forensic science, etc. Since DNA-based analysis has become an important part, region-specific issues related to gene sequences need to be addressed and resolved. This chapter has addressed the present nature of barcode regions derived from COI and the issues related to them so that the need for primer optimization [21] could be practiced at the earliest. Regardless of a single species (*M. rosenbergii*), identifying the nature of barcode regions on additional taxa in a broad spectrum could help to make the existing nucleotide database user-friendly, even to those who are beginning their research. “Error cascades” that occurred due to bad taxonomy in science made the research communities relying on advanced technologies like DNA barcoding for accurate species identification and taxonomic assignment. So, it could be beneficial if we are able to resolve or clarify these types of confusions so that DNA barcoding could be free from “error cascades” of molecular taxonomy. It is also recommended to have an integrative taxonomy in the case of morphologically recognizable organisms as suggested by Will et al. (2005) so that error-free results regarding a species could be drawn out only using both morphological and molecular approaches.

## Acknowledgements

Authors greatly acknowledge the Kerala State Council for Science, Technology and Environment (KSCSTE) for providing financial support. We gratefully acknowledge the director, School of Industrial Fisheries, Cochin University of Science and Technology, for providing necessary facilities and support for conducting this research.

## Author details

Deepak Jose* and Mahadevan Harikrishnan
School of Industrial Fisheries, Cochin University of Science and Technology, Kerala, India

*Address all correspondence to: deepak140887@gmail.com

## References

[1] Travers A, Muskhelishvili G. DNA structure and function. The FEBS Journal. 2015;**282**(12):2279-2295. DOI: 10.1111/febs.13307

[2] Kumar BR. DNA representation. In: Anjana M, editor. DNA Sequencing Methods and Applications. 2nd ed. ExLi4EvA; 2016. p. 3

[3] Stranneheim H, Lundeberg J. Stepping stones in DNA sequencing. Biotechnology Journal. 2012;**7**(9): 1063-1073

[4] Deshmukh VD. Principles of Crustacean Taxonomy. Manual on taxonomy and identification of commercially important crustaceans of India. 2013. pp. 28-39. Available from: http://eprints.cmfri.org.in/9645/

[5] Lee MS. The molecularisation of taxonomy. Invertebrate Systematics. 2004;**18**(1):1-6

[6] Tautz D, Arctander P, Minelli A, Thomas RH, Vogler AP. A plea for DNA taxonomy. Trends in Ecology & Evolution. 2003;**18**(2):70-74

[7] Deepak J, Harikrishnan M. Non-homologous COI barcode regions: A serious concern in decapod molecular taxonomy. Mitochondrial DNA Part A. 2016;**28**(4):482-492

[8] Deepak J, Nidhin B, Anil Kumar KP, Pradeep PJ, Harikrishnan M. A molecular approach towards the taxonomy of fresh water prawns *Macrobrachium striatum* and *M. equidens* (*Decapoda, Palaemonidae*) using mitochondrial markers. Mitochondrial DNA Part A. 2016;**27**(4):2585-2593

[9] Deepak J, Rozario JV, Benjamin D, Harikrishnan M. Morphological and molecular description for *Glyphocrangon investigatoris* Wood-Mason & Alcock, 1891 emphasizing its phylogenetic relationship. Mitochondrial DNA Part A. 2016;**27**(3):2053-2057

[10] Ficetola GF, Coissac E, Zundel S, Riaz T, Shehzad W, Bessière J, et al. An in silico approach for the evaluation of DNA barcodes. BMC Genomics. 2010;**11**(1):434

[11] Zhang J. Species identification of marine fishes in China with DNA barcoding. Evidence-based Complementary and Alternative Medicine. 2011;**10**:978253

[12] Hollingsworth PM, Forrest LL, Spouge JL, Hajibabaei M, Ratnasingham S, Bankvander M, et al. A DNA barcode for land plants. Proceedings of the National Academy of Sciences USA. 2009;**106**:12794-12797

[13] Folmer O, Black M, Hoeh W, Lutz R, Vrijenhoek R. DNA primers for amplification of mitochondrial cytochrome c oxidase subunit I from diverse metazoan invertebrates. Molecular Marine Biology and Biotechnology. 1994;**3**:294-299

[14] Vences M, Thomas M, Bonett RM, Vieites DR. Deciphering amphibian diversity through DNA barcoding: Chances and challenges. Philosophical Transactions of the Royal Society, B: Biological Sciences. 2005;**360**(1462):1859-1868

[15] Hebert PD, Cywinska A, Ball SL, Dewaard JR. Biological identifications through DNA barcodes. Proceedings of the Royal Society of London. Series B: Biological Sciences. 2003a;**270**(1512):313-321

[16] Hebert PD, Penton EH, Burns JM, Janzen DH, Hallwachs W. Ten species in one: DNA barcoding reveals cryptic species in the neotropical skipper butterfly *Astraptes fulgerator* . Proceedings of

the National Academy of Sciences. 2004a;**101**(41):14812-14817

[17] Deepak J, Harikrishnan M. Evolutionary history of genus *Macrobrachium* inferred from mitochondrial markers: A molecular clock approach. Mitochondrial DNA Part A. 2018;**30**(1):92-100

[18] Hajibabaei M, Singer GA, Hebert PD, Hickey DA. DNA barcoding: How it complements taxonomy, molecular phylogenetics and population genetics. Trends in Genetics. 2007;**23**(4):167-172

[19] Deepak J, Harikrishnan M, Jenson VR, Pradeep PJ, Saswata M, Sameera S. Targeted species substitution in giant freshwater prawn trade revealed by genotyping (submitted)

[20] Tedeschi R. DNA sequencing and crop protection. In: DNA Sequencing-Methods and Applications. IntechOpen; 2012. p. p63

[21] Schubart CD. Mitochondrial DNA and decapod phylogenies: The importance of pseudogenes and primer optimization. Decapod Crustacean Phylogenetics. 2009;**47**:65

[22] Hebert PD, Gregory TR. The promise of DNA barcoding for taxonomy. Systematic Biology. 2005;**54**(5):852-859

[23] Hebert PD, Ratnasingham S, de Waard JR. Barcoding animal life: Cytochrome c oxidase subunit 1 divergences among closely related species. Proceedings of the Royal Society of London. Series B: Biological Sciences. 2003b;**270**(suppl_1):S96-S99

[24] Hebert PD, Stoeckle MY, Zemlak TS, Francis CM. Identification of birds through DNA barcodes. PLoS Biology. 2004b;**2**(10):e312

[25] Rougerie R, Haxaire J, Kitching IJ, Hebert PD. DNA barcodes and morphology reveal a hybrid hawkmoth in Tahiti (*Lepidoptera*: *Sphingidae*). Invertebrate Systematics. 2012;**26**(6):445-450

[26] Collins RA, Cruickshank RH. The seven deadly sins of DNA barcoding. Molecular Ecology Resources. 2013;**13**(6):969-975

[27] Hickerson MJ, Meyer CP, Moritz C. DNA barcoding will often fail to discover new animal species over broad parameter space. Systematic Biology. 2006;**55**(5):729-739

[28] Will KW, Mishler BD, Wheeler QD. The perils of DNA barcoding and the need for integrative taxonomy. Systematic Biology. 2005;**54**(5):844-851

[29] Nass MM, Nass S. Intramitochondrial fibers with DNA characteristics: I. fixation and electron staining reactions. The Journal of Cell Biology. 1963;**19**(3):593-611

[30] Miller AD, Murphy NP, Burridge CP, Austin CM. Complete mitochondrial DNA sequences of the decapod crustaceans *Pseudocarcinus gigas* (*Menippidae*) and *Macrobrachium rosenbergii* (*Palaemonidae*). Marine Biotechnology. 2005;**7**(4):339-349

[31] Anderson S, Bankier AT, Barrell BG, de Bruijn MH, Coulson AR, Drouin J, et al. Sequence and organization of the human mitochondrial genome. Nature. 1981;**290**:457-465

[32] Smith DR. The past, present and future of mitochondrial genomics: Have we sequenced enough mtDNAs? Briefings in Functional Genomics. 2015;**15**(1):47-54

[33] Folmer O, Black M, Hoeh W, Lutz R, Vrijenhoek R. DNA primers for amplification of mitochondrial cytochrome c oxidase subunit I from diverse metazoan invertebrates. Molecular Marine Biology and Biotechnology. 1994;**3**:294-299

[34] Palumbi SR, Benzie J. Large mitochondrial DNA differences between morphologically similar penaeid shrimp. Molecular Marine Biology and Biotechnology. 1991;**1**(1):27-34

[35] Page TJ, Steinke D. No homology means there can be no analyses; a comment on Jose & Harikrishnan. Mitochondrial DNA. Part A, DNA Mapping, Sequencing, and Analysis. 2018;**29**(2):220-221

[36] Roe AD, Sperling FA. Patterns of evolution of mitochondrial cytochrome c oxidase I and II DNA and implications for DNA barcoding. Molecular Phylogenetics and Evolution. 2007;**44**(1):325-345

[37] Lefébure T, Douady CJ, Gouy M, Gibert J. Relationship between morphological taxonomy and molecular divergence within *Crustacea*: Proposal of a molecular threshold to help species delimitation. Molecular Phylogenetics and Evolution. 2006;**40**(2):435-447

[38] Wilson K, Cahill V, Ballment E, Benzie J. The complete sequence of the mitochondrial genome of the crustacean *Penaeus monodon*: Are malacostracan crustaceans more closely related to insects than to branchiopods? Molecular Biology and Evolution. 2000;**17**(6):863-874

[39] Batista AC, Negri M, Pileggi LG, Castilho AL, Costa RC, Mantelatto FL. Inferring population connectivity across the range of distribution of the stiletto shrimp *Artemesia longinaris* Spence Bate, 1888 (*Decapoda*, *Penaeidae*) from DNA barcoding: Implications for fishery management. ZooKeys. 2014;**457**:271

[40] Buhay JE. "COI-like" sequences are becoming problematic in molecular systematic and DNA barcoding studies. Journal of Crustacean Biology. 2009;**29**(1):96-110

[41] Liu MY, Cai YX, Tzeng CS. Molecular systematics of the freshwater prawn genus macrobrachium bate, 1868 (*Crustacea*: *Decapoda*: *Palaemonidae*) inferred from mtDNA sequences, with emphasis on east Asian species. Zoological Studies. 2007;**46**(3):272-289

[42] Costa FO, DeWaard JR, Boutillier J, Ratnasingham S, Dooh RT, Hajibabaei M, et al. Biological identifications through DNA barcodes: The case of the *Crustacea*. Canadian Journal of Fisheries and Aquatic Sciences. 2007;**64**(2):272-295

[43] Hogg ID, Hebert PD. Biological identification of springtails (*Hexapoda*: *Collembola*) from the Canadian Arctic, using mitochondrial DNA barcodes. Canadian Journal of Zoology. 2004;**82**(5):749-754

[44] Ward RD, Zemlak TS, Innes BH, Last PR, Hebert PD. DNA barcoding Australia's fish species. Philosophical Transactions of the Royal Society, B: Biological Sciences. 2005;**360**(1462):1847-1857

[45] Ratnasingham S, Hebert PD. BOLD: The barcode of life data system (http://www. Barcodinglife. Org). Molecular Ecology Notes. 2007;**7**(3):355-364

# 5

# Online Automated Micro Sample Preparation for High-Performance Liquid Chromatography

*Hiroyuki Kataoka, Atsushi Ishizaki and Keita Saito*

**Abstract**

Sample preparation is one of the most labor-intensive and time-consuming operations in sample analysis. Sample preparation strategies include the exhaustive or non-exhaustive extraction of analytes from matrices. Online coupling of sample preparation with the separation system is regarded as an important goal. In-tube solid-phase microextraction (SPME) is an effective sample preparation technique that uses an open tubular fused-silica capillary column as an extraction device. In-tube SPME is useful for trace enrichment, automated sample cleanup, and rapid online analysis. Moreover, this method can be used to determine the analytes in complex matrices by direct sample injection or merely by simple sample treatment such as filtration. In-tube SPME is frequently combined with high-performance liquid chromatography (HPLC) using online column-switching techniques. Various operating systems and new sorbent materials have been reported to improve extraction efficiency, such as sorption capacity and selectivity. This chapter discusses efficient micro sample preparation techniques for HPLC, especially online auto-mated in-tube SPME.

**Keywords:** sample preparation, online automated analysis, column switching, in-tube solid-phase microextraction, high-performance liquid chromatography

## 1. Introduction

Sample analysis consists of various analytical steps, including sampling, sample preparation, separation, detection and data analysis. One of the most important steps is sample preparation, which involves the extraction, isolation and concentration of target analytes from complex matrices. Sample preparation [1–18] is the most labor-intensive and error-prone process in analytical methodology and markedly influences the reliability and accuracy of analyte determination. In addition, sample preparation requires large amounts of sample and organic solvents, and is therefore difficult to automate. An ideal sample preparation technique should be simple and fast; be specific for analytes through the efficient removal of coexist-ing components; provide high sample throughput; utilize fewer operation steps to minimize analyte losses; and be solvent-free, inexpensive, and compatible with chromatography systems. Online automated sample preparation [19–29], in which sample preparation is directly connected to chromatographic separation systems, eliminates further sample handling between the trace-enrichment and separation steps. Online automated sample preparation methods usually improve data quality, increase sample throughput, reduce costs, and improve the productivity of personnel and instruments.

In-tube solid-phase microextraction (SPME), using a capillary tube as an extraction device, was introduced by Eisert and Pawliszyn [30] to overcome the problems inherent to conventional fiber SPME. These drawbacks included fragility, low sorption capacity, bleeding from thick-film coatings on fibers, limited effectiveness for extraction of weakly volatile or thermally labile compounds not amenable to gas chromatography (GC) or GC-mass spectrometry (MS), and reduced stability in solvents used in high performance liquid chromatography (HPLC). In-tube SPME was also developed to completely automate the sample preparation process and to enable direct online coupling of in-tube SPME with HPLC using capillary column switching systems [31].

This chapter reviews the configurations and characteristics of in-tube SPME technology and discusses current and future directions, including the strategies involved in extraction efficiency and method development. The details of in-tube SPME have been described in well documented reviews [27, 32–50].

## 2. Configurations of in-tube SPME

In-tube SPME is an efficient sample preparation technique for extraction in capillary columns using stationary phases coated on the inner wall of the capillary or on the surface of the packing material (**Figure 1**). Various in-tube SPME capillary devices have been developed, such as inner wall-coated fused-silica open tubular (**Figure 1A**), fiber-packed (**Figure 1B**), sorbent-packed (**Figure 1C**), and rod-type porous monolith (**Figure 1D**) capillaries [16, 31]. The capillaries are easily fixed with the autosampler injection system, and are generally reusable without plugging or breaking the column and without exfoliation of coating materials.

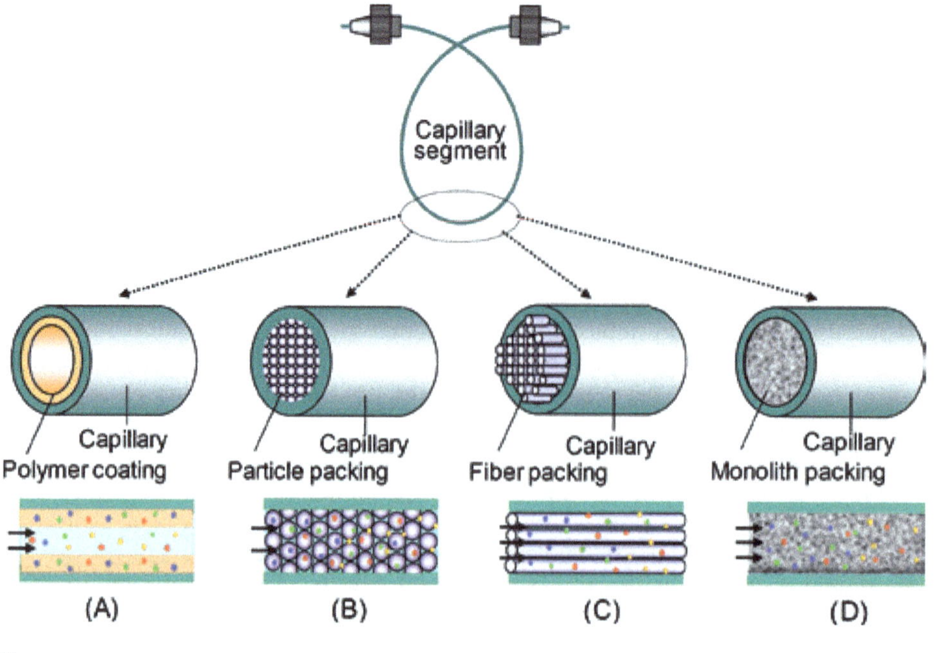


**Figure 1.**
*Capillary devices for in-tube SPME: (A) polymer coated, (B) sorbent-packed, (C) fiber-packed, and (D) monolith capillary tubes.*

## 2.1 Operating systems of in-tube SPME

Flow-through systems (**Figure 2**), in which sample solutions are continuously passed in one direction through a capillary column; or as repeated draw/ejection systems (**Figure 3**), in which sample solutions are repeatedly aspirated and dispensed from a capillary column, are used as an operating system of in-tube SPME [18]. These systems are operated by column switching techniques under computer control.

In flow-through systems, the complete analytical system consists of an automatic six-port valve, two pumps (a sample pump and a wash pump) and a liquid chromatography (LC) system. A capillary column is installed in the six-port valve or sometimes placed in the loop. Although one or two six-port valves are available, one valve mode is used more frequently than others. The procedure consists of four steps, conditioning, extracting, washing and desorbing. After conditioning of capillary column with water, the aqueous sample is pumped through the column under the load position (**Figure 2A**). Remaining matrix and residues in capillary are removed by washing with water. After switching the six-port valve to the injection position, the LC mobile phase is passed through the column (dynamic desorption), with the flow-rate of the LC pump (**Figure 2B**). The desorbed analytes are subsequently transferred to the analytical column for separation and detection. The flow-through extraction system, however, may include systematic troubles, such as contamination of the switching valve by sample matrix [18, 31, 37, 41].

Repeated draw/ejection systems include the placement of a capillary column for extraction between the injection loop and the injection needle of the autosampler. Since the sample solution moves only in the capillary, the metering pump and switching valve are not contaminated by sample matrix [18, 31, 37, 41]. A built-in UV diode array detector (DAD) or fluorescence detector (FLD) between the HPLC and the MS can enhance the multidimensional and simultaneous multi-detections, improving analyte identification. During the extraction and concentration step (**Figure 3A**), the injection syringe is programmed to repeatedly draw and eject sample solution from the vial until the concentration of the analyte reaches distribution equilibrium between the sample solution and the stationary phase. After switching the six-port valve to the injection position, the extracted analytes can be directly desorbed from the capillary coating by LC mobile phase flow (dynamic

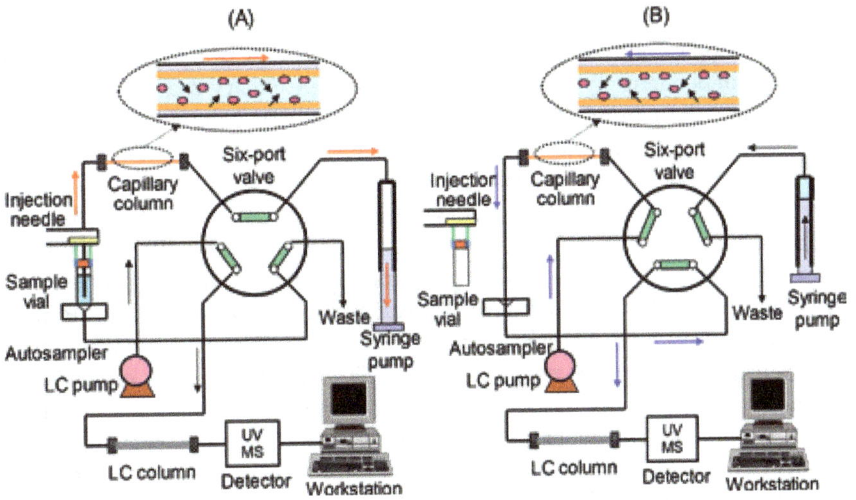


**Figure 2.**
*Schematic diagrams of a flow-through extraction system used for online in-tube SPME. (A) Load position (extraction), and (B) injection position (desorption).*

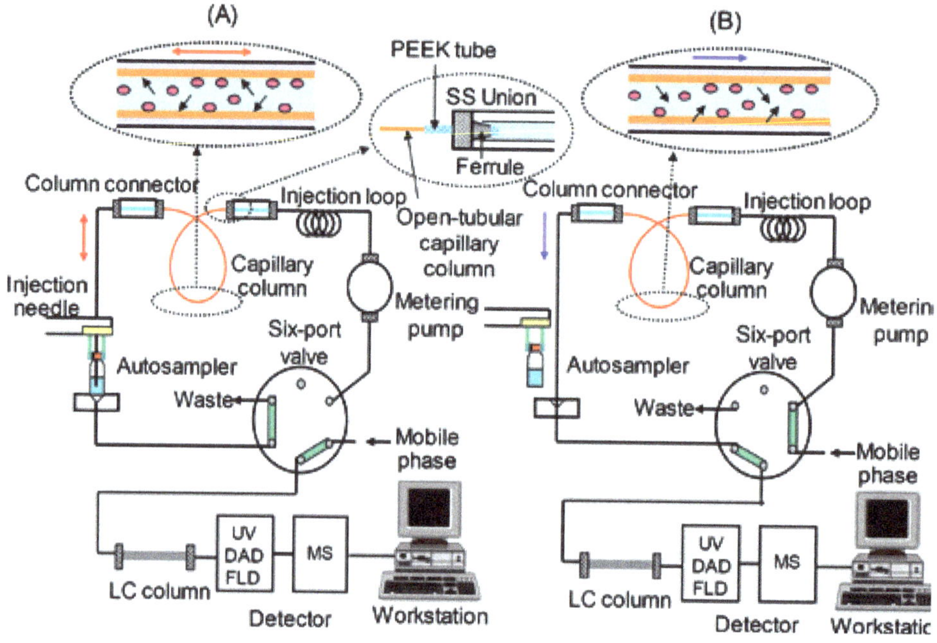


**Figure 3.**
*Schematic diagrams of a draw/eject extraction system used for online in-tube SPME (reproduced from Ref. [37]). (A) Extraction and concentration step, and (B) desorption and injection step.*

desorption) or by an aspirated desorption solvent (static desorption) (**Figure 3B**) [31]. The desorbed analytes are subsequently transferred to an LC column. The computer controls the drawing and ejection of sample solution; switching of the valves; control of peripheral equipment, such as the HPLC and MS; and analytical data processing, thus reducing labor and enhancing precision. In addition, the autosampler can automatically process a large number of samples without carry-over, because the injection needle and capillary column are washed in methanol and the mobile phase before the sample is extracted.

## 2.2 Extraction sorbent materials

The amount of analyte extracted into the stationary phase of the capillary during in-tube SPME is dependent on the characteristics of the capillary coating and the target analyte. Among the commercially available GC capillary columns, silica modi-fied columns have been found more suitable for the analysis of nonpolar compounds. Porous polymer type capillary columns such as Supel-Q PLOT (divinylbenzene polymer, film thickness 17 μm) have shown better extraction efficiencies due to their large surface area for most organic compounds than other liquid-phase type capillary columns, such as CP-Sil 5CB (100% polydimethylsiloxane, film thickness 5 μm), Quadrex 007–5 (5% phenyl polydimethylsiloxane, film thickness 12 μm), CP-Sil 19CB (14% cyanopropyl phenyl methylsilicone, film thickness 1.0 μm), and CP-Wax 52CB (polyethylene glycol, film thickness 1.2 μm). CP-Sil 19CB was superior for extraction of polyaromatic hydrocarbons, although the film layer was thin. In contrast, some compounds were effectively extracted with other PLOT type coatings, including Carboxen-1006 PLOT (carboxen molecularsives, film thickness 17 μm) and CP-Pora PLOT amine (basic modified styrene divinylbenzene polymer, film thickness 10 μm).

Several unique phases and technical solutions have been developed to improve extraction efficiency and selectivity when extended to microscale applications

[44, 51–53]. These include polypyrrole (PPY) coated capillaries; PEEK tube capillaries packed with molecularly imprinted polymer (MIP) particles [54–61]; and highly biocompatible SPME capillaries packed with alkyl-diol-silica (ADS) particles as restricted access media (RAM) [62, 63], immunosorbents [64], ionic liquids [65–67], monolithic materials [68–73], carbon nanomaterials [74–82], silica-coated magnetite ($SiO_2$-$Fe_3O_4$) [83–86], and temperature responsive polymers [87, 88]. Novel extraction sorbent materials for in-tube SPME are shown in **Figure 4**.

For example, chemically or electrochemically deposited PPY coatings have higher extraction efficiencies than commercial GC coatings due to the various types of interactions (e.g., π–π, polar, hydrogen bonding, and ionic interactions) between these multifunctional PPY coatings and the analytes. Capillary tubes have been coated with MIP, consisting of cross-linked synthetic polymers produced by copolymerizing a monomer with a cross-linker in the presence of a template molecule (**Figure 4A**), and PEEK tubes have been packed with MIP particles. By removing the template after polymerization, it is possible to leave open sites of a specific size and shape suitable for binding the same or similar chemicals in a sample.

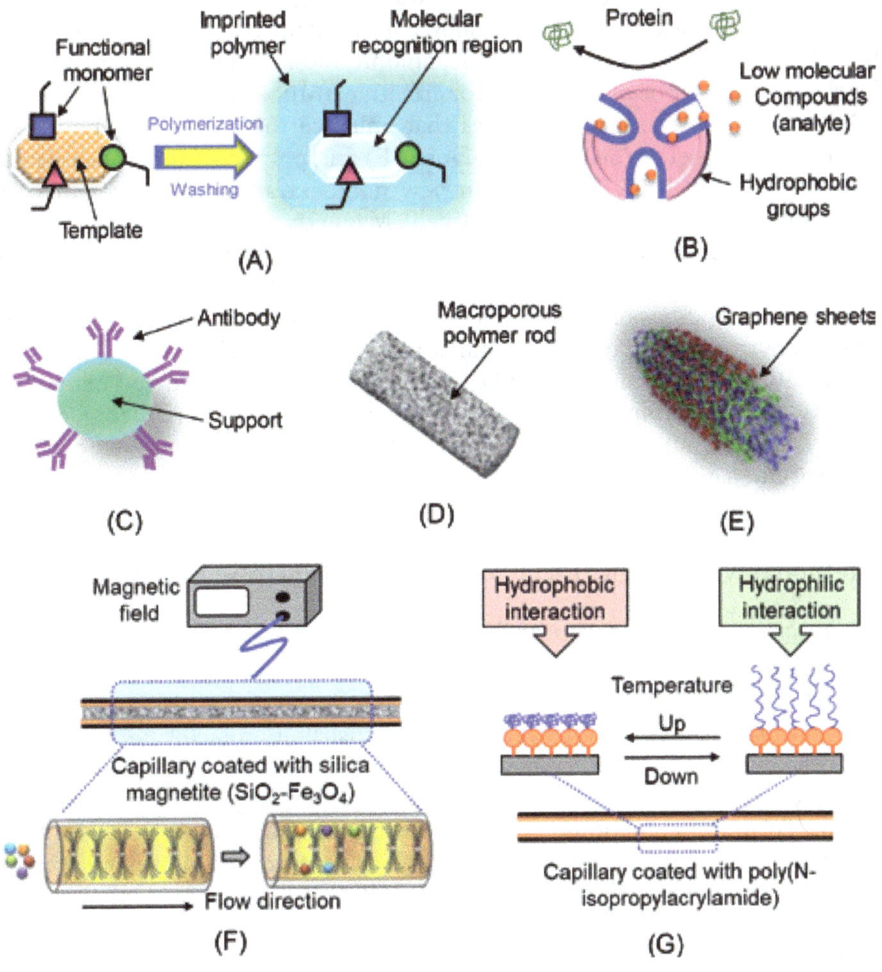


**Figure 4.**
*Novel extraction sorbent materials for in-tube SPME (eproduced from Ref. [37, 42, 84]). (A) Molecularly imprinted polymers, (B) restricted access media, (C) immunosorbents, (D) monolithic polymers, (E) carbon nanotubes, (F) silica-coated magnetite, and (G) temperature responsive polymers.*

MIPs recognize chemicals through combination of shape, hydrogen bonding, and hydrophobic and electrostatic interactions [16, 18, 31]. RAM materials possess defined diffusion barriers with small sized pores and biocompatible outer particle surfaces (**Figure 4B**). The bifunctionality of ADS particles used as a RAM SPME device can prevent fouling of the capillary by protein adsorption while simultaneously trapping the analytes in the hydrophobic porous interior. Furthermore, a simple SPME device has been fabricated for use in online immunoaffinity capillaries packed with immunosorbent materials, consisting of covalently immobilized antibodies (**Figure 4C**).

An alternative approach consists of in-tube SPME using monolithic capillary columns comprised of one piece of organic polymer or silica rods with a unique flow-through double-pore structure (**Figure 4D**). Monoliths are also highly permeable to liquids and biological samples, enabling reduced solvent use, varied support formats, and/or automation. Monolithic capillaries are especially suitable for in-tube SPME media due to the low pressure drop, allowing a high flow-rate to achieve high throughput and a total porosity greater than that of particle-packed capillaries. Hydrophobic main chains and acidic pendant groups of poly (methacrylic acid-ethylene glycol dimethacrylate) enhance the ability to extract basic analytes from aqueous matrices. The physicochemical properties of graphene-based sorbents and carbon nanotubes (**Figure 4E**) enable their use in extraction, with these combinations showing excellent results when used for in-tube SPME. In addition, various cationic, anionic and zwitterionic liquid-mediated sol–gel coatings have been developed for effective in-tube SPME.

Other innovative extractive phases that enhance the affinity of the analytes include silica magnetite ($SiO_2$-$Fe_3O_4$; **Figure 4F**) and poly (N-isopropylacrylamide; **Figure 4G**), which have been used in new microextraction processes involving magnetism and thermal energy, respectively. Magnetic and temperature controlled in-tube SPME are performed using flow-through systems, due to the need for additional equipment providing a magnetic or thermal field, which is easier to implement using flow-through devices. Other techniques include wire-in-tube SPME, using modified capillary columns with inserted stainless steel wires, and fiber-in-tube SPME, using PEEK tubes packed with fibrous rigid-rod heterocyclic polymers. These methods increase extraction efficiency by reducing capillary volume or increasing the extracting surface and have shown improved extraction efficiency when extended to microscale applications.

## 3. Method development and characteristics of in-tube SPME

### 3.1 Optimization of in-tube SPME

In-tube SPME depends on the distribution coefficient of each analyte. Extraction conditions may be optimized by increasing the distribution factor in the stationary phase. The selectivity and efficiency of extraction depend on the type of stationary phase and on the internal diameter, length, and film thickness of the capillary column. Sorption equilibrium is attained by optimizing various extrac-tion parameters for each type of analyte. These parameters include extraction rate, sample volume, sample pH, flow-rate, number of draw/eject cycles (only draw/eject system), and desorption conditions. As described in the preceding section, the choice of capillary coating is important for optimizing extraction selectivity and efficiency. Generally, low and high polarity columns selectively retain hydrophobic and hydrophilic compounds, respectively. Stationary phase consisting of a thicker film and longer column can extract larger amounts of compound, but quantitative desorption of compounds from capillary columns may be difficult. PLOT-type columns have a larger adsorption surface area and thicker film layer than liquid-phase-type columns, enabling more analytes to be extracted [16, 18].

Generally, the optimal length and internal diameter of a capillary column used in combination with HPLC is 20–80 cm and 0.25 or 0.32 mm, respectively. Although thick-film capillaries often show higher sample capacity and extraction sensitivity, it is extremely difficult to reliably bind thicker chemical coatings to the inner surfaces of fused-silica capillary tubes using conventional approaches. In contrast, thin-film capillaries can minimize the time to reach extraction equilibrium due to their low sample capacity. Capillary columns with chemically bonded or cross-linked liquid phases are very stable in water and organic solvents and can prevent loss of phase by LC mobile phase [18].

The volume of sample passed through a capillary is usually 0.2–2 mL in flow-through extraction systems, and their optimum extraction flow rates are 0.25–4 mL/min depending on the volume of the column. Although increases in the number and volume of draw/eject cycles can enhance extraction efficiency in draw/ejection systems, peak broadening is often observed [16]. Optimal conditions for a capillary column of inner diameter 0.25 mm and length 60 cm include a draw/ejection volume of 30–40 μL, a draw/ejection flow rate of 50–100 μL/min and 10–15 draw/ejection cycles. Below this rate, extractions require an inconveniently long time, and above this rate, bubbles form on the inside of the capillary, reducing extraction efficiency. Furthermore, the extraction efficiency of the analyte to the stationary phase varies with the pH of the sample solution. The presence of hydrophilic solvents such as methanol in the sample reduces the extraction efficiency. The analyte extracted on capillary coatings can be easily desorbed statically or dynamically without carryover [18].

### 3.2 Characteristics of the in-tube SPME technique

**Table 1** summarizes the characteristics of in-tube SPME. The main advantage is that the series of processes can be automated, which enables continuous extraction,

| Advantage | Disadvantage |
|---|---|
| • Minimal sample adjustment<br>• Large injection volume (flow-through system)<br>• Applicable to polar and thermolabile liquid samples<br>• Low solvent consumption<br>• Decreased handling of biohazardous samples<br>• Less sample loss due to online closed system<br>• Lower likelihood of carryover<br>• Higher mechanical stability of capillaries<br>• Reusability of capillaries without plugging or breaking<br>• Commercially available GC capillary columns<br>• Applicability of various unique adsorbents to specific and efficient extraction<br>• Easy on-line coupling with liquid chromatography<br>• Enabling of full automation by column switching<br>• Commercially available autosamplers<br>• Improvements in selectivity and sensitivity<br>• Better precision and accuracy | • Tendency of the capillary to clog<br>• Limited to particulate-free samples<br>• Stripping of non-bonding thick-film coatings<br>• Possible peak broadening<br>• Switching of valves, extraction columns, and pumps required<br>• Complicated switching system<br>• Relatively low enrichment factor<br>• Relatively long extraction time |

**Table 1.**
*Advantages and disadvantages of in-tube SPME.*

desorption and injection with column switching using a standard autosampler, and online coupling with the LC system [16, 18, 31]. In-tube SPME may be suit-able for the determination of polar and thermolabile compounds. Compared with manual techniques, automated sample-handling procedures not only shorten the total analysis time but are more accurate and precise. Automated techniques are also suitable for miniaturization, high-throughput performance, and online coupling with analytical instruments, and reduce the consumption of solvent. Online procedures can limit contact with dirty and hazardous samples, reducing sample contamination and loss. Online column-switching systems are highly sensitive due to pre-concentration resulting from the injection of large sample volumes into the extraction support without loss of chromatographic performance. The main disadvantage is that the capillaries tend to clog, which may be avoided by removing interfering phases such as particles or macromolecules by filtration or centrifugation before extraction. Although the absolute recovery rate of the in-tube SPME method is generally low, it can be extracted and concentrated reproducibly using an autosampler, and all extracts can be introduced into the LC column [16, 18, 31].

The online in-tube SPME method can be applied to polar and nonpolar compounds in liquid samples, and can be coupled with various analytical methods, such as HPLC and LC–MS. Early applications of online in-tube SPME have involved draw/eject extraction systems and commercially available open-tubular GC capillaries such as Supel Q PLOT and Carboxen 1006 PLOT capillaries. The subsequent development of various operating systems and new sorbent materials improved extraction efficiency, such as sorption capacity and selectivity, and extended the range of applications. Last decade, numerous applications of online in-tube SPME methods have been reported to many types of pharmaceutical and biomedical [86, 89–124], food [125–137], and environmental [138–178] analyses.

## 4. Conclusions and future directions

The online in-tube SPME techniques described in this chapter have many desirable features for automated separation of analytes, using column-switching techniques. These methods are especially well suited to the analysis of samples requiring significant cleanup and concentration to improve their selectivity and sensitivity, as well as being useful for high-throughput sampling. Since the in-tube SPME method using capillaries as an extraction device is useful for online sample preparation to extract and concentrate polar and non-polar compounds from aqueous solution, it has become an effective technique for convenient analysis of a wide variety of compounds in complex matrices such as biological, pharmaceutical, food and environmental samples [31]. Furthermore, various operating systems and new sorbent materials have been developed to improve extraction efficiency and sorption capacity and selectivity, and to extend the range of applications. These include MIPs, RAM, immunosorbents, monolithic materials, carbon nanoparticles, ionic liquids, temperature responsive polymers and magnetic hybrid adsorbents.

The main future direction in sample preparation is the development of more sensitive and selective extraction sorbents [31]. Chiral active phases, ionic liquids, dendrimers, aptamer modified sorbents, magnetic materials, temperature responsive materials may be available as new polymer devices for effective sample preparation. Furthermore, biomimetic coating materials including ultrasound and light responsive polymers may be available as a selective extraction device in the future. These customized coating materials, differing in type, shape, and size, are expected to result in highly efficient extraction of various samples. Biocompatible RAM and monolithic sorbents are useful for direct analysis, without pre-treatment other than dilution and centrifugation of biological samples. As another future direction, bet-ter integration of sampling/sample preparation and instrumental analysis will allow wider use of automated online analysis. Especially, the use of column-switching systems involving microextraction techniques and/or microdevices will offer convenient integration of sample preparation with various analytical instruments such as HPLC as well as other chromatographic systems, electrophoresis, direct MS, etc.

Finally, this chapter provides an overview of the configurations and characteristics of in-tube SPME technology for online automated micro sample preparation for HPLC. We hope that this chapter will serve as a guide to choosing the most effective sample preparation techniques for the analysis of various complex samples.

## Acknowledgements

This work was supported by a Grant-in-Aid for Basic Scientific Research (C, No. 17 K08259).

## Author details

Hiroyuki Kataoka*, Atsushi Ishizaki and Keita Saito
School of Pharmacy, Shujitsu University, Nishigawara, Okayama, Japan

*Address all correspondence to: hkataoka@shujitsu.ac.jp

## References

[1] Pawliszyn J, Lord H. Handbook of Sample Preparation. Hoboken: John Wiley & Sons; 2010

[2] Papadoyannis IN, Samanidou VF. Sample preparation for HPLC. In: Cazes J, editor. Encyclopedia of Chromatography. 3rd ed. Vol. III. Broken Sound Parkway: CRC Press; 2010. pp. 2090-2105

[3] Ashri NY, Abdel-Rehim M. Sample treatment based on extraction techniques in biological matrices. Bioanalysis. 2011;**3**:2003-2018

[4] Samanidou V, Kovatsi L, Fragou D, Rentifis K. Novel strategies for sample preparation in forensic toxicology. Bioanalysis. 2011;**3**:2019-2046

[5] Singleton C. Recent advances in bioanalytical sample preparation for LC-MS analysis. Bioanalysis. 2012;**4**:1123-1140

[6] Pawliszyn J. Comprehensive Sampling and Sample Preparation: Analytical Techniques for Scientists. Vol. 1-4. Burlington: Academic Press; 2012

[7] Namera A, Saito T. Recent advances in unique sample preparation techniques for bioanalysis. Bioanalysis. 2013;**5**:915-932

[8] Lundanes E, Reubsaet L, Greibrokk T. Chromatography: Basic Principles, Sample Preparations and Related Methods. Weinheim: Wiley-VCH; 2013

[9] Martíin-Esteban A. Sample preparation for chromatographic analysis. Advances in Chromatography. 2014;**51**:215-240

[10] Bylda C, Thiele R, Kobold U, Volmer DA. Recent advances in sample preparation techniques to overcome difficulties encountered during quantitative analysis of small molecules from biofluids using LC-MS/MS. Analyst. 2014;**139**:2265-2276

[11] Melnyk A, Wolska L, Namieśnik J. Coacervative extraction as a green technique for sample preparation for the analysis of organic compounds. Journal of Chromatography. A. 2014;**1339**:1-12

[12] Ribeiro C, Ribeiro AR, Maia AS, Gonçalves VM, Tiritan ME. New trends in sample preparation techniques for environmental analysis. Critical Reviews in Analytical Chemistry. 2014;**44**:142-185

[13] Vogliardi S, Tucci M, Stocchero G, Ferrara SD, Favretto D. Sample preparation methods for determination of drugs of abuse in hair samples: A review. Analytica Chimica Acta. 2015;**857**:1-27

[14] Rejczak T, Tuzimski T. Recent trends in sample preparation and liquid chromatography/mass spectrometry for pesticide residue analysis in food and related matrixes. Journal of AOAC International. 2015;**98**:1143-1162

[15] Raynie DE. Trends in sample preparation. LCGC North America. 2016;**34**:174-188

[16] Kataoka H. Sample preparation for liquid chromatography. In: Fanali S, Haddad PR, Poole CF, Riekkola ML, editors. Handbooks in Separation Science: Liquid Chromatography. 2nd ed. Vol. 2. Amsterdam: Elsevier; 2017. pp. 1-37. ISBN: 978-0-12-805392-8

[17] Barchanska H, Danek M, Sajdak M, Turek M. Review of sample preparation techniques for the analysis of selected

classes of pesticides in plant matrices. Critical Reviews in Analytical Chemistry. 2018;**48**:467-491

[18] Kataoka H. Pharmaceutical analysis/sample preparation. In: Worsfold P, Townshend A, Poole C, Miró M, editors. Encyclopedia of Analytical Science. 3rd ed. Vol. 8. Amsterdam: Elsevier; 2019. pp. 231-255. DOI: 10.1016/B978-0-12-409547-2.14358-6

[19] Wells DA. High throughput bioanalytical sample preparation: Methods and automation strategies. In: Progress in Pharmaceutical and Biomedical Analysis. Vol. 5. Amsterdam: Elsevier; 2003

[20] Zheng N, Jiang H, Zeng J. Current advances and strategies towards fully automated sample preparation for regulated LC-MS/MS bioanalysis. Bioanalysis. 2014;**6**:2441-2459

[21] Helle N, Baden M, Petersen K. Automated solid phase extraction. Methods in Molecular Biology. 2011;**747**:93-129

[22] Marquet P. LC-MS vs. GC-MS, online extraction systems, advantages of technology for drug screening assays. Methods in Molecular Biology. 2012;**902**:15-27

[23] Pan J, Zhang C, Zhang Z, Li G. Review of online coupling of sample preparation techniques with liquid chromatography. Analytica Chimica Acta. 2014;**815**:1-15

[24] Rogeberg M, Malerod H, Roberg-Larsen H, Aass C, Wilson SR. On-line solid phase extraction-liquid chromatography, with emphasis on modern bioanalysis and miniaturized systems. Journal of Pharmaceutical and Biomedical Analysis. 2014;**87**:120-129

[25] Mueller DM. The current role of on-line extraction approaches in clinical and forensic toxicology. Bioanalysis. 2014;**6**:2261-2274

[26] Wei D, Li M, King KW, Yang L. Online and automated sample extraction. Bioanalysis. 2015;**7**:2227-2233

[27] Barreiro JC, Luiz AL, Maciel SC, Maciel EV, Lanças FM. Recent approaches for on-line analysis of residues and contaminants in food matrices: A review. Journal of Separation Science. 2015;**38**:1721-1732

[28] Peng J, Tang F, Zhou R, Xie X, Li S, Xie F, et al. New techniques of on-line biological sample processing and their application in the field of biopharmaceutical analysis. Acta Pharmaceutica Sinica B. 2016;**6**:540-551

[29] Fumes BH, Andrade MA, Franco MS, Lanças FM. On-line approaches for the determination of residues and contaminants in complex samples. Journal of Separation Science. 2017;**40**:183-202

[30] Eisert R, Pawliszyn J. Automated in-tube solid-phase microextraction coupled to high-performance liquid chromatography. Analytical Chemistry. 1997;**69**:3140-3147

[31] Kataoka H. Recent advances in online column-switching sample preparation. In: Boyd R, Davies P, Decker M, et al., editors. Chemistry, Molecular Sciences and Chemical Engineering. Amsterdam: Elsevier; 2018. pp. 1-43. DOI: 10.1016/B978-0-12-409547-2.14547-0

[32] Kataoka H, Lord HL, Pawliszyn J. Applications of solid-phase microextraction in food analysis. Journal of Chromatography. A. 2000;**880**:35-62

[33] Lord H, Pawliszyn J. Microextraction of drugs. Journal of Chromatography. A. 2000;**902**:17-63

[34] Kataoka H. Automated sample preparation using in-tube solid-phase microextraction and its application—A review. Analytical and Bioanalytical Chemistry. 2002;**373**:31-45

[35] Lord HL. Strategies for interfacing solid-phase microextraction with liquid chromatography. Journal of Chromatography. A. 2007;**1152**:2-13

[36] Pragst F. Application of solid-phase microextraction in analytical toxicology. Analytical and Bioanalytical Chemistry. 2007;**388**:1393-1414

[37] Kataoka H, Ishizaki A, Nonaka Y, Saito K. Developments and applications of capillary microextraction techniques: A review. Analytica Chimica Acta. 2009;**655**:8-29

[38] Kataoka H. Recent developments and applications of microextraction techniques in drug analysis. Analytical and Bioanalytical Chemistry. 2010;**396**:339-364

[39] Kataoka H. New trends in sample preparation for analysis of plant-derived medicines. Current Organic Chemistry. 2010;**14**:1698-1713

[40] Campíns-Falcó P, Verdú-Andrés J, Sevillano-Cabeza A, Herráez-Hernández R, Molins-Legua C, Moliner-Martinez Y. In-tube solid-phase microextraction coupled by in valve mode to capillary LC-DAD: Improving detectability to multiresidue organic pollutants analysis in several whole waters. Journal of Chromatography. A. 2010;**1217**:2695-2702

[41] Kataoka H, Saito K. Recent advances in SPME techniques in biomedical analysis. Journal of Pharmaceutical and Biomedical Analysis. 2011;**54**:926-950

[42] Kataoka H. Current developments and future trends in solid-phase microextraction techniques for pharmaceutical and biomedical analyses. Analytical Sciences. 2011;**27**:893-905

[43] Kataoka H. Solid-phase microextraction: Biomedical applications. In: Jan Reedijk J, editor. Chemistry, Molecular Sciences and Chemical Engineering. Oxford: Elsevier, Academic Press; 2014. pp. 1-15

[44] Padrón ME, Afonso-Olivares C, Sosa-Ferrera Z, Santana-Rodríguez JJ. Microextraction techniques coupled to liquid chromatography with mass spectrometry for the determination of organic micropollutants in environmental water samples. Molecules. 2014;**19**:10320-10349

[45] Queiroz ME, Melo LP. Selective capillary coating materials for in-tube solid-phase microextraction coupled to liquid chromatography to determine drugs and biomarkers in biological samples: A review. Analytica Chimica Acta. 2014;**826**:1-11

[46] Kataoka H. SPME techniques for biomedical analysis. Bioanalysis. 2015;**7**:2135-2144

[47] Moliner-Martinez Y, Herráez-Hernández R, Verdú-Andrés J, Molins-Legua C, Campíns-Falcó P. Recent advances of in-tube solid-phase microextraction. Trends in Analytical Chemistry. 2015;**71**:205-213

[48] Kataoka H, Ishizaki A, Saito K. Solid-phase microextraction for pharmaceutical and biomedical analysis. Analytical Methods. 2016;**8**:5773-5788

[49] González-Fuenzalida RA, Moliner-Martínez Y, Molins-Legua C, Campíns-Falcó P. Miniaturized liquid chromatography coupled on-line to in-tube solid-phase microextraction for characterization of metallic

nanoparticles using plasmonic measurements. A tutorial. Analytica Chimica Acta. 2019;**1045**:23-41

[50] Eugenia M, Queiroz C, de Souza ID, Marchioni C. Current advances and applications of in-tube solid-phase microextraction. Trends in Analytical Chemistry. 2019;**111**:261-278

[51] Augusto F, Carasek E, Silva RG, Rivellino SR, Batista AD, Martendal E. New sorbents for extraction and microextraction techniques. Journal of Chromatography. A. 2010;**1217**:2533-2542

[52] Fumes BH, Silva MR, Andrade FN, Nazario CED, Lanças FM. Recent advances and future trends in new materials for sample preparation. Trends in Analytical Chemistry. 2015;**71**:9-25

[53] Maciel EVS, de Toffoli AL, Lanças FM. Current status and future trends on automated multidimensional separation techniques employing sorbent-based extraction columns. Journal of Separation Science. 2018;**42**:258-272

[54] Tamayo FG, Turiel E, Martín-Esteban A. Molecularly imprinted polymers for solid-phase extraction and solid-phase microextraction: Recent developments and future trends. Journal of Chromatography. A. 2007;**1152**:32-40

[55] Pichon V. Selective sample treatment using molecularly imprinted polymers. Journal of Chromatography. A. 2007;**1152**:41-53

[56] Haginaka J. Molecularly imprinted polymers as affinity-based separation media for sample preparation. Journal of Separation Science. 2009;**32**:1548-1565

[57] Turiel E, Martín-Esteban A. Molecularly imprinted polymers for sample preparation: A review. Analytica Chimica Acta. 2010;**668**:87-99

[58] Moein MM, Abdel-Rehim M. Molecularly imprinted polymers for on-line extraction techniques. Bioanalysis. 2015;**7**:2145-2153

[59] Sarafraz-Yazdi A, Razavi N. Application of molecularly-imprinted polymers in solid-phase microextraction techniques. Trends in Analytical Chemistry. 2015;**73**:81-90

[60] Speltini A, Scalabrini A, Maraschi F, Sturini M, Profumo A. Newest applications of molecularly imprinted polymers for extraction of contaminants from environmental and food matrices: A review. Analytica Chimica Acta. 2017;**974**:1-26

[61] Ansari S, Karimi M. Recent progress, challenges and trends in trace determination of drug analysis using molecularly imprinted solid-phase microextraction technology. Talanta. 2017;**164**:612-625

[62] Cassiano NM, Barreiro JC, Moraes MC, Oliveira RV, Cass QB. Restricted-access media supports for direct high-throughput analysis of biological fluid samples: Review of recent applications. Bioanalysis. 2009;**1**:577-594

[63] Yang SH, Fan H, Classon RJ, Schug KA. Restricted access media as a streamlined approach toward on-line sample preparation: Recent advancements and applications. Journal of Separation Science. 2013;**36**:2922-2938

[64] Hennion MC, Pichon V. Immuno-based sample preparation for trace analysis. Journal of Chromatography. A. 2003;**1000**:29-52

[65] Liu R, Liu JF, Yin YG, Hu XL, Jiang GB. Ionic liquids in sample preparation.

Analytical and Bioanalytical Chemistry. 2009;**393**:871-883

[66] Ho TD, Canestraro AJ, Anderson JL. Ionic liquids in solid-phase microextraction: A review. Analytica Chimica Acta. 2011;**695**:18-43

[67] Mei M, Huang X, Chen L. Recent development and applications of poly (ionic liquid)s in microextraction techniques. Trends in Analytical Chemistry. 2019;**112**:123-134

[68] Saunders KC, Ghanem A, Boon HW, Hilder EF, Haddad PR. Separation and sample pre-treatment in bioanalysis using monolithic phases: A review. Analytica Chimica Acta. 2009;**652**:22-31

[69] Xu L, Shi ZG, Feng YQ. Porous monoliths: Sorbents for miniaturized extraction in biological analysis. Analytical and Bioanalytical Chemistry. 2011;**399**:3345-3357

[70] Samanidou VF, Karageorgou EG. An overview of the use of monoliths in sample preparation and analysis of milk. Journal of Separation Science. 2011;**34**:2013-2025

[71] Bunch DR, Wang S. Applications of monolithic solid-phase extraction in chromatography-based clinical chemistry assays. Analytical and Bioanalytical Chemistry. 2013;**405**:3021-3033

[72] Nema T, Chan EC, Ho PC. Applications of monolithic materials for sample preparation. Journal of Pharmaceutical and Biomedical Analysis. 2014;**87**:130-141

[73] Fresco-Cala B, Cárdenas S. Potential of nanoparticle-based hybrid monoliths as sorbents in microextraction techniques. Analytica Chimica Acta. 2018;**1031**:15-27

[74] Lucena R, Simonet BM, Cárdenas S, Valcárcel M. Potential of nanoparticles in sample preparation. Journal of Chromatography. A. 2011;**1218**:620-637

[75] Tian J, Xu J, Zhu F, Lu T, Su C, Ouyang G. Application of nanomaterials in sample preparation. Journal of Chromatography. A. 2013;**1300**:2-16

[76] Scida K, Stege PW, Haby G, Messina GA, García CD. Recent applications of carbon-based nanomaterials in analytical chemistry: Critical review. Analytica Chimica Acta. 2011;**691**:6-17

[77] Xu L, Qi X, Li X, Bai Y, Liu H. Recent advances in applications of nanomaterials for sample preparation. Talanta. 2016;**146**:714-726

[78] Zhang BT, Zheng X, Li HF, Lin JM. Application of carbon-based nanomaterials in sample preparation: A review. Analytica Chimica Acta. 2013;**784**:1-17

[79] Jakubus A, Paszkiewicz M, Stepnowski P. Carbon nanotubes application in the extraction techniques of pesticides: A review. Critical Reviews in Analytical Chemistry. 2017;**47**:76-91

[80] Song XY, Chen J, Shi YP. Different configurations of carbon nanotubes reinforced solid-phase microextraction techniques and their applications in the environmental analysis. Trends in Analytical Chemistry. 2017;**86**:263-275

[81] Segro SS, Tran MP, Kesani S, Alhendal A, Turner EB, Malik A. Sol-gel microextraction phases for sample preconcentration in chromatographic analysis. Journal of Separation Science. 2010;**33**:3075-3096

[82] Zhao L, Qin H, Wu R, Zu H. Recent advances of mesoporous materials in sample preparation. Journal of Chromatography. A. 2012;**1228**:193-204

[83] de Toffoli AL, Maciel EVS, Fumes BH, Lanças FM. The role of

graphene-based sorbents in modern sample preparation techniques. Journal of Separation Science. 2018;**41**:288-302

[84] Moliner-Martínez Y, Prima-Garcia H, Ribera A, Coronado E, Campíns-Falcó P. Magnetic in-tube solid phase microextraction. Analytical Chemistry. 2012;**84**:7233-7240

[85] Herrero-Latorre C, Barciela-García J, García-Martín S, Peña-Crecente RM, Otárola-Jiménez J. Magnetic solid-phase extraction using carbon nanotubes as sorbents: A review. Analytica Chimica Acta. 2015;**892**:10-26

[86] Shamsayei M, Yamini Y, Asiabi H, Safari M. On-line packed magnetic in-tube solid phase microextraction of acidic drugs such as naproxen and indomethacin by using $Fe_3O_4$@ $SiO_2$@layered double hydroxide nanoparticles with high anion exchange capacity. Mikrochimica Acta. 2018;**185**:192

[87] Yu QW, Ma Q , Feng YQ. Temperature-response polymer coating for in-tube solid-phase microextraction coupled to high-performance liquid chromatography. Talanta. 2011;**84**:1019-1025

[88] Yang Y, Rodriguez-Lafuente A, Pawliszyn J. Thermoelectric-based temperature-controlling system for in-tube solid-phase microextraction. Journal of Separation Science. 2014;**37**:1617-1621

[89] Melo LP, Queiroz RH, Queiroz ME. Automated determination of rifampicin in plasma samples by in-tube solid-phase microextraction coupled with liquid chromatography. Journal of Chromatography. B, Analytical Technologies in the Biomedical and Life Sciences. 2011;**879**:2454-2458

[90] Chaves AR, Silva BJ, Lanças FM, Queiroz ME. Biocompatible in-tube solid phase microextraction coupled with liquid chromatography-fluorescence detection for determination of interferon α in plasma samples. Journal of Chromatography. A. 2011;**1218**:3376-3381

[91] Caris JA, Silva BJ, Moisés EC, Lanchote VL, Queiroz ME. Automated analysis of lidocaine and its metabolite in plasma by in-tube solid-phase microextraction coupled with LC-UV for pharmacokinetic study. Journal of Separation Science. 2012;**35**:734-741

[92] Chaves AR, Queiroz ME. Immunoaffinity in-tube solid phase microextraction coupled with liquid chromatography with fluorescence detection for determination of interferon α in plasma samples. Journal of Chromatography. B, Analytical Technologies in the Biomedical and Life Sciences. 2013;**928**:37-43

[93] Chaves AR, Queiroz MEC. In-tube solid-phase microextraction with molecularly imprinted polymer to determine interferon alpha 2a in plasma sample by high performance liquid chromatography. Journal of Chromatography. A. 2013;**1318**:43-48

[94] Santos MG, Tavares IMC, Barbosa AF, Bettini J, Figueiredo EC. Analysis oftricyclic antidepressants in human plasma using online-restricted access molecularly imprinted solid phase extraction followed by direct mass spectrometry identification/quantification. Talanta. 2017;**163**:8-16

[95] Wang R, Li W, Chen Z. Solid phase microextraction with poly(deep eutectic solvent) monolithic column online coupled to HPLC for determination of non-steroidal anti-inflammatory drugs. Analytica Chimica Acta. 2018;**1018**:111-118

[96] Souza ID, Hantao LW, Queiroz MEC. Polymeric ionic liquid open tubular capillary column for on-line in-tube SPME coupled

with UHPLC-MS/MS to determine endocannabinoids in plasma samples. Analytica Chimica Acta. 2019;**1045**:108-116

[97] Wang R, Chen Z. Boronate affinity monolithic column incorporated withgraphene oxide for the in-tube solid-phase microextraction of glycoproteins. Journal of Separation Science. 2018;**41**:2767-2773

[98] Wang C, Zhou W, Liao X, Zhang W, Chen Z. An etched polyether etherketone tube covered with immobilized graphene oxide for online solid phase microextraction of quaternary alkaloids prior to their quantitation by HPLC-MS/MS. Microchimica Acta. 2017;**184**:2715-2721

[99] Zheng MM, Wang ST, Hu WK, Feng YQ. In-tube solid-phase microextraction based on hybrid silica monolith coupled to liquid chromatography-mass spectrometry for automated analysis of ten antidepressants in human urine and plasma. Journal of Chromatography. A. 2010;**1217**:7493-7501

[100] Shamsayei M, Yamini Y, Asiabi H. Polythiophene/graphene oxide nanostructured electrodeposited coating for on-line electrochemically controlled in-tube solid-phase microextraction. Journal of Chromatography. A. 2016;**1475**:8-17

[101] Asiabi H, Yamini Y, Seidi S, Ghahramanifard F. Preparation and evaluation of a novel molecularly imprinted polymer coating for selective extraction of indomethacin from biological samples by electrochemically controlled in-tube solid phase microextraction. Analytica Chimica Acta. 2016;**913**:76-85

[102] Asiabi H, Yamini Y, Shamsayei M. Development of electrochemically controlled packed-in-tube solid phase microextraction method for sensitive analysis of acidic drugs in biological samples. Talanta. 2018;**185**:80-88

[103] Saito K, Yagi K, Ishizaki A, Kataoka H. Determination of anabolic steroids in human urine by automated in-tube solid-phase microextraction coupled with liquid chromatography-mass spectrometry. Journal of Pharmaceutical and Biomedical Analysis. 2010;**52**:727-733

[104] Kataoka H, Inoue T, Ikekita N, Saito K. Development of exposure assessment method based on the analysis of urinary heterocyclic amines as biomarkers by on-line in-tube solid-phase microextraction coupled with liquid chromatography-tandem mass spectrometry. Analytical and Bioanalytical Chemistry. 2014;**406**:2171-2178

[105] Mizuno K, Kataoka H. Analysis of urinary 8-isoprostane as an oxidative stress bio marker by stable isotope dilution using automated online in-tube solid-phase microextraction coupled with liquid chromatography-tandem mass spectrometry. Journal of Pharmaceutical and Biomedical Analysis. 2015;**112**:36-42

[106] Kataoka H, Mizuno K, Oda E, Saito A. Determination of the oxidative stress biomarker urinary 8-hydroxy-2′-deoxyguanosine by automated on-line in-tube solid-phase microextraction coupled with liquid chromatography-tandem mass spectrometry. Journal of Chromatography. B, Analytical Technologies in the Biomedical and Life Sciences. 2016;**1019**:140-146

[107] He J, Liu Z, Ren L, Liu Y, Dou P, Qian K, et al. On-line coupling of in-tube boronate affinity solid phase microextraction with high performance liquid chromatography-electrospray

ionization tandem mass spectrometry for the determination of cis-diol biomolecules. Talanta. 2010;**82**:270-276

[108] Ahmadi SH, Manbohi A, Heydar KT. Electrochemically controlled in-tube solid phase microextraction of naproxen from urine samples using an experimental design. Analyst. 2015;**140**:497-505

[109] Wang S, Xu H. Inorganic-organic hybrid coating material for the online in-tube solid-phase microextraction of monohydroxy polycyclic aromatic hydrocarbons in urine. Journal of Separation Science. 2016;**39**:4610-4620

[110] Luo X, Li G, Hu Y. In-tube solid-phase microextraction based on $NH_2$-MIL-53(Al)-polymer monolithic column for online coupling with high-performance liquid chromatography for directly sensitive analysis of estrogens in human urine. Talanta. 2017;**165**:377-383

[111] Saito A, Hamano M, Kataoka H. Simultaneous analysis of multiple urinary biomarkers for the evaluation of oxidative stress by automated online in-tube solid-phase microextraction coupled with negative/positive ion-switching mode liquid chromatography-tandem mass spectrometry. Journal of Separation Science. 2018;**41**:2743-2749

[112] Yasuhara R, Ehara K, Saito K, Kataoka H. Automated analysis of salivary stress-related steroid hormones by online in-tube solid-phase microextraction coupled with liquid chromatography–tandem mass spectrometry. Analytical Methods. 2012;**4**:3625-3630

[113] Kataoka H, Ehara K, Yasuhara R, Saito K. Simultaneous determination of testosterone, cortisol, and dehydroepiandrosterone in saliva by stable isotope dilution on-line in-tube solid-phase microextraction coupled with liquid chromatography-tandem mass spectrometry. Analytical and Bioanalytical Chemistry. 2013;**405**:331-340

[114] Moriyama E, Kataoka H. Automated analysis of oxytocin by on-line in-tube solid-phase microextraction coupled with liquid chromatography-tandem mass spectrometry. Chromatography. 2015;**2**:382-391

[115] Ishizaki A, Uemura A, Kataoka H. A sensitive method to determine melatonin in saliva by automated online in-tube solid-phase microextraction coupled with stable isotope-dilution liquid chromatography-tandem mass spectrometry. Analytical Methods. 2017;**9**:3134-3140

[116] Kataoka H, Inoue T, Saito K, Kato H, Masuda K. Analysis of heterocyclic amines in hair by on-line in-tube solid-phase microextraction coupled with liquid chromatography-tandem mass spectrometry. Analytica Chimica Acta. 2013;**786**:54-60

[117] Yamamoto Y, Ishizaki A, Kataoka H. Biomonitoring method for the determination of polycyclic aromatic hydrocarbons in hair by online in-tube solid-phase microextraction coupled with high performance liquid chromatography and fluorescence detection. Journal of Chromatography. B, Analytical Technologies in the Biomedical and Life Sciences. 2015;**1000**:187-191

[118] Inukai T, Kaji S, Kataoka H. Analysis of nicotine and cotinine in hair by on-line in-tube solid-phase microextraction coupled with liquid chromatography-tandem mass spectrometry as biomarkers of exposure to tobacco smoke. Journal of Pharmaceutical and Biomedical Analysis. 2018;**156**:272-277

[119] Chen D, Ding J, Wu MK, Zhang TY, Qi CB, Feng YQ. A liquid chromatography-mass spectrometry method based on post column derivatization for automated analysis of urinary hexanal and heptanal. Journal of Chromatography. A. 2017;**1493**:57-63

[120] Ying LL, Ma YC, Xu B, Wang XH, Dong LY, Wang DM, et al. Poly(glycidyl methacrylate) nanoparticle-coated capillary with orientedantibody immobilization for immunoaffinity in-tube solid phase micro-extraction: Preparation and characterization. Journal of Chromatography. A. 2017;**1509**:1-8

[121] Hakobyan L, Tolos JP, Moliner-Martinez Y, Molins-Legua C, Ramos JR, Gordon M, et al. Determination of meropenem in endotracheal tubes by in-tube solid phase microextraction coupled to capillary liquid chromatography with diode array detection. Journal of Pharmaceutical and Biomedical Analysis. 2018;**151**:170-177

[122] Wang S, Hu S, Xu H. Analysis of aldehydes in human exhaled breath condensates by in-tube SPME-HPLC. Analytica Chimica Acta. 2015;**900**:67-75

[123] Li Y, Xu H. Development of a novel graphene/polyaniline electrodeposited coating for on-line in-tube solid phase microextraction of aldehydes in human exhaled breath condensate. Journal of Chromatography. A. 2015;**1395**:23-31

[124] Hu Y, Song C, Li G. Fiber-in-tube solid-phase microextraction with molecularly imprinted coating for sensitive analysis of antibiotic drugs by high performance liquid chromatography. Journal of Chromatography. A. 2012;**1263**:21-27

[125] Wang J, Zhao Q , Jiang N, Li W, Chen L, Lin X, et al. Urea-formaldehyde monolithic column for hydrophilic in-tube solid-phase microextraction of aminoglycosides. The Journal of Chromatography A. 2017;**1485**:24-31

[126] Pang J, Yuan D, Huang X. On-line combining monolith-based in-tube solid phase microextraction and high-performance liquid chromatography-fluorescence detection for the sensitive monitoring of polycyclic aromatic hydrocarbons in complex samples. Journal of Chromatography. A. 2018;**1571**:29-37

[127] Wang J, Jiang N, Cai Z, Li W, Li J, Lin X, et al. Sodium hyaluronate-functionalized urea-formaldehyde monolithic column for hydrophilic in-tube solid-phase microextraction of melamine. Journal of Chromatography. A. 2017;**1515**:54-61

[128] Wu F, Wang J, Zhao Q , Jiang N, Lin X, Xie Z, et al. Detection of trans-fatty acids by high performance liquid chromatography coupled with in-tube solid-phase microextraction using hydrophobic polymeric monolith. Journal of Chromatography. B, Analytical Technologies in the Biomedical and Life Sciences. 2017;**1040**:214-221

[129] Wang TT, Chen YH, Ma JF, Hu MJ, Li Y, Fang JH, et al. A novel ionic liquid-modified organic-polymer monolith as the sorbent for in-tube solid-phase microextraction of acidic food additives. Analytical and Bioanalytical Chemistry. 2014;**406**:4955-4963

[130] Asiabi H, Yamini Y, Seidi S, Esrafili A, Rezaei F. Electroplating of nanostructured polyaniline-polypyrrole composite coating in a stainless-steel tube for on-line in-tube solid phase microextraction. Journal of Chromatography. A. 2015;**1397**:19-26

[131] Ishizaki A, Saito K, Hanioka N, Narimatsu S, Kataoka H. Determination

of polycyclic aromatic hydrocarbons in food samples by automated on-line in-tube solid-phase microextraction coupled with high-performance liquid chromatography-fluorescence detection. Journal of Chromatography. A. 2010;**1217**:5555-5563

[132] Ishizaki A, Saito K, Kataoka H. Analysis of polycyclic aromatic hydrocarbons contamination in tea products and crude drugs. Analytical Methods. 2011;**3**:299-305

[133] Ying LL, Wang DY, Yang HP, Deng XY, Peng C, Zheng C, et al. Synthesis of boronate-decorated polyethyleneimine-grafted porous layer open tubular capillaries for enrichment of polyphenols in fruit juices. Journal of Chromatography. A. 2018;**1544**:23-32

[134] Andrade MA, Lanças FM. Determination of ochratoxin A in wine by packed in-tube solid phase microextraction followed by high performance liquid chromatography coupled to tandem mass spectrometry. Journal of Chromatography. A. 2017;**1493**:41-48

[135] Saito K, Ikeuchi R, Kataoka H. Determination of ochratoxins in nuts and grain samples by in-tube solid-phase microextraction coupled with liquid chromatography-mass spectrometry. Journal of Chromatography. A. 2012;**1220**:1-6

[136] Wu F, Xu C, Jiang N, Wang J, Ding CF. Poly (methacrylic acid-co-diethenyl-benzene) monolithic microextraction column and its application to simultaneous enrichment and analysis of mycotoxins. Talanta. 2018;**178**:1-8

[137] Wang X, Ma Q, Li M, Chang C, Bai Y, Feng Y, et al. Automated and sensitive analysis of 28-epihomobrassinolide in Arabidopsis thaliana by on-line polymer monolith microextraction coupled to liquid chromatography-mass spectrometry. Journal of Chromatography. A. 2013;**1317**:121-128

[138] Poorahong S, Thammakhet C, Thavarungkul P, Kanatharana P. Online in-tube microextractor coupled with UV-vis spectrophotometer for bisphenol A detection. Journal of Environmental Science and Health. Part A, Toxic/Hazardous Substances and Environmental Engineering. 2013;**48**:242-250

[139] Feng J, Wang X, Tian Y, Luo C, Sun M. Basalt fibers grafted with a poly(ionic liquids) coating for in-tube solid-phase microextraction. Journal of Separation Science. 2018;**41**:3267-3274

[140] González-Fuenzalida RA, López-García E, Moliner-Martínez Y, Campíns-Falcó P. Adsorbent phases with nanomaterials for in-tube solid-phase microextraction coupled on-line to liquid nanochromatography. Journal of Chromatography. A. 2016;**1432**:17-25

[141] Saito K, Uemura E, Ishizaki A, Kataoka H. Determination of perfluorooctanoic acid and perfluorooctane sulfonate by automated in-tube solid-phase microextraction coupled with liquid chromatography-mass spectrometry. Analytica Chimica Acta. 2010;**658**:141-146

[142] Vitta Y, Moliner-Martínez Y, Campíns-Falcó P, Cuervo AF. An in-tube SPME device for the selective determination of chlorophyll a in aquatic systems. Talanta. 2010;**82**:952-956

[143] Bagheri H, Piri-Moghadam H, Es'haghi A. An unbreakable on-line approach towards sol-gel capillary microextraction. Journal of Chromatography. A. 2011;**1218**:3952-3957

[144] Moliner-Martínez Y, Molins-Legua C, Verdú-Andrés J, Herráez-Hernández R, Campíns-Falcó P. Advantages of monolithic over particulate columns for multiresidue analysis of organic pollutants by in-tube solid-phase microextraction coupled to capillary liquid chromatography. Journal of Chromatography. A. 2011;**1218**:6256-6262

[145] Prieto-Blanco MC, Moliner-Martínez Y, López-Mahía P, Campíns-Falcó P. Ion-pair in-tube solid-phase microextraction and capillary liquid chromatography using a titania-based column: Application to the specific lauralkonium chloride determination in water. Journal of Chromatography. A. 2012;**1248**:55-59

[146] Kataoka H, Shiba H, Saito K. Automated analysis of oseltamivir and oseltamivir carboxylate in environmental water samples by online in-tube solid-phase microextraction coupled with liquid chromatography–tandem mass spectrometry. Analytical Methods. 2012;**4**:1513-1518

[147] Prieto-Blanco MC, Moliner-Martinez Y, Campíns-Falcó P. Combining poly(dimethyl-diphenylsiloxane) and nitrile phases for improving the separation and quantitation of benzalkonium chloride homologues: In-tube solid phase microextraction-capillary liquid chromatography-diode array detection-mass spectrometry for analyzing industrial samples. Journal of Chromatography. A. 2013;**1297**:226-230

[148] Masiá A, Moliner-Martinez Y, Muñoz-Ortuño M, Pico Y, Campíns-Falcó P. Multiresidue analysis of organic pollutants by in-tube solid phase microextraction coupled to ultra-high performance liquid chromatography-electrospray-tandem mass spectrometry. Journal of Chromatography. A. 2013;**1306**:1-11

[149] Prieto-Blanco MC, Moliner-Martínez Y, López-Mahía P, Campíns-Falcó P. Determination of carbonyl compounds in particulate matter PM2.5 by in-tube solid-phase microextraction coupled to capillary liquid chromatography/mass spectrometry. Talanta. 2013;**115**:876-880

[150] Moliner-Martinez Y, Vitta Y, Prima-Garcia H, González-Fuenzalida RA, Ribera A, Campíns-Falcó P, et al. Silica supported $Fe_3O_4$ magnetic nanoparticles for magnetic solid-phase extraction and magnetic in-tube solid-phase microextraction: Application to organophosphorous compounds. Analytical and Bioanalytical Chemistry. 2014;**406**:2211-2215

[151] González-Fuenzalida RA, Moliner-Martínez Y, Prima-Garcia H, Ribera A, Campins-Falcó P, Zaragozá RJ. Evaluation of superparamagnetic silica nanoparticles for extraction of triazines in magnetic in-tube solid phase microextraction coupled to capillary liquid chromatography. Nanomaterials. 2014;**4**:242-255

[152] Ahmadi SH, Manbohi A, Heydar KT. Electrochemically controlled in-tube solid phase microextraction. Analytica Chimica Acta. 2015;**853**:335-341

[153] Moliner-Martínez Y, Serra-Mora P, Verdú-Andrés J, Herráez-Hernández R, Campíns-Falcó P. Analysis of polar triazines and degradation products in waters by in-tube solid-phase microextraction and capillary chromatography: An environmentally friendly method. Analytical and Bioanalytical Chemistry. 2015;**407**:1485-1497

[154] Zhang J, Zhang W, Bao T, Chen Z. Polydopamine-based immobilization of zeolitic imidazolate framework-8 for in-tube solid-phase microextraction.

Journal of Chromatography. A. 2015;**1388**:9-16

[155] Pla-Tolós J, Moliner-Martínez Y, Molins-Legua C, Herráez-Hernández R, Verdú-Andrés J, Campíns-Falcó P. Selective and sentivive method based on capillary liquid chromatography with in-tube solid phase microextraction for determination of monochloramine in water. Journal of Chromatography. A. 2015;**1388**:17-23

[156] Tan F, Zhao C, Li L, Liu M, He X, Gao J. Graphene oxide based in-tube solid-phase microextraction combined with liquid chromatography tandem mass spectrometry for the determination of triazine herbicides in water. Journal of Separation Science. 2015;**38**:2312-2319

[157] Sun M, Feng J, Bu Y, Luo C. Nanostructured-silver-coated polyetheretherketone tube for online in-tube solid-phase microextraction coupled with high-performance liquid chromatography. Journal of Separation Science. 2015;**38**:3239-3246

[158] Sun M, Feng J, Bu Y, Luo C. Highly sensitive copper fiber-in-tube solid-phase microextraction for online selective analysis of polycyclic aromatic hydrocarbons coupled with high performance liquid chromatography. Journal of Chromatography. A. 2015;**1408**:41-48

[159] Feng J, Sun M, Bu Y, Luo C. Development of a cheap and accessible carbon fibers-in-poly(ether ether ketone) tube with high stability for online in-tube solid-phase microextraction. Talanta. 2016;**148**:313-320

[160] Bu Y, Feng J, Sun M, Zhou C, Luo C. Facile and efficient poly(ethylene terephthalate) fibers-in-tube for online solid-phase microextraction towards polycyclic aromatic hydrocarbons. Analytical and Bioanalytical Chemistry. 2016;**408**:4871-4882

[161] Bu Y, Feng J, Sun M, Zhou C, Luo C. Gold-functionalized stainless-steel wire and tube for fiber-in-tube solid-phase microextraction coupled to high-performance liquid chromatography for the determination of polycyclic aromatic hydrocarbons. Journal of Separation Science. 2016;**39**:932-938

[162] Sun M, Feng J, Bu Y, Luo C. Ionic liquid coated copper wires and tubes for fiber-in-tube solid-phase microextraction. Journal of Chromatography. A. 2016;**1458**:1-8

[163] Bu Y, Feng J, Tian Y, Wang X, Sun M, Luo C. An organically modified silica aerogel for online in-tube solid-phase microextraction. Journal of Chromatography. A. 2017;**1517**:203-208

[164] Bu Y, Feng J, Wang X, Tian Y, Sun M, Luo C. In situ hydrothermal growth of polyaniline coating for in-tube solid-phase microextraction towards ultraviolet filters in environmental water samples. Journal of Chromatography. A. 2017;**1483**:48-55

[165] Mei M, Huang X. Online analysis of five organic ultraviolet filters in environmental water samples using magnetism-enhanced monolith-based in-tube solid phase microextraction coupled with high-performance liquid chromatography. Journal of Chromatography. A. 2017;**1525**:1-9

[166] Serra-Mora P, Jornet-Martinez N, Moliner-Martinez Y, Campíns-Falcó P. In tube-solid phase microextraction-nano liquid chromatography: Application to the determination of intact and degraded polar triazines in waters and recovered struvite. Journal of Chromatography. A. 2017;**1513**:51-58

[167] Wang X, Pan L, Feng J, Tian Y, Luo C, Sun M. Silk fiber for in-tube solid-phase microextraction to detect aldehydes by chemical derivatization.

Journal of Chromatography. A. 2017;**1522**:16-22

[168] Wang X, Feng J, Bu Y, Tian Y, Luo C, Sun M. Mesoporous titanium oxide with high-specific surface area as a coating for in-tube solid-phase microextraction combined with high-performance liquid chromatography for the analysis of polycyclic aromatic hydrocarbons. Journal of Separation Science. 2017;**40**:2474-2481

[169] Feng J, Wang X, Tian Y, Luo C, Sun M. Poly(ionic liquids)-coated stainless-steel wires packed into a polyether ether ketone tube for in-tube solid-phase microextraction. Journal of Separation Science. 2017;**40**:4773-4779

[170] Feng J, Tian Y, Wang X, Luo C, Sun M. Basalt fibers functionalized with gold nanoparticles for in-tube solid-phase microextraction. Journal of Separation Science. 2018;**41**:1149-1155

[171] Feng J, Mao H, Wang X, Tian Y, Luo C, Sun M. Ionic liquid chemically bonded basalt fibers for in-tube solid-phase microextraction. Journal of Separation Science. 2018;**41**:1839-1846

[172] Wang X, Feng J, Tian Y, Luo C, Sun M. Co-Al bimetallic hydroxide nanocomposites coating for online in-tube solid-phase microextraction. Journal of Chromatography. A. 2018;**1550**:1-7

[173] Feng J, Wang X, Tian Y, Bu Y, Luo C, Sun M. Electrophoretic deposition ofgraphene oxide onto carbonfibers for in-tube solid-phase microextraction. Journal of Chromatography. A. 2017;**1517**:209-214

[174] Jillani SMS, Alhooshani K. Urea functionalized surface-bonded sol-gel coating for on-line hyphenation of capillary microextraction with high-performance liquid chromatography. Journal of Chromatography. A. 2018;**1543**:14-22

[175] Pang J, Mei M, Yuan D, Huang X. Development of on-line monolith-based in-tube solid phase microextraction for the sensitive determination of triazoles in environmental waters. Talanta. 2018;**184**:411-417

[176] Prieto-Blanco MC, López-Mahía P, Campíns-Falcó P. On-line analysis of carbonyl compounds with derivatization in aqueous extracts of atmospheric particulate PM10 by in-tube solid-phase microextraction coupled to capillary liquid chromatography. Journal of Chromatography. A. 2011;**1218**:4834-4839

[177] Fernández-Amado M, Prieto-Blanco MC, López-Mahía P, Muniategui-Lorenzo S, Prada-Rodríguez D. Ion-pair in-tube solid phase microextraction for the simultaneous determination of phthalates and their degradation products in atmospheric particulate matter. Journal of Chromatography. A. 2017;**1520**:35-47

[178] Ishizaki A, Kataoka H. A sensitive method for the determination of tobacco-specific nitrosamines in main- and side-stream smoke samples by online in-tube solid-phase microextraction coupled with liquid chromatography-tandem mass spectrometry. Analytica Chimica Acta. 2019;**1075**:98-105

6

# DNA Sequencing Resolves Misdiagnosed and Rare Genetic Disorders

*Alice Abdel Aleem*

**Abstract**

This chapter focuses on the mandatory requirement of DNA sequencing approaches for genetic diagnosis and recurrence prevention of inherited diseases. Sequencing the DNA and coded transcripts has intensely promoted our understanding of functional genomics and the fundamental importance of non-coding genomic sequences in causing heritable diseases, when mutated. Though Sanger sequencing, the first employed approach in identifying genetic mutations has been replaced nowa-days in many laboratories with the highly robust massive parallel sequencing tech-niques, "Sanger" remains vital in countries with limited resources and also of essential importance in validating the results of large scale sequencing technologies. Next generation sequencing (NGS) enabled the parallel sequencing of the whole exome (WES) and whole genome (WGS) regions of human genome and has revolutionized the field of genetic and genomic research in human. WES and WGS have facilitated the identification of the role of previously unrecognized genes in causing neurologic phenotypes, brain structural malformation, and resolved the causal genes in puzzling and misdiagnosed genetic phenotypes. Role of fusion genes and non-coding RNA in causing neurogenetic recessive diseases has been uncovered by the application of NGS platforms, published examples are presented in this chapter. Extensive phenotypic variability that retained patients either as misdiagnosed or undiagnosed cases for years has been correctly diagnosed through NGS research applications.

**Keywords:** DNA sequencing in human genetic disorders, NGS platforms in rare diseases, leukodystrophies, neuromuscular disorders, muscle dystrophy, non-coding genetic mutation, puzzling phenotypes, NGS, WES, WGS

## 1. Introduction

Since the significant discovery made by Watson and Crick [1] delineating the DNA double helical structure of alternate units (nucleotides) composed of deoxyribose sugar phosphate backbone and nitrogen bases pyrimidines (Cytosine, C and Thiamine, T) and purines (Adenine, A and Guanine, G). And the following crucial findings, Chargaff's rules [2] informing that the quantity of nitrogen bases differs in between species and the numbers of A equal to T, same for C and G [concluding the pairing status), the field of genetics, genomics, and hereditary is magnificently progressed.

The biology of the genetic code "central dogma" describes the flow of heritable genetic information from the nuclear DNA through the transcription process into the mRNA that is further translated into proteins or families of proteins. Central dogma of noncoding regions of DNA has also its influences on the stability of mRNA, Exon-intron splicing machinery, and translational efficiency [3–6].

The order (sequence) of nucleotides within a known or yet undiscovered set of genes is the first check point that dictates the coded messenger message and translated proteins. DNA-regulatory sequences including promoters, untranslated regions, DNA-methylation related (epigenetic and posttranscriptional splicing modifications) interactively play in defining the transcriptome and proteome expression profiles in different tissues of the body. Newly developed sequencing technologies have enabled the discovery of these regulatory and expression-modifier sequences [7–9].

Changes in the sequence of DNA-nucleotides located at the coding, non-coding, or splicing regions of the genome are anticipated to amend, in different ways, the genetic message as well as properties of the coded proteins and hence its functions in the cell.

These sequence variations are either inherited (passing from a generation to the next through the germline's cells; ovum/sperm) or spontaneous (de novo) in a subject germ cells. Spontaneous mutations will be further potentially inherited, mostly in a dominant pattern, through the subject's descent when his/her reproduction ability is not affected by the mutation. Changes (polymorphic variations or disease underlying mutations) in the DNA sequence may arise through base substitution, small insertion or deletion of bases, structural variations (large deletions or complex rearrangements), dynamic mutations (expansion of repetitive elements of the genome). DNA, cDNA, or RNA sequencing tools are the evidence based investigations that help us as scientists or physicians to identify or "see in Sanger's chart" these nucleotide changes and accurately allocate its genomic position [10, 11].

Monogenetic (Mendelian) disorders caused by single gene defect(s) are regularly counted under rare (orphan) diseases. Population with high rate of consanguineous marriages described to have extended multiple generational families that harbor rare monogenetic diseases. The single gene defect can occur on the two copies (alleles) of a gene (homozygous mutant) or on one allele only (heterozygous). Inheritance of Mendelian disorders may be autosomal recessive (the two alleles of an autosomal gene should carry the causative mutation to produce the disease phenotype), dominant (one mutant allele will be enough to cause the genetic disease), or X-linked (the mutant gene is located on the X chromosome) with the disease transmission occurring mostly through the females who are obligate carrier of the X-linked mutation [12, 13].

Monogenetic diseases may affect various body systems; cardiovascular, central nervous, peripheral nerves, endocrine, renal, or pulmonary, etc. The clinical phenotypic spectrum of the different distinct categories of these diseases is likely heterogeneous or overlapping which harden the clinicians' decision in making a definitive diagnosis. Academic studies in the field of human genetic diseases as well as diagnostics has been complicated for a long period of time by the remarkable clinical and genetic heterogeneity that were evident for the subgroups of a bunch of familial recurrent diseases involving: congenital muscle dystrophies, limb girdle muscle dystrophies, cortical brain malformation, hereditary spastic paraplegias, hereditary sensory neuropathies, neurodevelopmental, or others. With the evolving NGS technologies progress and discovery has been promptly started.

DNA and cDNA high throughput and validation sequencing tools are fundamental approaches that should be implemented in laboratories to reach a correct genetic diagnosis and provide accurate genetic counseling for rare heritable diseases. Genetic diseases may remain for decades undiagnosed or incorrectly managed when sequencing technologies are either not available or not accessible to patients due to its high cost. Genes and mutations identification in patients services all family members; siblings, cousins, nephews, or other relatives allowing carrier detection, premarital planning when first cousin or relative marriage is considered, prenatal diagnosis or preimplantation genetics. These sequencing outcomes mark the long term goal of reducing the occurrence or recurrence of genetic disorders in the community achievable.

Discovery of new genes and novel genetic "mutations/etiologies" for rare diseases, has exposed the basis of genetic heterogeneity, increased depth of genomic investigations and been intensely empowered, starting 2005, by the emerged technologies of next generation sequencing (NGS) that enabled the massive parallel sequencing (MPS) of millions of DNA or RNA nucleotides at a time [11, 14].

Whole Exome Sequencing (WES), one of the NGS platforms, grew into a widely used genetic diagnostic test in certified diagnostic labs over the world as well as a research tool in academic studies. WES targets the variants located in the coding regions and splicing boundaries of genes simultaneously at a time. The protein coding genes have been estimated to constitute ~2% of the human genome. Though WES is a powerful tool for the identification of underlying genetic defect in Mendelian disorders, obviously it lacks the capacity to detect non-coding or regula-tory disease causing genetic variations [15–18].

Whole Genome Sequencing (WGS), most extensive NGS' platform, has the capacity to interrogate the whole genome of a subject; the promoters, the untranslated upstream and downstream genomic ends, intragenic and intergenic regions in addition to the coding and splicing parts. Its applications in monogenetic diseases are still mostly at the level academic research. Its value in discovering new causal roles of previously unrecognized genes in rare inherited diseases came from its nature in detecting non-coding, regulatory and large structural variations arise in subjects' genome [19].

Advances in NGS wet lab methodologies, improvements in informatics pipelines (read alignment, variants call), and the huge released data annotation and analysis platforms lead the new genes discovery, the identification of new etiologies for rare diseases, and new cellular mechanisms contributing genetic syndromes and disorders. The better understanding of molecular biology of gene's mutation constitutes the essentials for new therapeutics [20].

In this chapter we shed the light on live recent examples demonstrating the role of DNA sequencing tools in gene discovery and in resolving the dilemma of certain genetic phenotypes that were undiagnosed for years.

## 2. Sanger sequencing

### 2.1 Advances in diagnostics and research of Monogenetic diseases

To late nineties Sanger sequencing (Chain Terminator Method) was the tool we used to use both in service and research to identify gene mutations or recognize polymorphic sequence variations in particular gene(s). Sanger Sequence is named after Frederick Sanger and his colleagues who had developed the method in late seventies [21, 22]. This sequencing method enabled the identification of nucleotides sequence in a single DNA or RNA amplified fragments and hence the changes (variations) from the reference genomes. Sanger Sequencing was highly applicable in diagnostics when a particular gene or few alternative genes are in question.

### 2.2 Demonstrative example from author's experience: a well-defined genetic phenotype with two alternative claimed causative genes confirmed true by Sanger sequencing

Here, we show the value of Sanger sequencing in resolving, in a fairly good turn-around time, the genetic defect in a group of patients with a phenotype of abnormal cerebral white matter associated with subcortical cysts. The leukodystrophies are a group of diseases, collectively characterized by primarily white matter involvements at variable degrees of severity ranged from a change in signal intensity, on brain images, to cystic cavitation or vanishing of the brain white matter contents [23, 24]. This group of diseases is genetically heterogeneous, however with a good clinical history, examination and high resolution brain imaging, a differential diagnosis can be set and Sanger sequencing can be applied for the few differential genes. The association of distinctive clinical features of macrocephaly (large sized head) detected since birth or shortly thereafter, motor developmental delay, seizures and ataxia precipitated by trauma as well as brain images of diffusely swollen white matter with the very characteristic finding of subcortical cysts preferentially occurring in brain temporal or frontoparietal lobes (**Figure 1**) suggested a clinical diagnosis of megalencephalicleukoencephalopathy (MLC), an autosomal recessive disease [OMIM # 604004]. A long list of metabolic disorders can be listed for a differential diagnosis.

In 75% of these patients, MLC1 gene's mutations are causal for the disease phenotype, whereas in ~20% of cases it is another gene, the HEPACAM/Glia-CAM that contributes the MLC phenotype. Both MLC1 and Glia-CAM are of a reasonable coding regions' size. Application of direct Sanger sequencing had helped several of such patients to get a solid genetic diagnosis of their diseases and allowed their families to use the Sanger sequencing results in performing premarital counseling and preventive measures through the carrier detection and prenatal diagnosis. Thus in cases feature a rather defined phenotype, average sized coding region of genes are in claim, and few alternative candidate causative genes, application of Sanger sequencing empowers the genetic diagnosis in a fair short turnaround time and makes the disease primary prevention quite possible [25].

### 2.3 Immunohistochemistry-guided Sanger sequencing

In some other diseases due to a known contributing family of proteins coded by a subset of genes, the roundabout time may be quite consuming to resolve the specific causal gene and hence Sanger sequencing may not be the suitable diagnostic tool particularly when there is a large flow of samples. A good such example is the Limb Girdle muscle dystrophies (LGMDs) which constitute a large group of progressive muscle weakness and wasting. Each of the several main groups of LGMD possesses a list of several subtypes caused by genetic mutations in many of muscle proteins related genes.

Muscle biopsy (a specimen of muscle fibers) used in immunohistochemical staining is an invasive diagnostic approach applied in patients with LGMDs aiming to detect the specific missing (deficient) muscle protein, secondary to gene's alteration using mono- or polyclonal antibodies.

Sarcoglycanopathies is a known genetic group of LGMDs. It is comprised of a family of four proteins forming four subgroups of sarcoglycanopthies; alpha, beta, gamma, and delta annotated according to the encoded protein and the corresponding gene [26].

The antibodies implemented in the immunohistochemistry procedure are anticipated to have the capacity to confirm the diagnosis of sarcoglycanopathy-LGMD and the level of the specific protein expression in the muscles, or in the best case scenario may also suggest the specific type of deficient sarcoglycan, whether alpha or beta, etc. However, in order to confidently determine which of the four sarcoglycan genes, α, β, Y, or delta harbors a heritable causative pathogenic mutation, gene sequencing should follow the immunohistochemistry. In such cases, Sanger sequencing guided by the immunohistochemistry results possibly will be a valuable diagnostic approach in areas of limited resources, particularly in extended families with multiple affected subjects across successive generations (**Figure 2**). However, many of the times this is not the case since the antibodies cross react to its different proteins subtypes. In such situation,

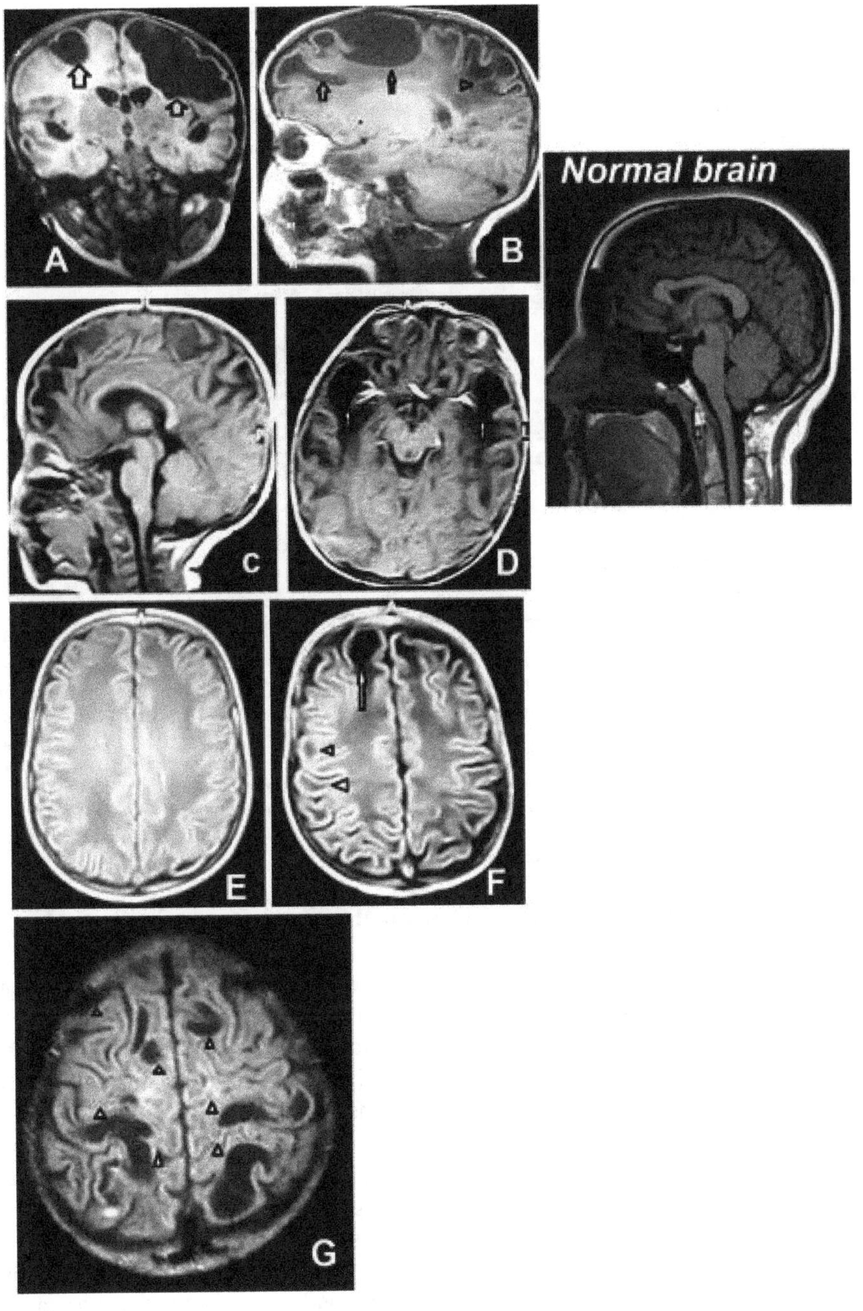


**Figure 1.**
*Brain magnetic resonance imaging (MRI) in Egyptian patients with MLC1 mutations [25]. Images are of different MLC patients captured at ages between 2 and 3 years old. Images T1-weighted sagittal (A and C), coronal (B), and axial (D): show extensive large vacuoles (A), widespread cystic changes involving frontal, temporal, most of the parietal, and occipital white matter (B), diffusely abnormal and swollen cerebral white matter (C and D). Cystic changes in frontal, parietal (C), and temporal subcortical regions (D). Images T1 weighted (E and F): display diffusely abnormal and swollen cerebral, subcortical, and periventricular, white matter. Cystic changes are visible in the frontal lobes (F). Axial T2 image at high level (G) demonstrates cystic changes (vacuolation) affecting different brain cortical regions. Sagittal T1 image of normal brain is included to support perception of the striking differences.*

though the time required to interrogate multiple related genes, each separately and release the results may be relatively long, however the sequencing outcomes' significance in disease's prevention and recurrence worth the time and efforts.

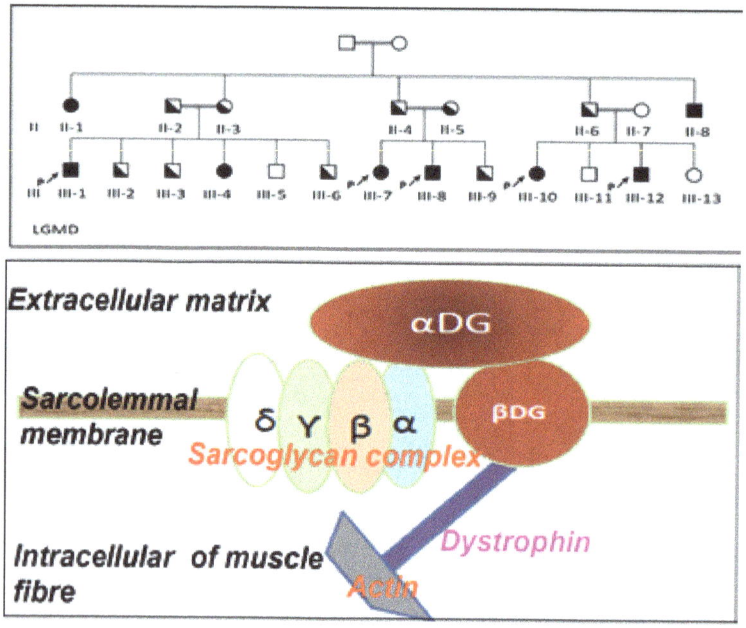


**Figure 2.**
*Extended pedigree with autosomal recessive LGMD alpha-Sarcoglycanopathy. Diagrammatic representation of sarcoglycans complex. The pedigree displayed the effect of consanguineous marriages in producing multiple generations with AR-limb girdle muscle dystrophy (LGMD): subtype alpha sarcoglycan that was confirmed only on DNA sequencing. Immunohistochemistry to the core family reported delta sarcoglycan [cross reactivity pitfall]. The diagram demonstrates the sarcoglycans complex (alpha, beta, gamma, and delta), a family of skeletal muscle sarcolemma proteins that are connected to the extracellular matrix through the alpha dystroglycans (αDG) which connects to the intracellular dystrophin and actin proteins (muscle cytoskeleton) via βDG forming the dystrophin-glycoprotein complex (DGC).*

## 2.4 Challenges for the diagnostic application of Sanger sequencing

Genes of extensively large coding regions like the FBN1, Titin, dystrophin, and many others constitute a challenge to use Sanger direct sequencing as a robust tool to characterize the underlying mutations. As a kind of solution, numerous commercial labs are limiting their molecular diagnostic service to specific gene's mutations' hot spots reported in the populations, when applicable. However, this approach is of a limited value when the case harbors a new or rare gene mutation.

In rather complex or non-specific clinical genetic presentations that are either of undetermined causative genes or of negative gene panel's results for a particular group of diseases, the Sanger sequencing remains unaccommodating.

The evolving roles of non-coding RNA and regulatory sequences alterations in causing heritable genetic diseases toughen the value of Sanger sequencing in diagnostics and human genetic research academic studies.

For all of these essentials new accommodating approaches were in need to satisfy the health care providers' goals to better serve patients with genetic diseases and the researchers need toward discovery of new genes and new etiologies for undiagnosed or misdiagnosed genetic disorders.

Targeted genes panel is a designed approach aiming to collectively sequencing a group of genes of a known causative relation to a particular inherited genetic disease or a group of closely related diseases. Examples involve panels for Limb girdle muscle dystrophy, hereditary spastic paraplegias (HSPs), inherited deafness, etc. This approach essentially and basically requires a continuous update of the designed panel to involve newly discovered genes aiming at avoiding false negative results. HSPs are a large group of diseases characterized by progressive lower limb spasticity, raised heal (tip toes) gait and associated in its complex phenotype with brain images abnormalities, developmental delay, ataxia, and other features. The list for HSPs associated gene defects is huge involving around 80 genes and continues to expand further [27]. Commercial HSP gene's panel are offered by various diagnostic laboratories, however pitfalls of negative results that falsely decline the diagnosis of HSPs is not uncommon.

Academic studies discover newly characterized HSP related genes yearly; this has to be regularly updating the diagnostic market. A proper alternative tool will be one of the cut edge NGS technologies.

## 3. Next Generation Sequencing (NGS)

### 3.1 NGS role in mapping genes and mutations to monogenetic diseases' phenotypes

WES and WGS yield a high throughput set of data. Of the interpretation process, these raw sequencing data/reads should be aligned to human reference nuclear genome. Differences between the subjects' sequencing reads and the reference genome are annotated as "variations" which may be counted either as common "polymorphic" or rare variants. The file contains all annotated variants of subject's sample is designated as the variants calling files (VCF).

The NGS' chemistry and nucleotide capture efficiency, depth of sequencing coverage, as well as bioinformatics pipelines employed in calling the variants of subjects' genome including the quality of mapping/alignment to the reference genome govern the potentials of the NGS' output [VCFs] in genes identification [28–30].

The key challenge in NGS data analysis is to identify the disease causal variants against the tremendous number of variants that are present at a low/rare frequency in genome or annotated, in-silico, as deleterious/pathogenic. Variants prioritization is the protocol employed to select the most potential disease causing variants. The diagram below (**Figure 3**) represents the number of variants originally called in WGS data of a subject and the filters sequentially applied aiming to highlight the most potential candidate disease related variants.

### 3.2 Gene discovery: identification of genes underlying a worldwide known clinical diagnosis

Kabuki syndrome (KS), OMIM # 147920 is a developmental, musculoskeletal, and intellectual disability with distinctive facial features genetic syndrome. This syndrome was first described, clinically, in families from Japan in 1981 [31] then described worldwide in patients from different ethnic groups. Intensive research has been made using the emerged high throughput sequencing technology to identify the KS causative gene, however unsuccessfully. The sporadic nature of KS (affected patients had negative family history and unaffected parent) harden the path of gene identification. The first Kabuki-associated gene (Lysine methyltransferase 2D, KMT2D, originally named as MLL2, a gene that regulates the expression of several downstream targets) was discovered only by late 2010 [32] along with the further developments made to WES and the process of variants identification and interpre-tation. KMT2D spontaneous gene mutations were found in over 75% of patients. A second X linked functionally related gene lysine demethylase 6A (KDM6A) contrib-utes 20% of KS cases [33].

This illustrates how it took about 30 years to identify the underlying gene(s) of a well-defined inherited genetic phenotype. Though the most modern high throughput technology was available for quite number of years, however refinement and

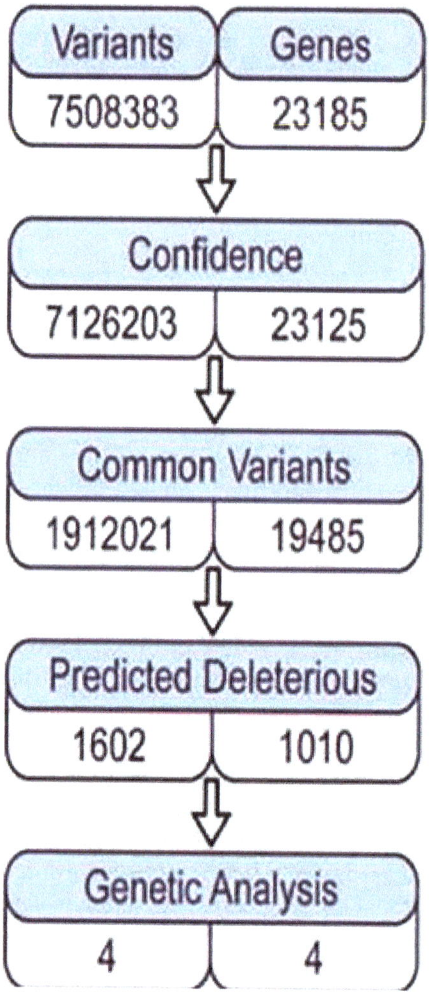


**Figure 3.**
*Diagrammatic representation of filters applied in WGS-VCF data analysis. The first line showed number of variants and corresponding number of genes called out of WGS in that particular sample. The confidence filter reveals the number of variants and corresponding genes passed the quality settings selected for this filter including the reads coverage/depth. The common variants filter presents the number of variants and genes after exclusion of "common" variants presented in public data base at frequency high ≥1%. The predicted deleterious filter has several options in its setting; the user can opt in the parameters wished to be considered in the analysis. The genetic filter has alternative options to select pattern of inheritance/transmission; homozygosity vs. heterozygosity in cases vs. controls among multiple other options.*

optimization of variants calling pipeline and variants analysis was recurrently visited to evolve into successful gene discovery for KS.

### 3.3 NGS approach resolves puzzling clinical phenotypes

With the author experience and the clinical examples discussed below we are aiming to outline the significance of NGS in driving research's discovery into clinical implementation and patients care.

Hereditary sensory and autonomic neuropathies, HSANs, are a genetically heterogeneous group of diseases, its phenotypic characteristics involve pain sensitivity (sensory loss) with its sequels, decreased sweating (hypohydrosis/autonomic function), plus mild motor weakness in a subset of patients [34]. Though the mechanism of development of disease pathology is not well understood, however; a known, short list of underlying genes were characterized and sequenced when the unique HSAN phenotype is suspected.

A consanguineous pedigree had two children, a boy and a girl aged 14 and 10 years respectively displayed a phenotype resembled that of hereditary sensory and autonomic neuropathies (HSANs). The clinical presentations characterized by two distinct features, sever pain insensitivity associated with hypohydrosis since birth along with the sequels of impaired pain sensation and severe aseptic destruction of large and small joints as well as the vertebrae (**Figure 4**). The two affected siblings had been examined by multiple local and international experts, the clinical diagnosis given was a general one describing an immune inflammatory disease (due to the joints destruction), however the association of the severely remarkable pain insensitivity remained unexplained in the context of immune-inflammation.

WGS revealed, unexpectedly, a homozygous mutation in LIFR. LIFR mutations have been associated with Stüve-Wiedemann syndrome (SWS), a lethal autosomal recessive skeletal dysplasia that may be associated with mild reduced pain sensation in atypical long survivors.

The complexity (overlapping phenotypes) as well as the striking severity of pain insensitivity phenotype, which phenocopy HSANs and atypically associated with extensive bone destruction challenges the diagnosis. The WGS had resolved this case dilemma, provided the family opportunities for preimplantation genetics as well as premarital counseling for other family members. Not only had that, but also reveals a new mechanism of LIFR's functional alteration (defective glycosylation of the mutant protein) [35]. WGS finding in these cases warrant the attention to consider LIFR testing in genetically unresolved phenotypes mimics HSAN.

### 3.4 NGS maps neurodevelopmental axonal guidance phenotype to a previously unrecognized gene

Neurodevelopmental disorders associated with brain malformation are the most extensively large group of neurological disorders. This group incorporates a broad spectrum of manifestations primarily involving the central nervous system and variably associated with motor and/or psychomotor delay, microcephaly, epilepsy, specific behavior, abnormal movements, eye symptoms, dysmorphic features, or hypotonia. Brain imaging is very helpful for the clinical diagnosis; however it remains challenging to reach a firm genetic diagnosis without NGS approaches. Each individual disease of this group is of the rare diseases. Some underlying genes have been identified and characterized; many others stay unknown or uncharacterized for its role in causing such diseases, waiting further research and discoveries.

We present here such example of a family with three affected siblings, a boy and a twin sister born to a consanguineous parent. The clinical phenotype of global developmental delay, learning difficulties associated with mild dysmorphism, hearing impairment was presented at variable severity between the older boy and the two affected female siblings. This clinical phenotype though can be categorized as neurodevelopmental disorder, however is very nonspecific. The older boy was given a provisional diagnosis of autistic spectrum hyperactivity due to some related features. The brain imaging of cortical malformation (polymicrogyria-cobblestone complex), central atrophy, and axonal guidance defects were variably shown in the three siblings. WGS applied for 8 members of this family (6 siblings: 3 affected and 3 unaf-fected plus parent) followed by bioinformatic variants analysis and genes functional reviews have successfully filtered the SNVs yield and identified a novel nonsense mutation in a previously unrecognized gene, Schwanomin-Interacting Protein1 (SCHIP1) (**Figure 5**) [36]. SCHIP1 was not previously associated to human neuro developmental disorders or brain malformation. However, mouse studies knocked out

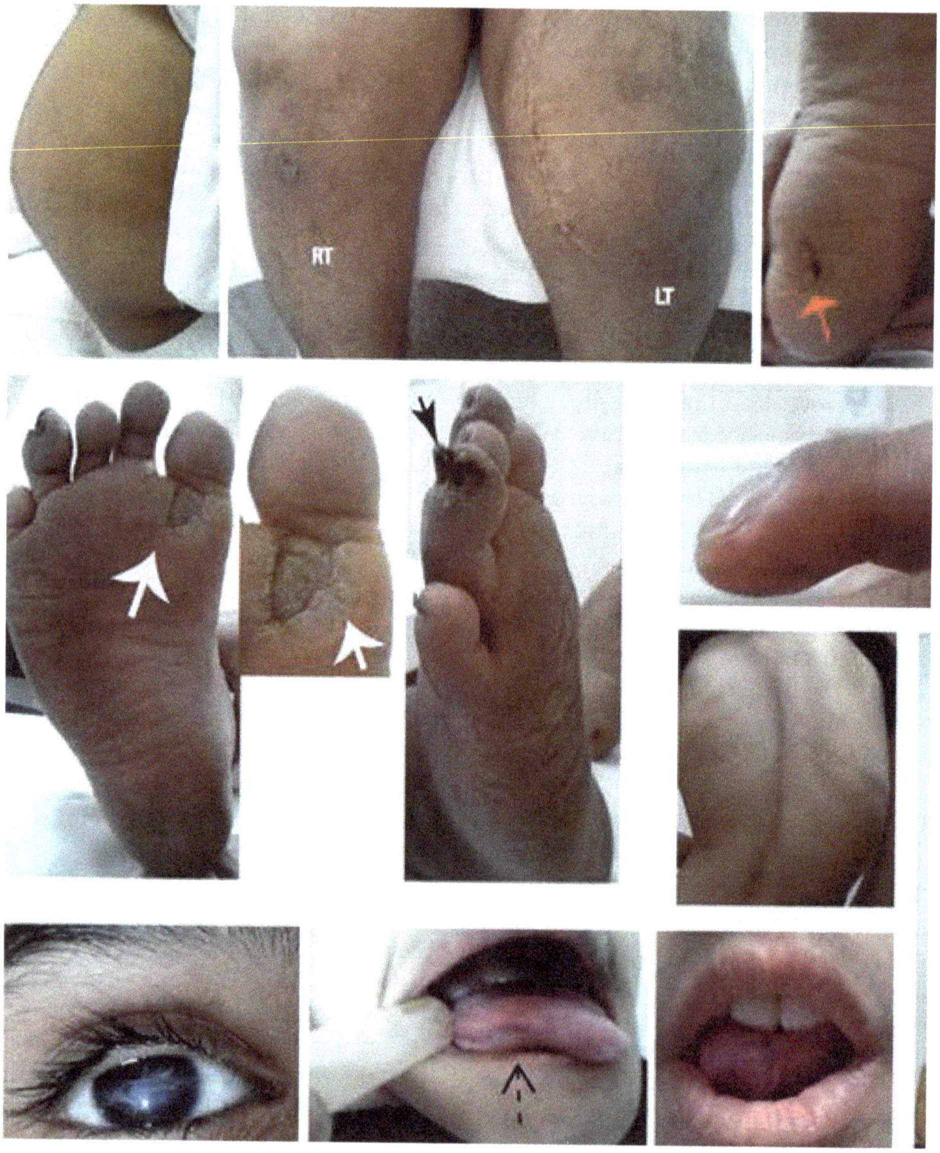


**Figure 4.**
*Phenotypic sequels of severe pain insensitivity, aseptic painless fractures and inflammation of large and small joints in patients with LIFR mutation [35]. Multiple images display: prominent spine kyphoscoliosis. Swollen knee joints; the scar in the left knee is due to surgical procedure treating joint's inflammation. Neuropathic chronic planter ulcerations: extensive and penetrating at the right heel, healing at the left big toe; evident hyperkeratosis of surrounding skin. Inflamed and painless distal inter-phalangeal joint of the left finger. Kyphoscoliosis, obvious corneal opacity (abnormal white band), tongue ulceration, and fissured lips (due to hypohydrosis-dryness) featured the pain insensitivity in the younger affected subject. This figure aims to demonstrate how the clinical picture mimics the diseases of pain insensitivity however the gene sequencing reveals a different disease category.*

schip1 isoforms produced a phenotype of brain axonal guidance defects, similarly to that detected in these patients. This gene has multiple isoforms including a fused gene (IQCJ-SCHIP1) isoform with variable tissue expression pattern and reported to have a role in axonogenesis during brain development. This example demonstrated the significant role of massive parallel sequencing approach as well as reviews of studies developed in mice with rather similar brain imaging phenotype in characterizing a new gene contributing neurodevelopmental-brain malformation phenotype.

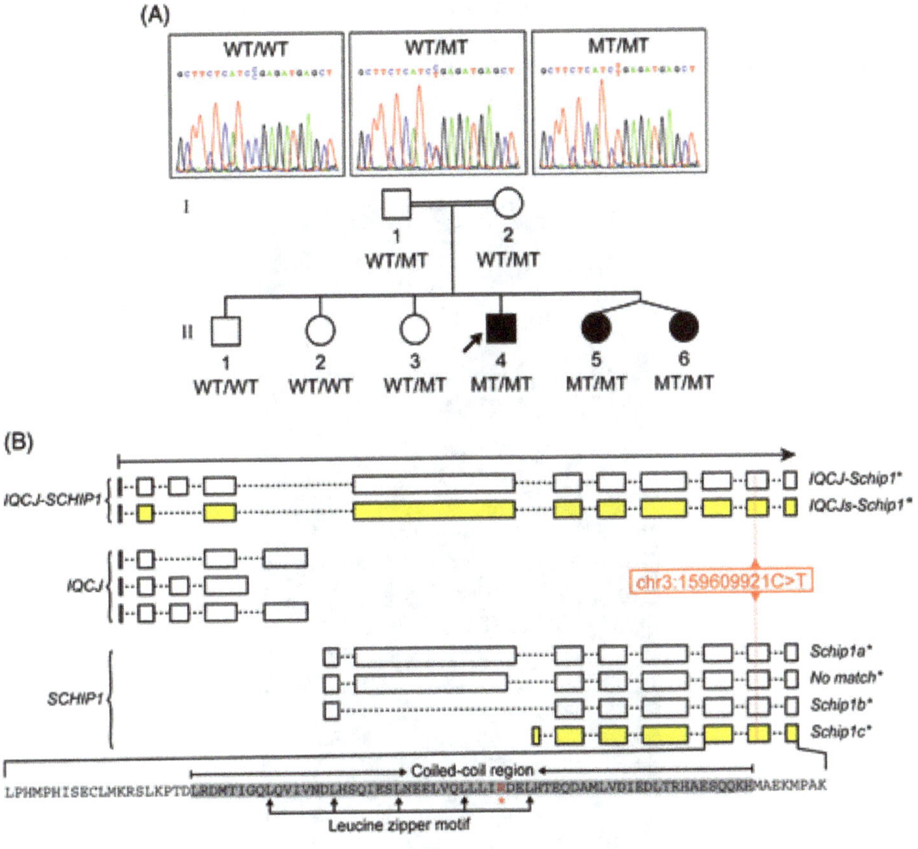


**Figure 5.**
*NGS maps neurodevelopmental axonal guidance phenotype in a consanguineous family to SCHIP1 [36]. Images showed (A) the family pedigree, the Sanger sequencing validating the nonsense mutation positioned in SCHIP1 and fused IQCJ-SCHIP1 isoform and its recessive segregation in family members. (B) Diagrammatic assembly of the IQCJ/SCHIP1 locus (exons illustrated as boxes, introns as lines) and the alternatively spliced transcriptional isoforms. The mutation position is marked in red. The asterisks indicate a mouse isoform of the human protein.*

### 3.5 WGS reveals new non-coding RNA minor splicing component's machinery that maps to a pure congenital cerebellar ataxia phenotype

Hereditary Cerebellar Ataxias (HCAs), the uncoordinated gait and body movements, can be inherited as autosomal dominant or recessive traits or in association with other neurological diseases. Hereditary ataxias are due to degeneration of cerebellar neurons or spinocerebellar tracts dysfunction [37]. Many several genes, its coding regions have been identified as causatives for the HCAs.

The emerging regulatory role of small non-coding RNA is evolving as a new mechanism leading human genetic diseases. WGS is particularly relevant to the identification of mutations in non-coding regions of the genome. An example, the 2nd worldwide of such condition was recently published [38]. In this referenced article, a large interrelated kindred had 6 patients with hereditary ataxias of unknown genetic etiology. Delayed speech and developmental milestones, congenital hypotonia, dysarthric speech, intention tremor, head nodding, and ataxic gait with a falling tendency were the main complains, however at variable severity among the affected patients. Brain images support the cerebellar involvements (**Figure 6**). Clinical diagnosis of an autosomal recessive cerebellar ataxia was suggested. Genetic investigations involving gene panel test and WES were performed; however results came back as negative.

WGS performed, on research basis, for 11 members of two branches of the extended family revealed interesting, nevertheless complex result that required functional testing to verify the causative gene and the biological impact of the

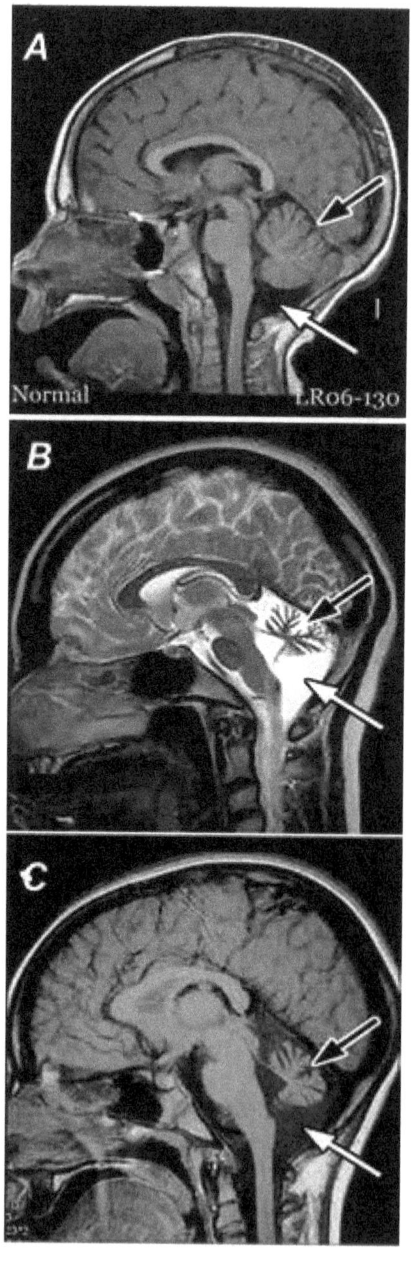


**Figure 6.**
*Brain MRI-sagittal views in normal and patients with ncRNU12 mutations [38]. Normal sagittal T1-weighted midline image (A) showing normal cerebellar foliation as well as normal brainstem proportions. A normal variant of prominent cisterna magna, the space at the posterior fossa (white arrow). Sagittal T2 weighted images (B and C) are of two affected female cousins, with variable degrees of clinical severity: pictures displayed dilatation of the cerebellar interfolia spaces indicating cerebellar atrophy or hypoplasia, reduction in the superior and inferior vermis' sizes (arrows). Brainstem is moderately affected (smaller in size). Widened CSF spaces around the posterior fossa, 2ndry to cerebellar and brain stem atrophy, more obvious in image B.*

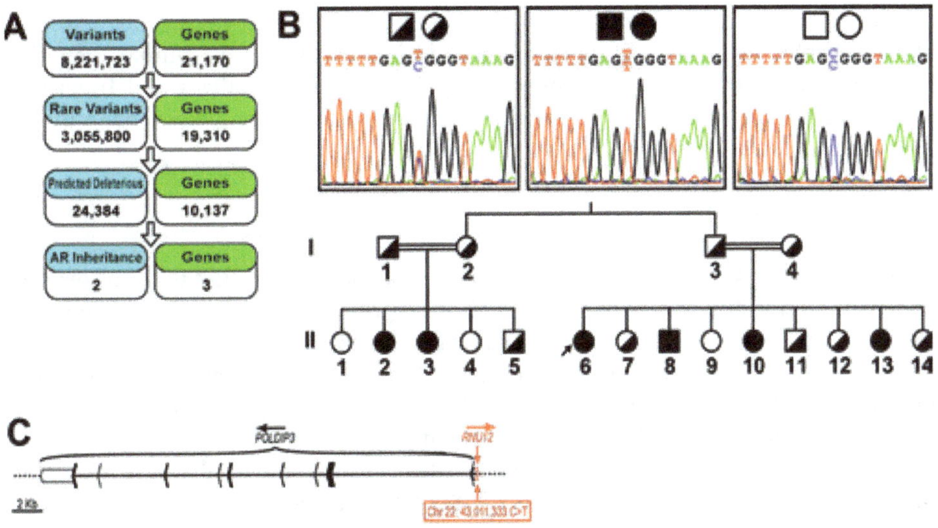

**Figure 7.**
*WGS variant identification, Sanger validation, and genomic organization of the region encompassing the RNU12 variant [38]. (A) Filters of ingenuity variant analysis (IVA) tool used to analyze the WGS data of 11 members of the extended family concluded two variants (one pathogenic, the 2nd was a kind of haplotype-linked polymorphism] passed all filters, corresponding to 3 genes, 2 that were recognized by IVA and a third gene identified by direct inspection of the locus. (B) Sanger sequencing of subjects' DNA validating the single nucleotide substitution on chromosome 22 and its recessive segregation with the phenotypes (normal/wild, normal/carrier, patient/mutant). (C) Diagrammatic representation of the chromosome 22 region, incorporating the variant that maps to the proximal promoter of POLDIP3 and a functional sequence of RNU12. POLDIP3 and RNU13 are transcribed from opposite strands (arrows). Multiple functional experiments were done and confirmed RNU12 contribution to the phenotype and disease pathomechanism.*

genomic mutation. WGS data analysis identified a variant (SNV) that was located in the promoter region of a protein coding gene POLDIP3 and fell as well in a small nuclear non-coding RNA gene (RNU12) that was transcribed from the opposite strand (**Figure** 7). Interestingly, RNU 12 was reported as a component of the U12-minor splicing machinery that functions in splicing of genes containing minor introns. Experimental investigations involved quantitative expression of the genes, RNA seq, semi-quantitative analysis of retention of minor introns containing genes (due to defective splicing machinery) established the causal relation of RNU12 to the disease phenotype in this large family. This story underscores the value of WGS in uncovering the unrecognized regulatory role of snRNU12 gene in human brain development and function. And the value in identifying the molecular gene defect in an example of monogenetic diseases that would have been remained uncovered when only WES was undertaken. This gene's result has been used by healthy family members in carrier detection, premarital counseling and prenatal diagnosis.

The ages at which patients of this kindred have getting the genetic diagnosis of their disease were as of 25 year old (for the female proband), 22 year old for her brother, 15 and 10 years old of her sisters (first branch), 19 and 13 years old of female siblings of second branch. This highlights how NGS empowered the diagnos-tic odyssey of monogenetic diseases translating research into clinic improving targeted patients care and prevention of diseases' recurrence in family and community.

## 4. Conclusion

Advancement of new therapeutics for genetic diseases is definitely influenced by research and technologies that support swift, reliable, and interpretable OMICs (genomic, transcriptomic and proteomic) research. DNA and RNA sequencing are of such technologies that greatly advanced the discoveries in human genetics. However, still further improvements of big-data pipeline analysis and functional investigations are mandatory to maximize and empower discoveries made by the "Sequencing."

## Acknowledgements

The author thanks the team and colleagues participated in the work of original images that are reused in this chapter following the permissions of the Copyright Owners. This chapter was made possible by funds received from the Qatar National Research Fund [grants: PPM1-1206-150013 and NPRP4-099-3-039; principal investigator: Alice Abdelaleem], a member of Qatar Foundation. The findings achieved herein are solely the responsibility of the author.

Author's appreciation extends to the institutes of Weill Cornell Medicine Qatar, Brain and Mind Research Institute-NY-USA, Hamad Medical Corporation Qatar, Pediatric Neurology-Cairo University Hospitals Egypt, and National Research Centre Egypt.

## Notes

Permissions for re-published figures have been obtained from the Copyright Owner(s).

## Author details

Alice Abdel Aleem
Weill Cornell Medicine Qatar, Brain and Mind Research Institute, Education City, Doha, Qatar

*Address all correspondence to: aka2005@qatar-med.cornell.edu

## References

[1] Watson JD, Crick FH. Molecular structure of nucleic acids: A structure for deoxyribose nucleic acid. Nature. 1953;**171**:737-738

[2] Chargaff E, Lipshitz R, Green C. Composition of the deoxypentose nucleic acids of four genera of sea-urchin. The Journal of Biological Chemistry. 1952;**195**:155-160. PMID 14938364

[3] Nirenberg M. Historical review: Deciphering the genetic code—A personal account. Trends in Biochemical Sciences. 2004;**29**:46-54. DOI: 10.1016/j. tibs.2003.11.0090

[4] Nirenberg M, Leder P. RNA code words and protein synthesis. Science. 1964;**145**:1399-1407. DOI: 10.1126/science.145.3639.1399

[5] Jimenez-Sanchez G, Childs B, Valle D. Human disease genes. Nature. 2001;**409**:853-855. DOI: 10.1038/35057050

[6] Stranger B, Dermitzakis E. From DNA to RNA to disease and back: The 'central dogma' of regulatory disease variation. Human Genomics. 2006;**2**:383-390. DOI: 10.1186/1479-7364-2-6-383

[7] 1000 Genomes Proj. Consort. An integrated map of genetic variation from 1,092 human genomes. Nature. 2012;**491**:56-65

[8] Goodwin S, Mcpherson J, Mccombie W. Coming of age: Ten years of next-generation sequencing technologies. Nature Reviews. Genetics. 2016;**17**:333-351

[9] Flusberg B, Webster D, Lee J, Travers K, Olivares E, et al. Direct detection of DNA methylation during single-molecule, real-time sequencing. Nature Methods. 2010;**7**:461-465

[10] Morey M, Fernandez-Marmiesse A, Castineiras D, Fraga JM, Couce ML, Cocho JA. A glimpse into past, present, and future DNA sequencing. Molecular Genetics and Metabolism. 2013;**110**:3-24

[11] Levy S, Myers R. Advancements in next-generation sequencing. Annual Review of Genomics and Human Genetics. 2016;**17**:95-115

[12] McKusick V. Mendelian Inheritance in Man: A Catalog of Human Genes and Genetic Disorders. 12th ed. USA: Baltimore Johns Hopkins University Press; 1998

[13] Amberger J, Bocchini C, Schiettecatte F, Scott A, Hamosh A. OMIM. org: Online Mendelian Inheritance in Man (OMIM®), an online catalog of human genes and genetic disorders. Nucleic Acids Research. 2015;**43**:D789-D798. DOI: 10.1093/nar/gku1205

[14] Shendure J, Balasubramanian S, Church G, Gilbert W, Rogers J, Schloss J, et al. DNA sequencing at 40: Past, present and future. Nature. 2017;**550**:345-353

[15] Voelkerding K, Dames S, Durtschi J. Next-generation sequencing: From basic research to diagnostics. Clinical Chemistry. 2009;**55**:4641-4658

[16] Bamshad M, Ng S, Bigham A, Tabor H, Emond M, Nickerson D, et al. Exome sequencing as a tool for Mendelian disease gene discovery. Nature Reviewes-Genetics. 2011;**12**:745-755

[17] Kaiser J. Human genetics. Affordable 'exomes' fill gaps in a catalogue of rare diseases. Science. 2010;**330**:903-911

[18] Ng SB, Buckingham KJ, Lee C, Bigham AW, Tabor HK, Dent KM, et al. Exome sequencing identifies the cause of a Mendelian disorder. Nature Genetics. 2010;**42**:30-35

[19] Mardis E, Whole-genome sequencing: New technologies, approaches, and applications. Genomic and Personalized Medicine. 2nd ed. Elsevier, UK: Ginsburg & Willard; 2013. pp. 87-93. Chapter 7. eBook ISBN: 9780123822284

[20] Salk J, Schmitt M, Loeb L. Enhancing the accuracy of next generation sequencing for detecting rare and subclonal mutations. Nature Reviews Genetics. 2018;**19**:269-285

[21] Sanger F, Nicklen S, Coulson A. DNA sequencing with chain-terminating inhibitors. Proceedings of the National academy of Sciences of the United States of America. 1977;**1077**(74):5463-5467

[22] Maxam AM, Gilbert W. A new method for sequencing DNA. PNAS. 1977;**74**:560-564

[23] van der Knaap M, Wolf N, Heine V. Leukodystrophies. Neurology clinical Practice. 2016;**6**:506-514. DOI: 10.1212/CPJ.0000000000000289

[24] van der Knaap M, Bugiani M. Leukodystrophies—Much more than just diseases of myelin. Nature Reviews | Neurology. 2018;**14**:747-748

[25] Mahmoud I, Mahmoud M, Refaat M, Girgis M, Waked N, El Badawy A, et al. Clinical, neuroimaging, and genetic characteristics of megalencephalic leukoencephalopathy with subcortical cysts in Egyptian patients. Pediatric Neurology. 2014;**50**:140-148

[26] Nalini A, Gayathri N, Thaha F, Das S, Shylashree S. Sarcoglycanopathy: Clinical and histochemical characteristics in 66 patients. Neurology India. 2010;**58**:691-696

[27] Bis-Brewer D, Züchner S. Perspectives on the genomics of HSP beyond mendelian inheritance. Frontiers in Neurology. 2018;**9**:1-11. DOI: 10.3389/fneur.2018.00958

[28] Choi M, Scholl UI, Ji W, Liu T, Tikhonova IR. Genetic diagnosis by whole exome capture and massively parallel DNA sequencing. PNAS. 2009;**106**:96-101

[29] Dewey FE, Grove ME, Pan C, Goldstein BA, Bernstein JA, et al. Clinical interpretation and implications of whole-genome sequencing. JAMA. 2014;**311**:1035-1045

[30] Lelieveld SH, Spielmann M, Mundlos S, Veltman JA, Gilissen C. Comparison of exome and genome sequencing technologies for the complete capture of protein-coding regions. Human Mutation. 2015;**36**:815-822

[31] Niikawa N, Matsuura N, Fukushima Y, Ohsawa T, Kajii T. Kabuki make-up syndrome: A syndrome of mental retardation, unusual facies, large and protruding ears, and postnatal growth deficiency. The Journal of Pediatrics. 1981;**99**:565-569

[32] Ng S, Bigham A, Buckingham K, Hannibal M, McMillin M, Gildersleeve H, et al. Exome sequencing identifies MLL2 mutations as a cause of Kabuki syndrome. Nature Genetics. 2010;**42**:790-793

[33] Lederer D, Grisart B, Digilio MC, et al. Deletion of KDM6A, a histone demethylase interacting with MLL2, in three patients with Kabuki syndrome. American Journal of Human Genetics. 2012;**90**:119-124

[34] Rotthier A, Baets J, Timmerman V, Janssens K. Mechanisms of disease in hereditary sensory and autonomic neuropathies. Nature Reviews Neurology. 2012;**8**:73-85

[35] Elsaid M, Chalhoub N, Kamel H, Ehlayel M, Ibrahim N, Elsaid A,

et al. Non-truncating LIFR mutation: Causal for prominent congenital pain insensitivity phenotype with progressive vertebral destruction? Clinical Genetics. 2016;**89**:210-216

[36] Elsaid M, Chalhoub N, Ben-Omran T, Kamel H, AlMureikhi M, Ibrahim K, et al. Homozygous nonsense mutation in SCHIP1/IQCJ-SCHIP1 causes a neurodevelopmental brain malformation syndrome. Clinical Genetics. 2018;**93**:387-391

[37] Anheim M, Tranchant C, Koenig M. The autosomal recessive cerebellar ataxias. The New England Journal of Medicine. 2012;**366**:636-646

[38] Elsaid M, Chalhoub N, Ben-Omran T, Kumar P, Kamel H, Ibrahim I, et al. Mutation in noncoding RNA RNU12 causes early onset cerebellar Ataxia. Annals of Neurology. 2017;**81**:68-78

# 7

# Principles of Chromatography Method Development

*Narasimha S. Lakka and Chandrasekar Kuppan*

### Abstract

This chapter aims to explain the key parameters of analytical method development using the chromatography techniques which are used for the identification, separation, purification, and quantitative estimation of complex mixtures of organic compounds. Mainly, the versatile techniques of ultra–/high-performance liquid chromatography (UPLC/HPLC) are in use for the analysis of assay and organic impurities/related substances/degradation products of a drug substance or drug product or intermediate or raw material of pharmaceuticals. A suitable analytical method is developed only after evaluating the major and critical separation parameters of chromatography (examples for UPLC/HPLC are selection of diluent, wavelength, detector, stationary phase, column temperature, flow rate, solvent system, elution mode, and injection volume, etc.). The analytical method development is a process of proving the developed analytical method is suitable for its intended use for the quantitative estimation of the targeted analyte present in pharmaceutical drugs. And it mostly plays a vital role in the development and manufacture of pharmaceuticals drugs.

**Keywords:** analytical method development, ultra performance liquid chromatography (UPLC), high-performance liquid chromatography (HPLC), assay, impurities, impurity profiling study, forced degradation study

### 1. Introduction

It is well known that chromatography is a laboratory technique used for separation and quantification of complex organic mixtures which cannot be separated effectively by other purification techniques. The constituents of a mixture dissolved in solvent get separated gradiently according to their affinities to the stationary phase with the help of mobile phase one after another. Chromatography is invented by *Mikhail Semenovich Tswett* in 1903 during his research on plant pigments such as chlorophylls, xanthophylls, and carotenoids [1] which got extended for analyzing organic molecules of different kinds especially pharmaceutical from the year 1920 [2]. Invention of chromatography made the jobs of organic chemist and the whole industry relying on them especially pharma industry easier. Keeping in mind the various fields where this technique has been used, this chapter focuses on the use of chromatography in pharmaceuticals for separating the drug (API) mixture in particular.

Nowadays, many different kinds of chromatography techniques, such as thin-layer chromatography (TLC), paper chromatography, and liquid chromatography (e.g., HPLC, UPLC, and preparative HPLC), supercritical fluid chromatography, and gas chromatography (GC)) have been designed and utilized for the separation and purification of pharmaceutical drugs [3]. In this chapter, the authors discuss the principles for chromatography method development using ultra/high-perfor-mance liquid chromatography (UPLC/HPLC) techniques for the analysis of assay and organic impurities/related substances/degradation products of pharmaceuticals (any drug product/drug substance/intermediate/raw material of pharmaceuticals). These techniques are developed substantially as a result of the work of *Archer John Porter Martin* and *Richard Laurence Millington Synge* during the 1940s and 1950s, for which they won the 1952 Nobel Prize in Chemistry [4]. Commonly used character-izing technique in pharma industry is liquid chromatography (e.g., HPLC, UPLC, and LC–MS). Each one varies in the stationary phase and operational conditions. HPLC and UPLC can be used as a quantitative technique if coupled with a mass detector (MS) to elucidate the structure of the molecule and quantification.

In pharma industry specific, stability-indicating HPLC/UPLC methods have to be developed to estimate the assay and to quantitatively determine the impurities of new drug substances and drug products [5]. Assay is a quantitative test of a substance to determine the amount of an individual components present in it. Impurity is an unknown component of drug substance that is not the chemical entity. Assay and impurity tests are major and critical quality attributes of the pharmaceutical dosage forms which help to check and ensure the quality, safety, and efficacy of drug substances and drug products. This chapter will discuss the various parameters that have to be chosen to run the chromatography in order to have a better separation and maximum purity. The process of changing the conditions in order to design a best method run for a particular drug mixture or compound is called the analytical method development.

## 2. Analytical method development

Analytical method development is a process of proving that the developed chromatography method is suitable for its intended use in the development and manufacturing of the pharmaceutical drug substance and drug product. The basic separation techniques and principles involved in the analytical method development using the HPLC and UPLC are listed as follows:

- Selection of chromatography mode
- Selection of detector
- Selection of column (stationary phase)
- Selection and optimization of mobile phase
    - Buffer and its strength
    - pH of buffer
    - Mobile-phase composition
- Selection of organic modifiers
- Selection of ion-pair reagents
- Selection of flow rate

- Selection of solvent delivery system (elution mode)
- Selection of diluent
- Methods of extraction
- Samples to be used
- Experimentation to finalize the method
- Selection of test concentration and injection volume
- Forced degradation studies (stress resting)
- Evaluation of stress testing
- Mass balance study
- Finalization of wavelengths
- Stability of solution
- System suitability
- Robustness of the method
- Relative response factor
- Quantification methods

### 2.1 Literature search

i. Before starting an analytical method development, literature on some of the column characteristics as mentioned below has to be referred for the target molecules or similar molecules or precursors from open resources like articles, books, pharmacopeia reports, etc. This will give a tentative choice in designing a method for initial or test experiments, which will be further modified or updated to develop a method which fits the separation process for better results in terms of reproducibility, quantification, etc. *Solubility profile* of drug substance in different solvents at different pH conditions is useful while selecting the diluents for standard solutions and extraction solvents for test solutions.

ii. *Analytical profile* is useful in understanding the physicochemical properties (e.g., *pKa*, melting point, degradation pathways, etc.) and absorption characteristics of drug in selecting the detector wavelength for analysis.

iii. *Stability profile* of the drug substance with respect to storage conditions (sensitivity of the drug towards light, heat, moisture etc.) is useful as it helps in adopting the suitable/adequate precautions while handling drug and its formulated products.

iv. *Impurity profile* collects the information of impurities and degradation profile of the drug substance during their formation pathways. This helps a lot in developing the method for separation of all possible impurities and degradation products of targeted analyte. It should be borne in mind that impurity profile may vary depending on the manufacturing process (which uses different methods, precursors, and conditions), which makes it clear that not all manufacturing processes yield the same impurity profile.

v. *Metabolic pathway* is a chemical reaction which occurs within a cell when the drug molecule reacts with an enzyme and forms a metabolite [6]. Metabolic pathway gives the information on oxidation, reduction, and hydrolysis products which gives critical inputs on the possible degradation products.

vi. Stability-indicating method is to identify the closely related structures by collecting the structures of the molecule and its impurities and degradation products. This helps to develop a specific and stability-indication method with a good resolution between the closely related structures.

vii. *Checking the polarity* of the drug molecule using the functional groups as elucidated from structural analysis techniques. By comparing the structures of impurities and degradation products with the structure of drug molecule, it will help in understanding the polarity based on the nature of functional groups. This makes the scientists' job easy in choosing the right solvents with either lesser or higher in polarity than the compound of interest.

viii. *Estimation of maximum daily dose (MDD).* Calculate the reporting, identification, and qualification thresholds of drug substance and drug product based on the maximum daily dose as per ICH Q3A guideline [7, 8]. MDD info can also be obtained from physical desk reference (PDR), innovator product information leaflet (PIL), and the website of RX-list.

## 2.2 Selection of chromatography mode

Chromatography can be operated by two ways, normal mode and reverse phase modes. The choice of the mode is very important, which is dependent on the type of sample which has to be separated. In general, the usage of reversed-phase chromatography (in which the mobile phase is polar and stationary phase is nonpolar in nature) is the preferred mode for most of the molecules, except in the case of isomer (enantiomers) separation where the normal-phase chromatography (in which the mobile phase is nonpolar and stationary phase is polar in nature) is used. Revered-phase chromatography separates the components with a good resolution based on their hydrophobicity. A compound with a greater polarity elutes earlier, and those with the least polarity elute later.

## 2.3 Selection of detector

Detector plays an important role in the finalization of any analytical method. Generally most of the organic/drug molecules are aromatic or unsaturated in nature, which has an absorption in the UV–vis region. This comes as an advantage in quantifying and analyzing the molecules and its associated impurities. The absorbance maxima of the compound shall be collected by analyzing the UV–vis spectrophotometer or diode array detector (DAD) of HPLC/UPLC. From the area intensity of the test compound using calibration curves, the quantification of the test compound can be done [9–10].

If the compounds do not absorb and if they do not have chromophores, other detectors like refractive index detector (RID) and evaporative light scattering detector (ELSD)/corona-charged aerosol detector (CAD) can be used for the quantitative determination of assay and impurities [11]. If the compounds of interest contain a part, which is non-chromophoric, which may likely be cleaved and produce a non-chromophoric impurity, then both UV and other detectors like RI/ELSD/CAD can be coupled in order not to miss any impurity.

Alternatively, non-chromophoric compounds can also be analyzed by UV after converting it into a derivative which will be active. But the usage of derivatives has to be carefully assessed keeping in view the functional group involved in the derivatization reaction [12, 13]. In case the molecule of interest is having fluorescence properties, a fluorescence detector (FLD) can be used for compounds for which structural information is available [14]. But when FLD is to be used for estimation of unknowns, it needs to be carefully assessed whether fluorescence properties are available in all possible impurities and degradation products.

## 2.4 Selection of column stationary phase

The choice of the right column (stationary phase) is the basis of the whole technology. Most chromatographic separations are achieved due to a wide variety of columns available in the market and due to their flexibility in changing and controlling the parameters. A widely used choice of column material is silica either as neat or modified depending on the nature of the solute mixture in normal-phase chromatography, wherein the eluent (mobile phase) is nonpolar an organic solvent. The silanol groups on the surface of the silica give it a polar character.

Though silica remains the most common support for liquid chromatography (LC) columns, other commonly used materials are cross-linked organic polymers, zirconia, etc. The silica support for columns was gradually modified for the betterment through the years by three different manufacturing technologies commonly described as "evolution through three generations." The initial process started with type A silica where the raw material used is from inorganic sols. A slightly modified type A silica by performing a chemical treatment to remove the metal impurities is termed as a second-generation material which is called as base-deactivated silica. Third generation silica (type B) is an altogether new process which uses organic sols instead of inorganic sols. These materials are similar in properties to the second-generation silica because both have a minimum level of metal impurities. Silica-based liquid chromatography columns with a different percent of cross-linking and functionalization of silanol groups with substituted aliphatic and aromatic moieties were designed for varying polarities of the separating medium. An increasing order of functionalized silica is represented below with alkyl groups at the nonpolar end, phenyl and amino functionalized in the moderate polar region, and cyano and silica groups at the polar end.

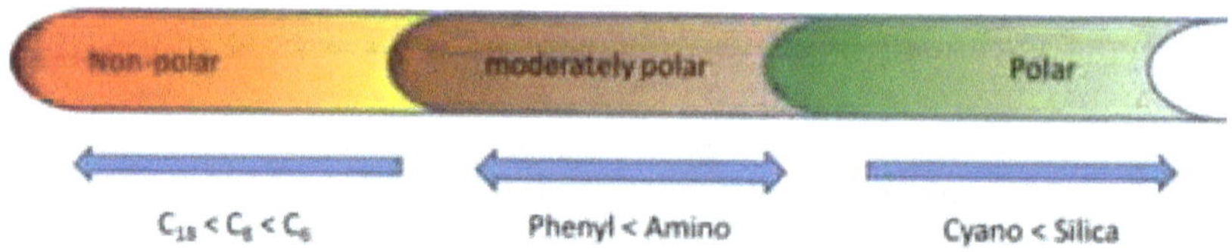

The following are the parameters of a chromatographic column which need to be considered while choosing a column (stationary phase) for separation of assay, impurities, and degradation products:

i. Length and diameter of column

ii. Packing material

iii. Shape of particles

iv. Size of particles

v. Percent (%) of carbon loading

**Column dimension**: Length and internal diameter of packing bed.

- Short (30–50 mm)—can result in short run times and low back pressure

- Long (250–300 mm)—can result in higher-resolution and long run times

A column with a diameter of 2.1 mm leads to a high resolution.

**Particle size**: Decrease in particle size leads to increase in resolution but with a corresponding increase in back pressure. In general smaller particles offer higher efficiency, but there is a chance to get high back pressure limiting the separation efficiency. Less (3 µm) particles are usually used for resolving complex and multicomponent samples, where the lesser surface area induces better resolution and separation characteristics.

**Pore size and surface area**: Larger pores allow larger solute molecules to be retained for a longer time through maximum surface area exposure. High surface area generally provides greater retention, capacity, and resolution for multicomponent samples. Low surface area materials generally equilibrate quickly and provide lesser separation efficiency but can be highly preferred and important in gradient analyses.

**Carbon loading**: Higher carbon loads generally offer greater resolution and longer run times. Low carbon loads shorten run times, and many show a different selectivity. A pictorial representation of difference in carbon loading is as shown below.

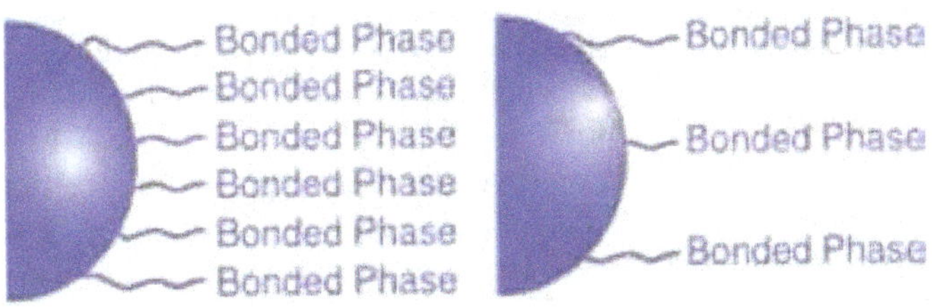


**End capping**: End capping reduces peak tailing of polar compounds that interact excessively with the otherwise exposed, mostly acidic silanols. Non-end capped packing provides a different selectivity than do end-capped packing, especially for polar compounds. A pictorial representation of difference in end capping is shown below.

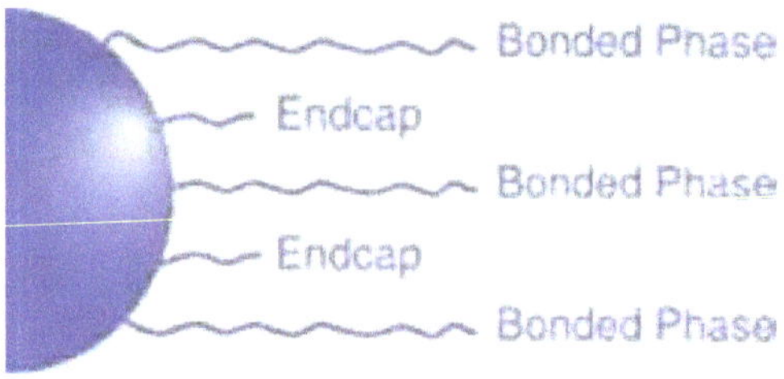


## 2.5 Selection and optimization of mobile phase

Though adsorption is the principle behind chromatography, real separation happens only when the adsorbed compound is eluted using a mobile phase of the required polarity. The selection of mobile phase is done always in combination with the selection of column (stationary phase). The following are the parameters which shall be taken into consideration while selecting and optimizing the mobile phase.

- The right choice of buffer and its eluting efficiency
- pH of the buffer or pH of the mobile phase
- Mobile-phase composition inclusive of binary and tertiary solvent mixture

### *2.5.1 Buffer and its strength*

Buffer and its efficiency play an important role in deciding the peak symmetries (shapes) and peak separation. Various types of organic/inorganic buffers are employed for achieving the required separation. The most commonly used buffers are:

- Phosphate buffers—$KH_2PO_4$, $K_2HPO_4$, $NaH_2PO_4$, $Na_2HPO_4$, $H_3PO_4$, etc.
- Acetate buffers—Ammonium acetate, sodium acetate, etc.
- Triethylamine/diethylamine buffers
- Buffers with various ion-pair reagents like tetrabutyl ammonium hydrogen sulfate, butane sulfonic acid, hexane sulfonic acid, heptane sulfonic acids, etc.

The choice of buffer is to reduce the tailing factor for each peak separated which occurs due to varying ionic strength. The retention time of analyte(s) is delayed and got separated well when more concentrated buffer is used [15]. Better separation happens when the molarity of buffer used is in the range of 0.05 to 0.20 M. The concentration of buffer is chosen by carefully choosing the composition of organic mobile phase.

Depending on the need of the chosen mixture of separation, the strength of the buffer can be increased or decreased if necessary to achieve the required separa-tion, and it can be varied between 10 and 20%, and the effect of variation has to be studied in detail before using. But it should be ensured that increased or decreased buffer strength should not result in precipitation or turbidity either in mobile phase during operation or during storage in refrigerator. Before using the chosen buffer of specific strength to run a column, test experiments have to be done in optimizing the separation to avoid peak tailing, better separation, and reproducibility.

### *2.5.2 pH of buffer*

pH plays an important role in achieving the chromatographic separations as it controls the elution properties by controlling the ionization characteristics. The pH of buffer or mobile phase should be selected based on the *pKa* of analyte or test mixture, which is based on the structure of the molecule. Depending on the *pKa*, drug molecules change retentions, e.g., acids show an increase in retention as the pH is reduced, while the base shows a decrease. If the *pKa* of the compound is high, lower pH or acidic mobile phase has to be chosen as it will stop unwanted association with the stationary phase. For basic compounds, the use of high pH or basic mobile phase and, for neutral compound, neutral mobile phase is highly preferable for better separation.

It is important to maintain the pH of the mobile phase in the range of 2.0 ~ 8.0 as most columns do not withstand the pH which is outside this range. This is due to the fact that the mostly used silica column gets deactivated at high pH (<2) and at low pH (>8) due to cleavage of siloxane linkages. If a pH outside the range of 2.0 ~ 8.0 is found to be necessary, stationary phase which can withstand the range shall be chosen [16–18].

### *2.5.3 Mobile-phase composition*

It is well reported in literature that to achieve better efficiency, binary and tertiary solvent mixtures are used along with other components like buffer and acids or bases. The ratio of the organic versus (vs.) aqueous or polar vs. nonpolar solvents is varied accordingly to get better separation. This is due to the fact that a fairly large amount of selectivity can be achieved by choosing the qualitative and quantitative composition of aqueous and organic portions. Experiments shall be conducted with mobile phases having buffers of different pH and different organic phases to check for the best separations between the impurities. Most chromatographic separations can be achieved by choosing the optimum mobile phase composition [18].

## 2.6 Selection of organic modifiers

Most widely used solvents in reverse-phase chromatography are methanol and acetonitrile. Tetrahydrofuran (THF) is also used but to a lesser extent [19, 20]. In most of the systems, acetonitrile is used as the default organic modifier because of favorable UV transmittance and low viscosity. It is recommended to mix acetonitrile with 5–10% of the aqueous solution(s) to avoid the pumping problems associated with a higher percent (%) of acetonitrile usage. Methanol is also the second most widely used solvent in liquid chromatography, but it gives the back pressure to LC column. Though THF has some disadvantages like higher UV absorbance, reactivity with oxygen, and slower column equilibration, sometimes it gives very unique selectivity for closely eluting peaks. Intermediate selectivity (if needed for a particular sample) can be obtained by blending appropriate amounts of each of these solvents.

Order of polarity: methanol > acetonitrile > ethanol > THF > propanol. Order of solvent strength: propanol > THF > ethanol > acetonitrile > methanol.

## 2.7 Selection of ion-pair reagents

Ion pair reagents are necessary as a mobile-phase additive when structurally or chemically or polarity wise inseparable closely related compounds are to be separated [21, 22]. For example, if a mixture of ionic and nonionic analyte(s) having the same polarity and same retention time is required to be separated, start by optimizing for one of the analytes by adding an ion pair reagent in a mobile phase which reduces or increases the polarity of component and helps in increasing the elution time difference. Careful choice of an appropriate ion-pair reagent is required in such cases to get the necessary selectivity. A dedicated LC column is used when an ion pair reagent (0.0005 M to 0.02 M) is intended to employ for specific analysis, but an appropriate cleaning procedure has to be established to enhance the lifetime of the column material. Alkyl ammonium salts (tertiary or quaternary) and alkyl sulfonate salts are the most useful in the separation of acidic and basic compounds, respectively. Sodium perchlorate can also be used for acidic components.

### 2.8 Selection of flow rate

Separation of mixtures is highly influenced by the flow of mobile phase inside the column [23, 24]. The flow rate is highly crucial in having well-separated peaks with no tailing. The flow rate of the mobile phase can be optimized based on the retention time, column back pressure, and separation of closely eluting adjacent peaks or impurities and peak symmetries from the test run. Preferably the flow rate is fixed not more than 2.0 mL/minute. The flow which gives the least retention times, good peak symmetries, least back pressures, and better separation of adjacent peaks/ impurities could be the chosen as an optimized flow rate for the analysis.

### 2.9 Selection of column temperature

Temperature is another criterion which has to be optimized for any sample, as the flow rate and the rate of adsorption vary with temperature. It is generally believed that with increasing temperature, it can help to improve the resolution between the adjacent/closely eluting peaks and peak merging. So a careful choice of the temperature is a must which might change the pressure of the column and ultimately the elution and resolution [25–28].

Choosing ambient temperature for the analysis is always preferred as it will minimize the degradation of the test sample; however, higher temperatures are also advisable under unavoidable conditions after confirming the stability of the compound. The temperature range which is usually allowed in liquid chromatography is 25 and 60° C. Higher temperatures above 60°C are preferred if the peak symmetry is not good and to increase the retention time for closely occurring peaks.

### 2.10 Selection of solvent delivery system (elution mode)

Chromatographic separations with a single eluent (isocratic elution: all the constituents of the mobile phase are mixed and pumped together as a single eluent) are always preferable. However, the gradient elution is a powerful tool in achieving separation between closely eluting compounds or compounds having narrow polarity difference [29–31]. An important feature of the gradient elution mode which makes it a powerful tool is that the polarity and ionic strength of the mobile phase are changed (increased or decreased) during the run. Experiments using different mobile-phase combinations and different gradient programs have to be performed prior to achieving better separation.

In a gradient run, two mobile phases which have different compositions of polar and nonpolar solvents are premixed using a single pump before introducing to the column which is called as *low pressure gradient (LPG)*, and when the mobile phases are pumped at different flow rate and mixed in a chamber, then introduced into the column is known as *high pressure gradient (HPG)*. It is better to select the gradient run, whether *LPG* or *HPG*, while optimizing the chromatography method. HPG can be only preferred for use when more than 80% organic phase is pumped. To avoid the pumping problems due to the low viscous solvents like acetonitrile in mobile phase, at least 10% aqueous portion could be added to the organic phase.

While optimizing the gradient program, it is important to monitor the following. Pressure graph is needed to be monitored so as to ensure that the overall system pressure will not cross 400 bar or 6000 psi at any point during the run. Flow rate has to be physically cross-checked by collecting the output from the detector during the run at different time intervals, especially when the gradient is running with higher organic-phase composition so as to ensure that there were no pumping problems during the run when mobile phases of different compositions are pumped. It is also important to optimize the program for initialization after each run and before going for the next injection. The program for initialization shall be optimized such that there shall be no carry-over to the next run and the system stabilizes with initial composition before the next injection.

One standard program which can be used for optimizing is discussed below. For starting a method development, a solvent gradient system is always preferred. Initially, start with a gradient of 50:50 buffer and mobile phase, and change the program linearly up to 5:95, and retain the ratio for at least 30 minutes. Then try with a gradient of 95:5 and the program linearly changed up to 5: 95, and retain for at least 30 minutes. The typical gradient program is as follows:

## 2.11 Selection of diluent

Diluent is an aqueous solution or a solvent used to dissolve and extract the drug moiety for analysis. Select a diluent in which impurities, starting material, by-product, intermediates, degradation products, and the analyte are soluble. It is advisable to check first in the mobile phase. All the analytes should be completely soluble and the solution should be clear [32]. Diluent should be compatible with the mobile phase to obtain the good peak shape.

- Selection of diluent based on extraction efficiency and peak shapes: Select the diluent for finished dosage forms, in which the analyte should be extracted at least 95% for assay and 90% for organic impurities. Calculate the % extraction against pure standard compound in the concentration of linear range, (preferably <1 AU) by diluting the test preparation.

- The peak shapes of all compounds should be good in the selected diluent: Select an initial flow rate of 1.0 mL/min or 1.5 mL/min and select column temperature as ambient (25–30°C).

Diluent is selected initially based on solubility of the substance. However, the finalization of diluent is based on its extraction efficiency, peak symmetries, resolution of impurities, and diluent blank injection interference. Inject the diluent blank and test solution spiked with known impurities into the chromatographic system, and establish the noninterference of blank in estimation of the drug and the effect of diluent on resolution of impurities from drug peak and peak symmetry.

## 2.12 Methods of extraction

General methods followed for extraction are sonication, rotary shaking, or seldom both. In some cases where the analyte cannot be extracted by the above procedures, heating can be adapted if the substance is stable and should not precipitate upon cooling to room temperature [33–34].

### 2.13 Samples to be used for analysis

- Use mixture of impurities, starting material, by-product, intermediates, and degradation products to establish the separations.
- Use the reaction mass/mother liquor/what if study samples for the above study if all the impurities samples are not available in the beginning for method development.
- Use forced degradation samples, if degradation products are not available in the case of drug API.
- Prepare a mixture of known impurities spiked on API at a test concentration of about 0.5 or 1.0 mg/mL.
- Prepare a placebo (mixture of excipients to be used in formulation) solution at a concentration equivalent to test concentration (of about 0.5 or 1.0 mg/mL) for the dose in which higher placebo content is expected.

### 2.14 Experimentation to finalize the method

Inject individual solution of standard and impurities to confirm the retention times. Check for the interference from blank. Check for the interference from placebo components in the case of the formulation.

Gradient program will provide an assessment of the elution pattern of polar and nonpolar impurities. Also, run an isocratic run with a mobile phase of a buffer with a suitable pH and acetonitrile in the ratio of acetonitrile: buffer (90:10) using a 250 × 4.6 mm, 5 µm silica column. This will help to know whether any highly nonpolar impurities are still uneluted from the column.

### 2.15 Selection of test concentration and injection volume

The test concentration and injection volume are generally chosen based upon the response of API peak at the selected detector wavelength [35]. However, the test concentration shall be finalized after it is proven that drug (API) is completely extractable at the selected test concentration. After finalizing the test concentration and diluent, prepare a test solution, and keep the filtered solution in closed condition on a bench top, and check whether the solution has any precipitation or turbidity after 24 hours. Generally, the test solution must be clear and should not show any turbidity or precipitation.

### 2.16 Forced degradation studies (stress testing)

It's a method of subjecting the drug substance or drug product to stress with varied strengths of stressing agents to obtain the degradation. The stressed samples were analyzed using an LC system equipped with a PDA detector and monitored for the separation of degradation products formed under the stressed conditions and the peak purity of the analyte peak. The method is considered as stability-indicating for the estimation of the drug if it meets the peak purity requirement [36, 37].

Forced degradation studies are conducted basically to meet the following objectives:

- To investigate the likely degradation products; this, in turn, helps to establish the degradation pathways and the intrinsic stability of the drug molecule.
- To provide a foundation for developing a suitable stability-indicating method.
- Ensure the force degradation limit of 2–20%.

The major forced degradation studies which are to be carried out are as follows:

a. Thermolytic degradation

This stress testing method studies the degradation that is caused by exposure to temperature high enough to induce bond breakage. Solid-state reactions often proceed in an autocatalytic pathway involving an induction period (lag), followed by a period of rapidly increasing degradation and then slowing down of the degradation rate as the compound is consumed. Thus, solid-state reaction kinetics will often follow an S-shaped curve when degradation vs. time is plotted. Thus, before conducting thermolytic degradation, determine the melting point of the compounds of interest. Then, choose a temperature of 70°C for all the drugs for which melting point is <100°C, or choose a temperature which is 40°C below the melting point.

For the compounds for which melting point is >150°C, stress the samples at 105°C. Keep the samples directly exposed in the oven for 1 week or until about 2–20% degradation is achieved, whichever is earlier. Stress the drug substance, placebo, and drug product separately. In the case of the multicomponent drug products, stress testing of placebo with other actives excluding the one at a time shall be performed additionally.

b. Hydrolytic degradation

Drug degradation that involves hydrolysis reaction is called hydrolytic degradation. Hydrolysis reactions are typically acid or base catalyzed. Acidic, neutral, and basic conditions should therefore be employed in order to induce potential hydrolytic reactions. As these hydrolytic stress studies are to be conducted in aqueous solutions, solubility of the drug molecule of interest in water has to be estimated first. Many small molecule drugs are not soluble in water at the concentrations typi-cally used for analytical evaluations (0.1 to 1 mg/mL); in those cases either a slurry or suspension must be used to examine the hydrolytic stability of a compound, or a cosolvent must be added to facilitate the dissolution under the conditions of low solubility. Two most commonly used cosolvents are acetonitrile and methanol. Methanol has the potential of participating in the degradation chemistry which has to be used with caution especially under acidic conditions when the compound being tested contains a carboxylic acid, ester, or amide.

Acetonitrile is generally regarded as inert solvent and is typically preferable to methanol in hydrolytic stress testing studies. However, acetonitrile is not completely inert and can participate in the degradation reactions, leading to art factual degradation results.

The other cosolvents that are recommended for the hydrolytic stress testing studies are shown below.

| Acidic pH | Neutral pH | Basic pH |
|---|---|---|
| Acetonitrile | Acetonitrile | Acetonitrile |
| DMSO | N-methyl pyrrolidine | DMSO |
| Acetic acid | | Diglyme |
| Propionic acid | | p-Dioxane |

The hydrolytic degradations (using water/0.1 M HCl/0.1 M NaOH with or without cosolvent) are recommended to be performed at a temperature of about 70°C with a reflux condenser installed to avoid the loss of evaporation. Reflux until about 2–20% degradation is achieved. Stress the drug substance, placebo, and drug product separately. Neutralize the stressed solutions before injection. Prepare a stressed solution at a higher concentration than that of test concentration. In the case of the multicomponent drug products, stress testing of placebo with other actives excluding the one at a time shall be performed additionally.

c. Humidity stress

Stress the samples to 90% humidity for 1 week. Stress the drug substance, placebo, and drug product separately. In the case of the multicomponent drug products, stress testing of placebo with other actives excluding the one at a time shall be performed additionally.

d. Oxidative degradation

Oxidative degradation is one of the most common mechanisms of drug degradation. Oxidative drug degradation reactions are typically autoxidative, that is, the reaction is radical initiated. Radical initiated reactions start with an initiation phase involving the formation of radicals followed by propagation phase and eventually a termination phase. Thus, the reaction kinetics will often follow S-shaped curve when the degradation vs. time is plotted and will not follow Arrhenius kinetics.

In oxidative stress study, the use of temperature > 30°C is not recommended because the reaction rate in solution may reduce at higher temp due to the decrease in oxygen content of the solvent. Thus, it is always suggested to perform the degradation with 3% hydrogen peroxide at room temperature (25–30°C) with constant stirring in the dark. Stress for 24 hours or until about 1–20% degradation is achieved or whichever is earlier. Stress the drug substance, placebo, and drug product separately.

In the case of the multicomponent drug products, stress testing of placebo with other actives excluding the one at a time shall be performed additionally.

e. Photolytic degradation

Photolytic degradation is the degradation that results from exposure to UV or visible light. Expose the samples to 3 times to 1.2 million lux-hr visible and 200 W-hr/$m^2$ UVA. Stress the drug substance, placebo, and drug product separately. In the case of the multicomponent drug products, stress testing of placebo with other actives excluding the one at a time shall be performed additionally.

## 2.17 Evaluation of stress testing

Peak purity can be evaluated for the main peak and the major degradants which have the peak heights less than 1 AU. Identify the degradation products by co-injection, in case of known impurities and have comparable spectra.

If any known impurity is observed to be increased in stress, it can be examined properly. If process impurity is found to be increased in stress study, it needs to be assessed whether there is any secondary pathway of formation of this impurity via some other degradant route.

After conducting these studies, verify the chromatograms, and observe any peaks merging with respect to main peak and any critical pairs. If any situations were arrived, adjust the mobile-phase compositions, column parameters, etc. and conclude the method parameters.

After method finalization, check the method using different detectors (RI/ELSD/CE/LC–MS), and compare the data with other detectors like UV, fluorescence, etc. The UV inactive components can be found with these experiments. Identify the mass of major degradant which may be formed greater than 1.0% in stress studies, and try to establish the structures.

## 2.18 Mass balance study

Mass balance is a process of adding together the assay value and levels of degradation products to see how closely these add up to 100% of the initial value. It is important to have methods that detect all major degradation products. This is generally done by performing the assay of forced degraded samples and assesses the mass balance. Mass balance has to be achieved at least up to 95% level. If it is less than the required criteria, investigation has to be done and justified. The following are some of the reasons for not achieving the mass balance.:

- Degradation products are:
    - Not eluted from the LC column
    - Not detected by the detector used
    - Lost from the ample matrix, due to insolubility, volatility, or adsorption losses
    - Co-eluted with the parent compound
    - Not integrated due to poor chromatography
- Parent compound may be lost from the sample matrix, due to insolubility, volatility, or adsorption losses
- Inaccurate quantification due to differences in response factors

## 2.19 Detector wavelengths

After separation of all impurities and degradation products, absorption spectra of all the compounds are recorded and compared by taking overlay spectra of all known impurities along with the main analyte in each stress condition and finalizing a wavelength where all impurities are detected and quantified and have the maximum absorbance. In case this is not feasible, select different wavelengths to estimate all impurities. It is also recommended to extract the chromatograms at lower wavelengths like 210 nm–220 nm to see if there is any additional impurities found, which are found to be missing at higher wavelengths; this is likely the case when parent compound breaks into two parts during forced degradation study with one part highly UV active and second part an alkyl chain where alkyl chain will have poor UV character.

## 2.20 Stability of analytical solutions

The stability of analytical solutions (sample or standard) can be established on auto-injector for at least 12 hours continuously in a sequence mode to know the stability of all components and ruggedness of the method (peak shapes, column back pressure over the period of time). To get better results, choose a diluent in which a test solution is stable for at least 12 hours. If the solution is found to be unstable by its nature, then incorporate the stability of solution in test method.

## 2.21 System suitability

System suitability tests verify and ensure whether the system's performance is acceptable at the time of analysis in accordance with the criteria set forth in the procedure or not. System suitability parameters are chosen based on the criticality of separation. In general, resolution factor for the two adjacent peaks or closely eluting peaks is selected as a system suitability requirement. If the separation of impurities from each other and from API peak is found to be satisfactory, there is no need to keep a resolution factor as a system suitability parameter. In such a case, only a diluted standard reproducibility can be adopted as a system suitability requirement. Before finalizing the system suitability parameters, the separation needs to be studied during the robustness study to understand its behavior during the various deliberate changes in method.

System suitability checking must be performed on two different make of HPLC systems whenever the separation of any impurities is critical. For in-process-related impurity issues, the quantification limit (QL) concentration is to be injected, and signal to noise ratio (S/N) must be kept as a system suitability parameter.

## 2.22 Robustness of the method

Robustness by definition means the reliability of an analysis with respect to deliberate variations in method parameters. After finalizing all chromatographic conditions, robustness study with regard to mobile phase composition (±10%), pH (±0.2), gradient (±0.2%/min), flow rate (±0.2 mL/min), and temperature (±5°C) can be carried out to ensure that the developed method is stability-indicating. If the method of analysis is in a gradient mode, it needs to be checked on two different brands of HPLC or different HPLC to check the effect of the system volumes on separations.

## 2.23 Relative response factor

The relative response factor is used to correct the difference in the detector response of impurities with respect to the main analyte peak. It is mainly used to control the impurities or degradation products in a drug substance or drug product. RRF is established for all the known impurities using any of the slope methods. The standard solutions of API and all impurity can be prepared in at least five different concentrations in the range of 0.1–1.0% (e.g., 0.1, 0.3, 0.5, 0.7, and 1.0%) and analyzed using the liquid chromatography. RRF is calculated by using the slope of the respective impurity and slope of the main drug (API) [38, 39].

### 2.24 Quantification methods

The following methods can be used for the quantitative determination of assay and organic impurities [40, 41]:

a. *External standard method:* This method is used for the assay and impurity estimation in a given sample, where the impurities are estimated using the respective impurity standard and without the API standard peak. It's possible to estimate the concentration from calibration curve.

b. *Area normalization:* If the RRF value of known impurity is close to the API (analyte), i.e., 0.9–1.1, the area normalization method is chosen for quantification. The recovery needs to be established without using the response factors.

c. *Diluted standard method:* If the RRF values of impurities are different from the analyte, the diluted standard method can be chosen.

d. *Internal standard method:* If the sample preparation procedure involves different extraction steps to avoid the error in the extraction procedure, internal standard procedure shall be chosen (normally for derivatization techniques and bioanalytical methods).

## 3. Conclusion

Principles involved in chromatography method development, especially for the analytical method development for the separation, identification, purification, and quantitative estimation of organic compounds using the liquid chromatography techniques (HPLC, UPLC, LC–MS, preparative HPLC, etc.), were emphasized in this chapter. Though many different types of chromatography techniques are currently in use, the liquid chromatographic methods HPLC, UPLC, and LC–MS are most widely utilized for the separation and quantitative determination of organic compounds. This chapter mainly focused on and explained the major and critical parameters of the liquid chromatography for the method development and optimization of a suitable stability-indicating LC method and impurity profiling studies. Each and every parameter which controls the purification of most of the organic compounds inclusive of drug, its precursors, and degraded products has been explained in detail in this chapter. The information given in this chapter will help the reader in choosing the right conditions for a particular compound to quantitatively separate from the reaction mixture or drug composition.

## Author details

Narasimha S. Lakka[1,2] and Chandrasekar Kuppan[1*]

1 Department of Science and Humanities, VIGNAN'S Foundation for Science, Technology and Research (VFSTR), Guntur, Andhra Pradesh, India

2 Department of Analytical Research and Development, Jodas Expoim Private Limited, Hyderabad, India

*Address all correspondence to: ramachan16@gmail.com

## References

[1] Ettre LS. Milestones in Chromatography. LCGC North America. 2003. Available from: http://files.pharmtech.com/alfresco_images/pharma/2014/08/22/e5b0c0c2-2e1c-463d-8d4f-4993f7245f47/article-56954.pdf [Accessed: 10 July 2019]

[2] Blomberg L. Stationary phases for capillary gas chromatography. Trends in Analytical Chemistry. 1987;**6**(2):41-45

[3] Zlatkis A, editors. 75 Years of chromatography: A historical dialogue. Journal of Chromatography Library. Elsevier; 1979:7. ISBN 0-444-41754-0 (Vol. 17); ISBN 0-444-41616-1 (Series)

[4] The Nobel Prize in Chemistry. 1952. Available from: nobelprize.org [Retrieved: 25 August 2016]

[5] FDA, United States-Published in the Federal Register. Q6A Specifications: Test Procedures and Acceptance Criteria for New Drug Substances and New Drug Products: Chemical Substances. 2000. Vol. 65; 83041-83063. Available from: https://database.ich.org/sites/default/files/Q6A_Guideline.pdf

[6] Nicholson DE. An Introduction to Metabolic Pathways by S. Dagley. 59, No. 2 ed. Sigma Xi. The Scientific Research Society; 1971. 266p. Available from: https://onlinelibrary.wiley.com/doi/pdf/10.1002/food.19710150432

[7] Q3A(R2) Impurities in New Drug Substances: FDA, United States-Published in the Federal Register on June 2008. Available from: https://www.ich.org/fileadmin/Public_Web_Site/ICH_Products/Guidelines/Quality/Q3A_R2/Step4/Q3A_R2__Guideline.pdf [Accessed: 10 July 2019]

[8] Q3B(R2) Impurities in New Drug Products. FDA, United States-Published in the Federal Register. 14 November 2003; 68; 220; 64628-9. Available from: https://database.ich.org/sites/default/files/Q3B_R2__Guideline.pdf; https://www.ich.org/fileadmin/Public_Web_Site/ICH_Products/Guidelines/Quality/Q3B_R2/Step4/Q3B_R2__Guideline.pdf [Accessed: 10 July 2019]

[9] Liljefors T, Allinger NL. Conformational analysis. 128. The Woodward-Fieser rules and α,β-unsaturated ketones. Journal of the American Chemical Society. 1978;**100**(4):1073. DOI: 10.1021/ja00472a008

[10] Silverstein B. Morrill: Spectroscopic Determination of Organic Compounds. 5th ed1991

[11] Swartz M. HPLC detectors: A brief review. Journal of Liquid Chromatography & Related Technologies. 2010;**33**:1130-1150. DOI: 10.1080/10826076.2010.484356

[12] Rigas PG. Review: Liquid chromatography—post-column derivatization for amino acid analysis: Strategies, instrumentation, and applications. Instrumentation Science & Technology. 2012;**40**(2-3): 161-193. DOI: 10.1080/10739149.2011.651669

[13] Santa T, Al-Dirbashi OY, Fukushima T. Derivatization reagents in liquid chromatography/electrospray ionization tandem mass spectrometry for biomedical analysis. Drug Discoveries & Therapeutics. 2007;**1**(2):108-118

[14] Scott RPW. liquid chromatography detectors. Chapter 3. The fluorometric detector. Journal of Chromatography Library. 1977;**11**:121-130. DOI: 10.1016/S0301-4770(08)61018-0

[15] McCalley DV. Choice of Buffer for the Analysis of Basic Peptides in Reversed-Phase HPLC. LCGC

ASIA PACIFIC VOLUME 8. 2005. Available from: http://alfresco. ubm-us.net/alfresco_images/pharma/2014/08/22/806d27ce-71f0-4239-b9a6-f0541666d136/article-254604.pdf [Accessed: 10 July 2019]

[16] Chemical Analysis. ZORBAX HPLC Columns. Available from: http://quimica.udea.edu.co/~carlopez/cromatogc/ph/95121tb.html [Accessed: 10 July 2019]

[17] Zapala W. Influence of mobile phase composition on retention factors in different HPLC systems with chemically bonded stationary phases. Journal of Chromatographic Science. 2003;**41**. Available from: https://pdfs.semanticscholar.org/2310/bbddc ddcf462b99666f1d7cc351fcab 48952.pdf [Accessed: 10 July 2019]

[18] Janeček M, Šlais K. Design of mobile phase composition for liquid chromatography with an internal pH gradient. Chromatographia. 1993;**36**(1):246-250. DOI: 10.1007/BF02263872

[19] Joshi VS, Kumar V, Rathore AS. Role of organic modifier and gradient shape in RP-HPLC separation: Analysis of GCSF variants. Journal of Chromatographic Science. 2015;**53**(3):417-423. DOI: 10.1093/chromsci/bmu222

[20] Shimada K, Yoshida H, Komine Y. The effect of organic modifier in the mobile phase on the separation of bile acids and its fluorescent derivatives in inclusion chromatography. Journal of Liquid Chromatography. 1991;**14**(4):605-617. DOI: 10.1080/01483919108049274

[21] Ion-pair Reagents for HPLC. TCI/A1084E 20170303. Available from: https://www.tcichemicals.com/a-cmn/en/common/support-download/brochure/ion-pair_reagents_for_hplc.pdf [Accessed: 10 July 2019]

[22] Shibue M, Mant CT, Hodges RS. Effect of anionic ion-pairing reagent hydrophobicity on selectivity of peptide separations by reversed-phase liquid chromatography. Journal of Chromatography A. 2005;**1080**(1):68-75. DOI: 10.1016/j.chroma.2005.03.035

[23] 1.14.4 High-Performance Liquid Chromatography: The International Pharmacopoeia. 8th ed2018. Available from: http://apps.who.int/phint/pdf/b/7.1.14.4.1.14.4-High-performance-liquid-chromatography.pdf [Accessed: 10 July 2019]

[24] Du Q, CaijuanWu GQ, Wu P, Ito Y. Relationship between the flow-rate of the mobile phase and retention of the stationary phase in counter-current chromatography. Journal of Chromatography A. 1999;**835**(1-2):231-235

[25] Heidorn M. The Role of Temperature and Column Thermostatting in Liquid Chromatography. White Paper. WP71499-EN 09/16S. Thermo Scientific. Available from: http://tools.thermofisher.com/content/sfs/brochures/WP-71499-LC-Temperature-Column-Thermostatting-WP71499-EN.pdf [Accessed: 10 July 2019]

[26] Thermo Scientific Product Spotlight SP71195: Best in UHPLC Column Thermostatting to Fit All Needs. 2014. Available from: http://www.thermoscientific.com/content/dam/tfs/ATG/CMD/cmd-documents/bro/bro/chrom/lc/sys/SP-71195-Vanquish-Column-Thermostatting-SP71195-EN.pdf [Accessed: 10 July 2019]

[27] Thermo Scientific Poster Note PN71314. Thermostatting in UHPLC: Forced Air Mode, Still Air Mode, and Method Transfer. Germering, Germany. 2014. Available from: http://www.thermoscientific.com/content/dam/tfs/ATG/CMD/cmd-documents/sci-res/posters/chrom/lc/sys/ PN-71314-

ISC2014- UHPLC- Thermostat -ting-PN71314- E N.pdf [Accessed: 10 July 2019]

[28] Dolan J. Separation Science. HPLC Solutions # 53. Temperature and Retention. Available from: https://owl.english.purdue.edu/owl/resource/747/08/ [Accessed: 10 July 2019]

[29] Gradient HPLC Solvent Delivery System: WPP10. Waters. Available from: https://gimitec.com/file/wpp10.pdf [Accessed: 10 July 2019]

[30] Gradient Design and Development, Breaking the Bad Gradient Cycle: Agilent Technologies. Available from: https://www.agilent.com/cs/library/slidepresentation/public/Gradient%20Design%20and%20Development.pdf [Accessed: 10 July 2019]

[31] Good Habits for Successful Gradient Separations: Getting the Most From Your Method. Agilent Technologies. Available from: https://www.agilent.com/cs/library/slidepresentation/Public/Successful%20Gradient%20Separations.pdf [Accessed: 10 July 2019]

[32] Wrezel PW, Chion I, Pakula R. Key factors in sample diluent selection for HPLC assays of active pharmaceutical ingredients. LCGC North America. 2005;**23**(7):682-686

[33] Slack GC, Snow NH. 8 HPLC sample preparation. Separation Science and Technology. 2007;**8**:237-268. DOI: 10.1016/S0149-6395(07)80014-6

[34] LCGC Editors. Overview of sample preparation. 2015; **33** (11):46-51. Available from: http://www.chromatographyonline.com/overview-sample-preparation [Accessed: 10 July 2019]

[35] John W. Dolan: How much can i inject? Part I: Injecting in mobile phase. LCGC North America. 2014; **32** (10): 780-785. Available from: http://www.chromatographyonline.com/how-much-can-i-inject-part-i-injecting-mobile-phase [Accessed: 10 July 2019]

[36] Singh S, Junwal M, Modhe G, Tiwari H, Kurmi M, Parashar N, et al. Forced degradation studies to assess the stability of drugs and products. Trends in Analytical Chemistry. 2013;**49**:71-88

[37] Jain D, Basniwal PK. Forced degradation and impurity profiling: Recent trends in analytical perspectives. Journal of Pharmaceutical and Biomedical Analysis. 2013;**86**:11-35

[38] Kalyana Chakravarthy VV, Kishore Babu G, Lakshmana Dasu R, Prathyusha P, Aparna Kiran G. The role of relative response factor in related substances method development by high performance liquid chromatography (HPLC). Rasayan Journal of Chemistry. 2011;**4**(4):919-943

[39] Bhattacharyya L, Pappa H, Russo KA, Sheinin E, Roger L. Williams: U.S. pharmacopeia. The use of relative response factors to determine impurities. Pharmacopeial Forum. 2005;**31**(3). Available from: https://www.researchgate.net/publication/295423011_The_use_of_Relative_Response_Factors_to_determine_impurities

[40] Altria KD. Essential peak area normalisation for quantitative impurity content determination by capillary electrophoresis. Chromatographia. 1993;**35**(3-4):177-182

[41] de Oliveira EC, Muller EI, Abad F, Dallarosa J, Adriano C. Internal standard versus external standard calibration: an uncertainty case study of a liquid chromatography analysis. Química Nova. 2010;**33**:4. DOI: 10.1590/S0100-40422010000400041

# 8

# Biological Evidence Analysis in Cases of Sexual Assault

*Benito Ramos González, Miranda Córdova Mercado, Orlando Salas Salas, Juan Carlos Hernández Reyes, Martín Guardiola Ramos, Elton Solis Esquivel, Gerardo Castellanos Aguilar and Porfirio Diaz Torres*

**Abstract**

Sexual assault (SA) is a crime of violence against a person's body resulting in a physical trauma, mental anguish, and suffering for victims generating expenses for government intended criminal investigation, medical care, and psychological attention. During the crime scene investigation, the identification and recovery of biological evidence (BE) are utmost important, since sometimes these are the only way to prove sexual contact and the perpetrator's identity. The examiner, with the help of specific technologies and techniques, must be able to find evidence that otherwise could go unnoticed. Forensic laboratories identify biological evidence with systemized protocols and use molecular methods to generate DNA profiles based on the amplification and DNA sequencing. Before the arrival of the new-generation sequencers, the application of other markers (single nucleotide polymorphisms (SNPs), insertion-deletion of nucleotides (INDEL), or microhaplotypes (MHs)) was laborious, expensive, and not very informative for forensic purposes; however, now they are useful in this field. Next-generation sequencing (NGS) brought a new series of applications like epigenetics, microbiota, messenger RNA, and microRNA analysis and the inferences in the ancestry and phenotyping of individuals. In the end, the results obtained from such analyses and stored in databases are very useful for the identification of sexual aggressors.

**Keywords:** sexual assault, biological evidence, DNA analysis, next generation sequencing, human identification

## 1. Introduction

Sexual assault is considered a serious offense all around the world due to the impact it has on the victims, their relatives, and society in general. The investigation of sex crimes requires a group of multidisciplinary forensic professionals focused on the identification, recovery, packing, and analysis of evidence.

This document describes general recommendations and the decision-making process is necessary for the recovery and analysis of the collected evidence in sexual assault especially that of biological nature coming from victims, perpetrators, and crime scenes. The use of appropriate tools for identifying biological evidence (BE) is a key element in the success of the investigation since it allows forensic investigators to make decisions and utilize presumptive or confirmatory methods to recover and forward evidence to specialized laboratories. Additionally, the utilization of microscopic techniques and genetic fluorescence hybridization allows accurate work while selecting and isolating components on a cellular level thus increasing the possibilities of obtaining a genetic profile that identifies the perpetrator.

Obtaining genetic profiles out of BE in sexual assault cases requires the use of DNA extraction techniques designed for the separation of cells (sperm cells from the aggressor and epithelial cells from the victim) which contribute to the acquiring of differentiated genetic profiles of the contributors.

The use of short tandem repeats (STRs) in forensic investigation has been, for many years, a key element in human identification. Other techniques such as mitochondrial DNA (mtDNA) and single nucleotide polymorphism (SNP) analysis and its variants broaden the possibility of obtaining a profile by providing additional information when other methods fall short. DNA methylation analysis, microRNAs, and genome sequencing of microorganisms provide scientific information for criminal investigation.

The development of new-generation sequencing has set the perspective of analysis on establishing geographical origin of individuals, estimating marker frequency of different population groups around the world just as genetic markers of phenotypic expression allow acquiring information of visible external characteristics (height, baldness, eye color, skin and hair), they provide help for criminal investigation.

The implementation and use of databases which register the information acquired from sexual assault investigations are a necessary tool to facilitate the comparison of the resulting information; hence, establishing the parameters to enter or delete perpetrator profiles and other genetic profiles, different in nature, from such databases must be contained in every country's legislation to be used in criminal investigation processes.

This chapter focuses on the location, sampling, molecular analysis and management of biological evidence in sexual crimes, as well as on specific aspects and a panorama of their molecular analysis.

## 2. Crime scene investigation and recovery of biological evidence

### 2.1 General inspection

Before the investigators begin examining the scene of the crime, they should gather as much information as possible about the setting to prevent lost or destruc-tion of valuable and/or fragile evidence such as shoeprints, trace evidence, etc. The main areas of inspection are the floor, rugs, bathroom, bedding, and trash receptacles where other elements could be discarded by the aggressor during cleaning such as condoms; the inspection should be extended to the neighborhood if necessary [1, 2].

In the search for signs of sexual contact, the investigator can identify evidence through naked eye observation; however, it is convenient to emphasize that evidence of contact is frequently not visible. These elements of BE require the use of forensic light sources for detection due to their natural characteristics, such as light absorption (blood) or fluorescence emissions (semen, saliva, and urine). This method is a simple, presumptive, and nondestructive test [3–5].

In cases where evidence is not detected with the use of forensic light, it is necessary to use other techniques such as Bluestar®to detect washed blood stains, low light or magnifying glasses to observe fibers, and the use of vacuum machines that retain material in filters which could be analyzed in a criminal laboratory [2, 4, 5].

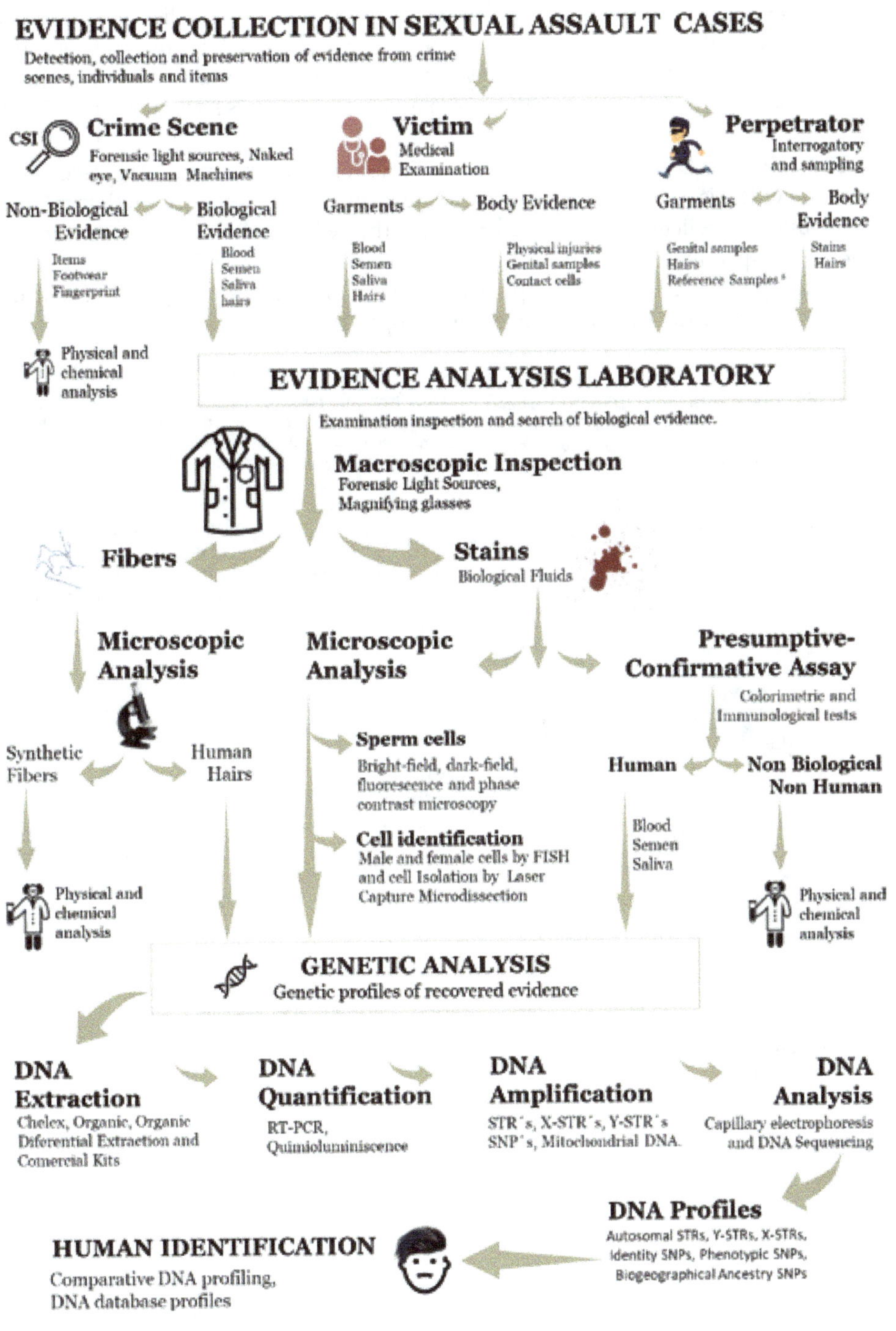


**Figure 1.**
*Evidence collection in sexual assault cases. Workflow of inspection, recovery and analysis of evidence from the crime scene, victims and perpetrator up to the genetic profile and human identification. Crime scene investigator (CSI) is in charge of collecting and analyzing crime scene evidence; medical examination: forensic medical examiner, medical technician or nurse; interrogation and sampling is done by a law enforcement agent and by the scientific police investigator; evidence analysis laboratory is done by the research scientists specialized in evidence analysis.*

## 2.2 Recovery of evidence at the crime scene

In a SA investigation, it is necessary to identify any possible source of BE left on the victim or at the crime site (e.g., condoms, body fluids on objects or textiles, bottles, cigarette butts, and hair). Transportable evidences will be packed and sent to the laboratory. When BE are in non-transportable objects, the use of a dry or lightly moistened swab passed gently through and rotated in the same spot (swabbing method) is sufficient for recovery. In the case of wet evidence, care should be taken to dry them to avoid damage of BE, by the growth of microorganisms that cause degradation of DNA [6].

The success of DNA typing is related to the amount of target material recovered from an evidentiary item. Absorption and adsorption are two features that related to capability to collect BE and to later release the cells/DNA during the extraction process, respectively. Synthetic swabs release more cells/DNA during the extraction process and yielded up to 2.5 times more alleles compared to cotton swabs because portions of DNA remain entrapped in the fibers [7].

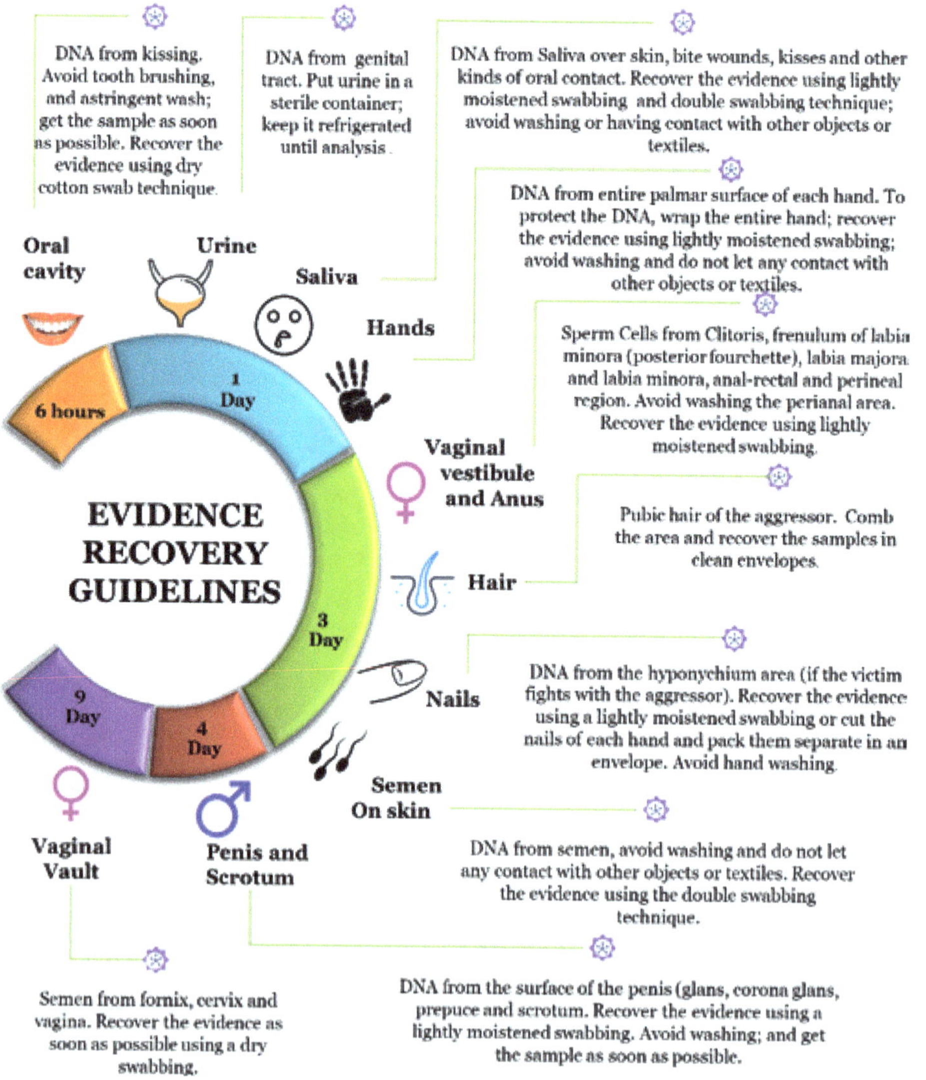


**Figure 2.**
*Evidence recovery guidelines. Recommended time frame for evidence collection from different anatomical areas according to DNA persistence and its sampling methods in SA cases.*

Swabs of different design, shape, and size used for evidence recovery are commercially available (X-Swab™ Diomics Corporation and Copan 4N6FLOQSwab™); all of them with highly absorptive properties. The use of double swabbing method are recomended to recovery of touched (trace) evidence; this technique increases the possibility of obtaining DNA profiles; however the use of cotton swabs is not recommended for trace evidence [7–10]. **Figure 1** shows the workflow of evidence recovery from the crime scene.

### 2.3 Recovery of evidence on victim and perpetrator

When a SA is reported, authorities order a medical interview and examination for evidence recovery; during the interview, the expert needs to document the type of sexual aggression (penile-vaginal rape, oral, copulation, sodomy, penetration with foreign objects, or digital penetration), Personal hygiene, and the elapsed time after the incident are crucial; these information will indicate the type of sampling to be performed. Additionally, the examiner will look for elements that are associated with aggression (e.g., bites and body fluids), and these will be obtained from anatomical regions that show signs of injury or attack [8, 11–13] (**Figure 1**).

One source of evidence in SA investigation is the suspect or perpetrator. It is known that the evidence could potentially be transferred from the suspect to the victim and vice versa. Therefore, depending on the type of contact involved in a SA, the suspect's body may actually be a better source of probative evidence. The biological evidence deposited on the victim and perpetrator deteriorates rapidly; therefore, it needs to be collected as soon as possible [14]. **Figure 2** shows the evidence recovery guidelines.

The sperm cells are resistant to biological degradation compared to somatic cells; this rationale is supported by the knowledge that the protein composition of the sperm nucleus (protamine) acts as a protector of the damage caused by the nucle-ases, delaying the degradation process [15].

## 3. Identification of biological evidence in the laboratory

The evidence/garments collected (from the victim, corpse, aggressor, and crime scene) are inspected in the laboratory in order to perform a search for blood, semen, hair, saliva, sweat, tissues, fibers, and other elements. One of the first interventions is the macroscopic analysis that consists of evaluating evidence through meticulous and sequential observation, evaluating and establishing strategies to find biological spots. When BE is not visible to the naked eye, it is then necessary to use technologi-cal help: the forensic light sources with specific wavelengths for its detection [3–5] (**Figure 1**).

In daily forensic practice, the latent spots of some biological fluids such as semen, saliva, urine, and sweat require the application of light radiation with specific wavelengths for detection by fluorescence depending on their emission properties or absorption of light; although fibers and hairs are elements that can be observed without instruments, the lack of contrast in the background makes their visibility difficult; in such cases, the use of magnifying glasses or lights helps to generate shadows that can help to locate them.

Once identified, the BE on the area—depending of surface or support of the fluid—is taken with moistened swabs with sterile water, or a portion (of support) is cut to perform a presumptive or confirmatory analysis of the evidence. In the case of trace evidence, it should be kept in its original support (textile) and analyzed ensuring sufficient evidence is left for subsequent trials [10].

The applications of presumptive chromatic reaction tests are useful for orientation in the identification of its nature and its selection of confirmatory test for determination of human origin through immunological tests. It is important to consider the amount of BE for the destructive processes for some test and to apply necessary measures for its preservation or greater use for subsequent studies.

Some forensic laboratories analyze semen through optic microscopes, aiming to identify the sperm cells. There is controversy regarding this procedure since a portion of the sample is separated from the original support, making it difficult to apply other analyses, even though it is important to consider it as minimal evidence for obtaining genetic profiles. On the other hand, laboratories use fluorescence microscopy for cytological preparations to apply fluorescent techniques that allow increasing the sensitivity in the detection of spermatozoa, confirming the presence of these cells in the analyzed fluids [16, 17]. **Figure 3** describes the advantages and disadvantages of presumptive and confirmatory forensic tests.

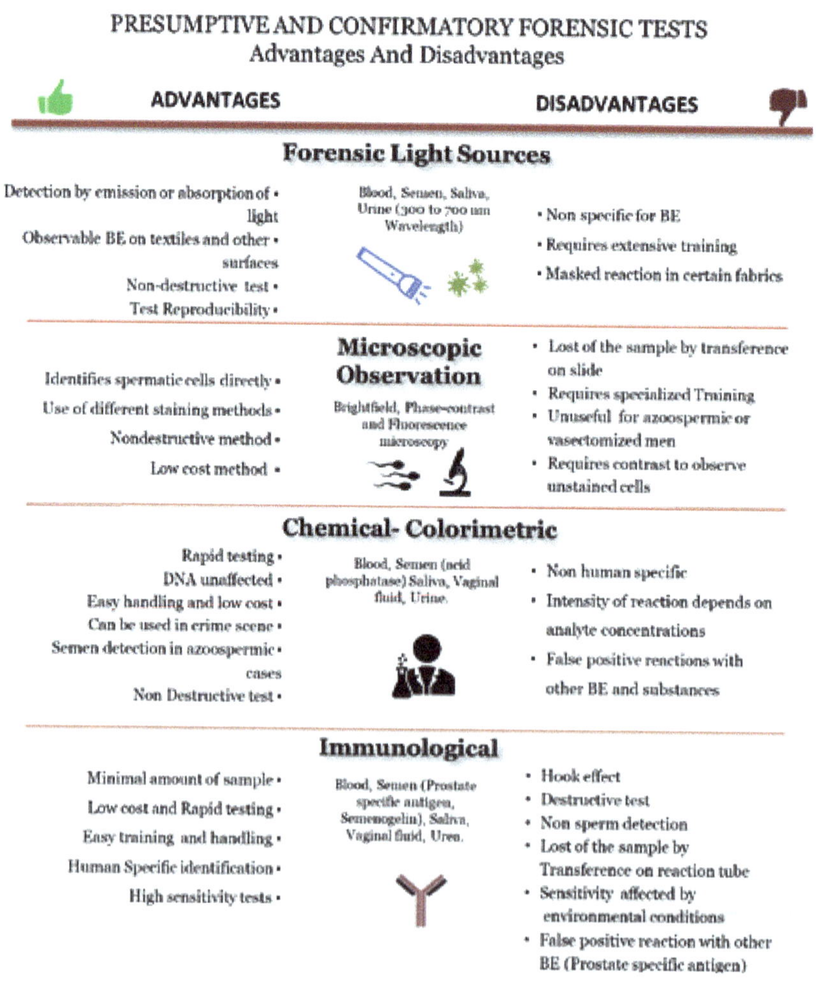


**Figure 3.**
*The advantages and disadvantages of presumptive and confirmatory tests used in the laboratory to locate and identify the type of BE. Its application use goes from general to particular fluid identification, considering its destructive nature based on the analyst's criteria. BE: biological evidence; Bf: bright field microscopy; Ph: phase contrast microscopy.*

## 4. Cell isolation from biological evidence

Biological cell mixtures represent one of the major challenges in forensic genetics. In principle, when more individuals contribute to a mixture with different biological fluids, their single genetic profiles can be obtained by separating the distinct cell types [18, 19]. There are standard DNA extraction methods developed to separate the sperms (male fraction) from the epithelial cells (female fraction) as preferential lysis; however, these methods are incapable of separating single-source sperm from multiple male donors [20].

There has been a recent use of modern tools to reach that goal. Laser microdissection (LMD) is a technology that has been around for more than 40 years; it combines the amplification power of a microscope with the precision cut of objects allowed by the laser technology. Only in the last decade has LMD been used for forensic purposes, mainly in SA for isolating sperm cells from vaginal swabs [18, 21–23].

### 4.1 Laser microdissection (LMD)

The use of LMD in the forensic field was first described in 2003 as a way of recovering sperm cells from slide smears of SA cases. LMD allows the selection of individual cells based on morphologic analysis (e.g., sperm and epithelial cells) or on labeling with specific fluorescent dyes. The microscopic search for sperm in cases where there is a limited number of cells can be exhaustive and prolonged [24]. However, this technology includes an automatic searching function module as introduced by the manufacturers [20, 24].

Until today two variants of this technique are noted: laser capture microdissection (harvesting cells by melting a thermoplastic membrane) and laser cutting microdissection (harvesting cells by catapulting). The operating principles of these types of LMD are the identification of cells, using the laser to perform clean cuts in the supporting layer around them and not requiring physical manipulation of the cells eliminating the risk to foreign contamination [19, 22, 23]. The cell analysis in a mixture with an azoospermic or oligospermic contributor is more difficult. This is because in the absence of sperm cells, the male and female cells are indistinguishable; therefore, the use of specific fluorescent dyes is required [20].

### 4.2 Fluorescence in situ hybridization and laser microdissection

The use of LMD does not always allow distinguishing the sperms in the microscopic bright field for several reasons: they can lose the tail; few sperms; or azoospermic cases. However, non-sperm cells can be found in semen, such as leukocytes and epithelial cells from the ejaculatory duct and urethra [18, 25].

Fluorescence in situ hybridization (FISH) method allows distinguishing male cells from female ones in cellular mixtures. The DNA is hybridized with DNA probes for the "X" and "Y" chromosomes (marked with fluorophores) and then observed in fluorescence microscopy, enabling individual identification [18, 25, 26]. The LMD in combination with the FISH technology can greatly improve the identification and later separation of male non-spermic cells from epithelial female cells.

This technique (FISH with LMD) has been shown to be capable of producing autosomal STR profiles from samples that previously would have proved difficult or impossible to separate; additionally, it has applications in numerous other sample types where the ratio of female cells to male cells is large, including cases involving penetration without ejaculation, digital penetration, or oral sex [18, 27]. On the other hand, other separation methods [28] were developed which consisted of separating sperms from epithelial cells taking the difference in size and shape; this gave mixed genotypes in the results. Other new methods have also been proposed for cell separation, such as low-volume polymerase chain reaction (LV-PCR) used for single sperm isolation and detection, aspiration capillaries, microfluidic devices, the mDip technique, and fluorescence-activated cell sorting with flow cytometry, based on immunolabeling only applicable on fresh vaginal lavages and not on vaginal smears or archived material [20].

## 5. DNA analysis

### 5.1 DNA extraction methods

There are many extraction methods available, and they vary in their ability to extract the DNA in an efficient way; some of the factors to consider are the kind of sample to be analyzed, the time it takes to process, the operator intervention, the risk of contamination, and the difficulty or ease of use. This is the basis for successful forensic DNA profiling [6, 29].

The method of preference has the task to not only ensure that the DNA is efficiently extracted from each sample, but it must also remove possible inhibitors which may interfere with other processes like the amplification [29].

#### *5.1.1 Techniques for DNA extraction*

One of the most common techniques used in DNA extraction is Chelex, which is a chelating resin that uses ion exchange to bind transition metal ions protecting the DNA from degradation. The advantage of the Chelex® method is that it is quick, it does not require multiple tube transfers, and it does not use toxic organic solvents; the main disadvantage is that it is unable to remove inhibitors that interfere with the amplification process [6, 30–32].

When processing samples with inhibitors, it is advisable to use the organic extraction method, which requires lysis of cells carried out in a salt solution containing detergents and proteases to denature proteins and release the DNA from the cell. This cocktail can be separated by using a mixture of phenol-chloroform-isoamyl alcohol, which leaves the DNA in the aqueous phase. The extracted DNA can be concentrated from the aqueous phase by ethanol precipitation or with a centrifugal filter unit, which allows for additional purification and concentration of the DNA in the samples [6, 29, 31].

The advantage of the organic extraction method is that it can obtain genetic material from difficult samples (degraded and/or low amount of DNA) and can successfully remove the presence of inhibitors for the PCR. While this method remains one of the most reliable and efficient, it is also very time-consuming, uses hazardous chemicals, and, because of the greater hands-on effort and multiple tube transfers involved, introduces increased risks for contamination and sample mishandling [6, 31].

#### *5.1.2 Differential lysis in DNA mixtures*

The genetic analysis of the evidence collected in sexual crimes commonly includes genetic profiles of two or more contributors; in this kind of mixtures, the genetic contribution of the individuals is generally unbalanced. In some circumstances, the biological mixture presents a minimal level of one contributor, usually the perpetrator in cases of SA. The genetic rate of this donor is likely not to be detected because of the sensitivity limits or the reaction saturation by the component that has more quantity. In most cases, the minor contributor in the DNA mixture cannot be detected when ratios exceed 1:20 [29].

The recovery of evidence in cases of SA is a great challenge for the DNA forensic analysts, because it requires the separation of DNA from epithelial (the victim) and sperm (perpetrator) cells. The differential extraction was first described in 1986 by Gill and coworkers [33], as a modification of the organic phenol-chloroform extraction, and it is called differential lysis because the non-sperm cells are selectively lysed with detergent and proteases, while the sperm cells are not lysed due to the heavily disulfide cross-linked proteins in the sperm head that resist protease treatment [6, 29, 33, 34].

In DNA forensic labs, the differential lysis method has long been the standard for separating spermatozoa from epithelial cells. Although this technique can theoretically provide two fractions, as pointed out earlier (one comprising the offender's DNA and the other containing the victim's DNA), the separation is not always complete, resulting in mixed genotypes [29, 33].

#### *5.1.3 Other DNA extraction methods*

There are other methods to separate sperm and epithelial cells from sexual assault samples. The Differex™ System method involves a proteinase K-selective digestion of epithelial cells, followed by differential centrifugation and phase separation. The use of this method in DNA laboratories indicates it offers efficiency equal to the two-step method for extracting sperm DNA from mixed stains [6, 35, 36].

## 5.2 Molecular methods for human identification

The first use of DNA testing in a forensic setting came in 1986; two girls were sexually assaulted and then brutally murdered in 1983 and 1986, in Leicestershire, England. This case showed an innocent being accused and 1 year later the guilty responsible one being found and processed [37].

In the last 30 years, DNA molecular analysis has become an important tool in forensic investigations. Currently, DNA profiling is based on polymerase chain reaction (PCR) analyses. This method includes the autosomal STRs, Y and X chromosomes. The PCR is a process of replicating a specific region on the genome, over and over again to yield many copies of a region [29, 38].

Before the PCR, the DNA has to be quantified. This is essential in order to ensure its correct amplification; its primary purpose is to determine the amount of DNA template, resulting from the isolation. There are many methods with different accuracy, but knowing the DNA concentration present in the samples allows the forensic scientist to establish the ideal amount of DNA required for its amplification in order to make it possible to obtain a genetic profile that falls within the quality parameters set by the laboratory [29, 39].

The genetic analysis of the evidence collected in sexual crimes commonly includes genetic profiles of two or more contributors; in this kind of mixtures, the genetic contribution of the individuals is generally unbalanced. This will further impair the identification process through a series of stochastic effects, such as preferential amplification, which it is known to possibly affect PCR [29, 40].

### 5.2.1 Short tandem repeat (STR) analysis

Short tandem repeat (STR), also called microsatellites, or simple sequence repeats (SSRs) contain a core of nucleotides (length) that are tandemly repeated, and their use in forensic science opened a new path in human identification [29, 40, 41].

It is well known that STRs have a high degree of discrimination due to their hypervariable markers, which are useful when it is intended to involve the perpetrator in the crime scene or in the victim. Artifacts are a common challenge in forensic cases; biological ones (stutter products, incomplete adenylation, etc.) and instrumental ones (arise from voltage spikes, dye blobs, etc.) must often be sorted through in order to generate a complete and accurate STR profile [29, 41, 42].

Biological evidence showing fragmented DNA is commonly found in SA cases and can be recovered more effectively when the PCR products are smaller. By moving the PCR primers closer to the STR region, the product sizes can be reduced while retaining the same information [43–45]. In practice, the success rates in recovering information from compromised DNA samples improve with mini STR systems compared with conventional STR kits.

The sex chromosomal STR indicates biological lineage of a person, obtaining a low power of exclusion between relatives. Y-STR markers can play a role when mixed profiles of opposite sexes are involved, in cases where differential extraction is not possible, in an azoospermic male or in aged sexual stains [46, 47]. The X-STR markers have a wide range of forensic applications and can be used for establishing the relationship between distant relatives, such as aunt, niece, and cousins [48, 49].

Furthermore, theoretical and the first empirical evidence was provided to show that a set of 13 RM Y-STRs (rapidly mutating Y-STRs) is able to achieve an order of magnitude higher than male relative differentiation. The effects of this near-complete male individualization will be of great benefit to forensic applications (e.g., to reduce the inclusion of innocent individuals in sexual investigations due to adventitious haplotype matches) [50, 51].

### 5.2.2 Single nucleotide polymorphisms (SNPs) analysis

Single nucleotide polymorphisms (SNPs) are a single-base sequence variation between individuals at a particular point and take place in millions of sites in the human genome which means they could differentiate individuals from one another. SNPs are able to recover information from degraded DNA samples that show no stochastic phenomena, the sample processing and data analysis can be more automated because a size-based separation is not needed, and it has the ability to predict ethnic origin and certain physical traits with a careful selection of markers [6, 52].

One of the biggest challenges of using SNPs in forensic DNA typing applications is the inability to simultaneously amplify enough SNPs in robust PCR multiplexes, from small amounts of DNA. Because a single biallelic SNP yields less information than a multiallelic STR marker, it is necessary to analyze a larger number of SNPs in order to obtain a reasonable power of discrimination to define a unique profile. Formerly, high-density SNP arrays allow hundreds of thousands or even millions of SNPs to be analyzed in parallel.

The basic principles of SNP array are the convergence of DNA hybridization, fluorescence microscopy, and solid surface DNA capture. The three mandatory components of the SNP arrays are an array containing immobilized allele-specific oligonucleotide (ASO) probes; fragmented nucleic acid sequences of target, labeled with fluorescent dyes; and a detection system that records and interprets the hybridization signal. However, these arrays typically require hundreds of nano-grams of DNA, which are usually not available from forensic casework samples arising from minute biological stains, and for this reason it is more often used in ancestry studies [6, 29, 53, 54].

Another form of a biallelic (or di-allelic) polymorphism is insertion-deletion of nucleotides or INDEL which can be a DNA segment. Most INDELs exhibit allele of few nucleotides length differences. The PCR amplicons were designed to be less than 160 bp, and with this a complete profile could be obtained down to approximately 300 pg of DNA template [29, 55]. However, not all INDELs are highly informative in all populations, and the exact number of INDELs in the human genome remains unknown [56].

Both SNPs and INDELs can now be typed using multiplexes based on fragment length analysis on instruments available in all routine forensic laboratories, thus making it possible to extend the range of markers beyond the currently used STRs. In recent years haplotype systems based on multiple SNPs are being tried as optimal markers for the forensic area due to their discriminating power nearing that of STRs which provides a powerful alternative for the analysis. The microhaplotypes (MHs) have 2 or more SNPs in a span of less than 200 nucleotides (creating a multi-allelic locus), with extremely low recombination rates and discriminating power similar to STRs useful in cases with fragmented DNA and mixture sample analysis [57–59].

### 5.2.3 Mitochondrial DNA analysis

Mitochondrial DNA (mtDNA) analysis is commonly performed using the Sanger sequencing chemistry [60–63]. This DNA sequencing is performed in both the forward and reverse directions so that the complementary strands can be compared to one another for quality control purposes. The focus of most forensic DNA studies to date has involved two hypervariable regions within the control region commonly referred to as HVI (HV1) and HVII (HV2). Occasionally a third portion of the control region, known as HV3, is examined to provide more information regarding a tested sample [29, 64, 65].

Human mitochondrial DNA is considered to be inherited strictly from our mothers and is commonly used in parental linkage. The middle piece of sperm cells contains mtDNA, and this DNA is more resistant than autosomal DNA because in small circular genomes, the double membrane of the mitochondrion and the circular structure (without open ends) act as protective agents against the degradation processes [35, 66].

The forensic applications for mtDNA include analysis of samples that are degraded or with low amount of DNA (e.g., stains, hairs, bones), and it was used for the identification of Tsar Nicholas II and his brother Georgij Romanov [67]. In recent approach, it has been demonstrated that the mtDNA could be used for the identification of sperm cells in the vaginal tract through a micromanipulation technique [68, 69]. Besides the physical separation, sequence-specific primers (SSP) for the man were used to ensure that the woman's mtDNA would not be co-amplified. The primer design was based on the mtDNA haplotype differences between contributors determined after mtDNA analysis of buccal swabs. This procedure allows the characterization of the male mitotype from a single sperm cell present in a vaginal swab [70].

### 5.2.4 Next-generation sequencing for forensics

There are several next-generation sequencing (NGS) platforms using different sequencing technologies. All of them perform sequencing of millions of small fragments of DNA in parallel; they use the bioinformatic analyses to piece together these fragments by mapping the individual reads to the human reference genome, providing to deliver accurate data and an insight into unexpected DNA variations [70–72].

The bases of the method consist in DNA polymerase catalyzing the incorporation of fluorescently labeled deoxyribonucleotide triphosphates into a DNA template strand during sequential cycles of DNA synthesis. During each cycle, at the point of incorporation, the nucleotides are identified by fluorophore excitation. The critical difference is that, instead of sequencing a single DNA fragment, NGS extends this process across millions of fragments in a massively parallel fashion [73, 74].

The NGS analysis allows to find differences in the ordering of nucleotides in the DNA in cases where the alleles are of the same size. It also allows us to analyze multiple polymorphisms simultaneously in a single workflow (Autosomal STRs, Y-STRs, X-STRs, Identity SNPs, Phenotypic SNPs and Biogeographical ancestry SNPs) [74]. NGS reveals substantial sequence variation in addition to repeat length, thereby increasing the discriminatory power of STRs compared to conventional fragment analysis; it also allows for the analysis of large panels of SNPs when severely degraded DNA is involved [72].

On the other hand, the information obtained from multiple analyses in NGS is not needed in all forensic cases and can take up large portions of the sequencing capacity which will eventually result in fewer samples per sequencing run and a higher cost of the investigation. In our experience, a reliable quality control platform for the sizing and quantification of the libraries is necessary.

NGS can also be used for the detection and identification of microorganisms found in biological evidence (on victims or perpetrators) and the sexually transmitted infections (STIs) with the aim of tracing the source of microbes besides estimating the postmortem interval (PMI) related to changes in microbial community profiles or "microbial clocks" [75, 76]. NGS has the advantage of high throughput and multiplexing capability and accuracy, which makes it suitable for rapid whole-genome typing of polymorphisms detected by analyzing every base of the genome, thus giving forensic data higher resolution and greater accuracy.

Edaphic, necrobiomic microorganisms at the cadaver-soil interface construct multi-species communities that change when the host body dies and begins to decompose. Characterization of these dynamic changes has been made possible by metagenomic technologies [71, 72, 75]. It is expected that a high-quality forensic microbial database will soon become a reality and aid in the fast and accurate identification of criminals and biological terrorists.

Even nonhuman species identification is an important component of forensic practice: The species that range from domestic animals (common in the urban areas) to insects that were present in crime scene [29, 77]. Entomological evidence is used to define the PMI, and it is essentially based on the morphological recognition of the insect and an estimation of its insect life cycle stage; however, molecular genotyping methods can also provide an important support for forensic entomological investigations when the identification of species or human genetic material in their digestive tract is required [71, 72, 75, 76].

Epigenetic approaches based on NGS technology include whole-genome bisulfite sequencing and methylated DNA sequencing. Interestingly, extremely low amounts of starting DNA (100 pg) were successfully analyzed through genome-wide amplification of a bisulfite-modified DNA template, followed by quantitative methylation detection using pyrosequencing. Additionally, another encouraging study performed bisulfite genomic DNA sequencing with micro-volume blood spot samples. This can also be used to predict tissue type and associations with diseases and determine the sex and age of a DNA donor [71, 72, 75].

Furthermore distinguishing monozygotic twins has been a limitation in forensic genetics, since they exhibit identical STR profiles; the high number of readings of a single sequence that is able to reach NGS, allows to see the variations of methylated DNA and mitochondrial SNPs, giving us a way to distinguish them [78, 79].

MicroRNAs (miRNAs) have only recently been introduced to forensic science; they are a class of endogenous small RNA molecules with 18–24 nucleotides in length. There small size, resistance to degradation, and tissue-specific or highly tissue-divergent expression plays an essential regulative role for many cellular processes. They are suitable for forensic body fluid identification making it possible to conclusively link a DNA profile to a particular body fluid, species identification, different disease states, and PMI [71].

NGS technology in forensic science will increase the field of applications which contribute to the resolution of criminal cases. The standardization of procedures among laboratories will lead to the acceptance before the court, as well as to the understanding of their uses and limitations (see **Figure 3**).

### 5.3 Ancestry and phenotypic expression

Humans are 99.9% identical in their DNA. The difference between each human genome is small. Yet, in analyzing these small differences, we can begin to understand what makes us unique. The variation between human genomes is not randomly distributed across the globe. Humans are more likely to have descendants with people that live nearby; the closer geographically two individuals or populations are, the more genetically similar to they tend to be. If we were to gather DNA from across the globe, we could connect certain genetic signatures to geographic spaces. Population-specific alleles have been found in both STR and SNP markers. The genetic patterns of human population variation arose from a series of sequen-tial migrations and bottleneck events [80, 81].

#### 5.3.1 Analysis of genetic markers of ancestry

SNPs are more convenient to become "fixed" in a population than are STRs, because of their lower mutation rate. SNPs change on the order of once every hundred generations, while STR mutation rates are approximately one in a thousand. Ancestry informative markers (AIMs) possess alleles with large frequency differences between populations that can help distinguish them. A small proportion of SNP variants have emerged as particularly informative for ancestry, inferred by comparing a sample's genetic diversity with the patterns of variation in contemporary populations. When selecting suitable ancestry informative markers, the degree of divergence between populations and the number of populations that a test seeks to differentiate have both a bearing on the selection process [80, 82].

Ancestry inference offers many other applications, including aiding cold case reviews with additional data on linked profiles; achieving more complete identifications of missing persons or disaster victims; assessing atypical combinations of physical characteristics in individuals with admixed parentage; and enhancing genetic studies where forensic sensitivity is necessary, e.g., testing medical archive material or archaeological DNA [82].

AIMs, however, are not 100% accurate for predicting ancestral background of samples; for example, individuals with mixed ancestral backgrounds may not possess the expected phenotypic characteristics. Thus, results from genetic tests attempting to predict ethnic origin or ancestry should always be interpreted with caution and only in the context of other reliable evidence. In countries like the United States where movement of the population is more fluid, greater levels of admixture are expected, and thus genetic testing results would not be as likely to correlate strongly with geographic location. However, the possibility of admixed ancestry raises a warning in the use of any statistic with any panel of AIMs. Admixed ancestry cannot be estimated accurately unless the ancestral populations are represented among the reference populations [83].

AIMs are limited, identification of the optimal SNPs could change between group of samples, and some panels are based on very large numbers of SNPs, thereby limiting the ability of others to test different populations. AIMs in forensic genetic investigations of crime scene can be performed on very small amounts of DNA, less than 1 ng. The strategy for interpretation of the result of AIM investigations can be explorative. The likelihoods of the AIM profiles in various populations may be calculated, and the one with the highest likelihood may be considered the population of origin. When two populations are identified a priori, the likelihood ratios of the populations are calculated.

The likelihood that one population is greater than another does not prove that any of the two populations are relevant to the AIM profile, due to the fact that even though the populations may be exclusive, they are not exhaustive in the sense that covers all possible human populations [84, 85].

Due to continuous migrations, AIM alleles are shared across all human groups; it is not the absolute presence/absence of an allele, rather its frequency in the population that is usually analyzed when inferring ancestry. The recombination of autosomal markers can provide additional information about the admixed nature of an individual. Y-chromosome markers and mitochondrial DNA (mtDNA) sequence variation have benefits and limitations for ancestry inference that relate to their maternal and paternal lineages [82]. INDELs may also be valuable AIMs, but the number of markers and the informative value are less than those of SNPs [84].

### 5.3.2 Analysis of genetic markers of phenotypic expression

Forensic phenotyping can provide useful intelligence regarding the ancestry and externally visible characteristics (EVCs) of the donor of an evidentiary sample. Currently, SNPs base inference of externally visible characteristics. This may substitute and support eyewitness testimony when descriptions are unavailable or uncertain, in which DNA from the perpetrator is available but no suspect is identified [80, 86].

The predicting phenotypes of EVCs from DNA genotypes have the final aim of concentrating police investigations to find persons completely unknown, without database matches or low quality/quantity of DNA available and finally requesting standard forensic STR profiling only for the reduced number of EVC matching suspects aiming DNA individualization for courtroom use [86, 87].

The ability to predict the physical appearance of an individual directly from crime scene material can in principle help police investigations by limiting a large number of potential suspects where unknown perpetrators are involved, where STR profiling could not provide a hit within the DNA (profile) database or could not provide a match with a suspect singled-out by authorities or cases where an STR profile could simply not be generated due to low quality and/or quantity of DNA available.

In the case of an unidentified body being found in an advance state of decomposition with no visible physical characteristics, EVCs are expected to provide leads for human identification. However, work is still being done to identify predictive DNA markers for several other EVCs such as skin color, hair color, body height, male baldness, and hair morphology [84, 87–89].

Numerous global studies describe correlations between population geographical distribution and variations in the allele frequencies that are linked to several human phenotypes, including the skin, hair, and iris pigmentation, biological metabolism, biological modification variants, disease susceptibility, and morphology, because these variations are expected to display great population diversity. The investigators and juries may have trouble understanding probabilities from ancestry or phenotyp ing predictions using DNA results. Telling a detective that the individual donor of a biological sample at a crime scene has an 80% chance of having blue eyes is new territory when he or she typically associates a DNA result as being irrefutable evidence.

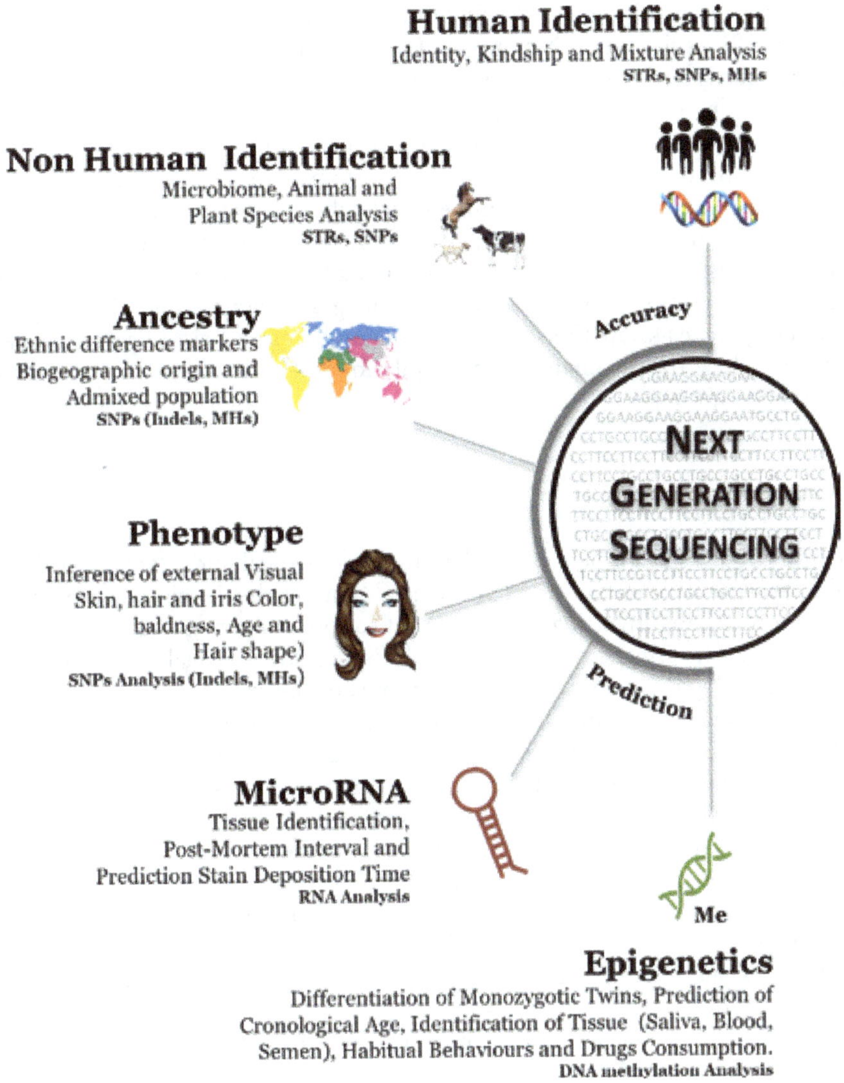


**Figure 4.**
*Next generation sequencing applications. NGS analysis in forensic science provides ample information showing the highest level of precision on individual identity (profiling) and the lowest on prediction of the habitudes of a person (epigenetic biomarkers).*

If ancestry prediction and forensic phenotyping are pursued, then expectations of individuals using the information will need to be managed [89, 90]. **Figure 4** shows next-generation sequencing applications and its usefulness in human identification.

## 6. Genetic DNA database

Forensic genetics has become a key test in multiple criminal and civil proceedings for its ability to confirm or eliminate a suspect. In the criminal field, it allows to analyze criminal strategies and identify authors, improving judicial and police management [5, 6, 12, 29, 91]. The DNA databases pursue the resolution of criminal cases allowing the automated comparison of DNA profiles from the crime scene, of suspects or convicts and sometimes of the victims. The usefulness of this type of database is indisputable in all the countries in which it exists [6, 14, 29, 92].

Currently genetic database CODIS (Combined DNA Index System) developed by the US FBI exchange and compare DNA profiles electronically from crime scenes and convicted offenders are stored. CODIS can be searched to determine if a DNA profile pulled from biological evidence in a crime matches the DNA of a known offender or DNA from evidence in another crime.

The legislations of each country vary in certain points that affect these issues. Another important point is to determine which laboratories can generate DNA profiles that are included in the database. It is likely that in the near future, developed countries will establish collaboration agreements for the exchange of genetic data, which could be a fundamental tool for the fight against some crimes. It is important that public agencies know the scope of these databases and establish collaboration agreements for the exchange and collation of information for criminal investigation purposes.

## 7. Conclusions

Sexual assault is a complex crime that involves medical and psychological attention for the victim and generates high financial cost per the development of forensic investigation. During investigation the identification, collecting and packing of biological fluids in the crime scene and the analysis of evidence in labs are fundamental since errors during this stage would affect the rest of the investigation [6]. The use of protocols of interventions in crime scene decreases the possibility of loss of data that could clarify the crime, and even the protocol must be complemented with the interview of witnesses and/or victims in order to make decisions in broadening the area of evidence search. The standardization and quality control of procedures guarantee that all personnel manage a crime scene in the same way.

For the correct and successful investigation of sexual crimes, it is necessary to recover evidence in three principal areas: crime scene, victims, and perpetrator. Evidence recovery must be completed during the first hours after the crime; this is crucial for the success of the investigation, although it does not always happen for some investigation units [8, 11, 13, 14].

The analysis of evidence in the laboratory continues with the macroscopic examination of biological spots. The methods used by crime laboratories are presumptive screening tests, and some of them have confirmatory tests that will conclusively identify their presence. A disadvantage of most of these current methods is that they are designed to detect a specific body fluid (**Figure 3**); the investigator needs to decide which test to perform based on the fluid that is most likely present [6]. It is necessary to develop a universal confirmatory test that can be applied to an unknown stain and which will be able to identify any of the body fluids. However, in 2016 Scientific Working Group on DNA Analysis Methods (SWGDAM) recommends the SA-targeted testing approach: direct to DNA. The serology test employed by laboratories is less sensitive than modern DNA typing kits; However, DNA typing only the swabs which screen positive in the serology test enables the possibility of missing elegible profiles [42, 71, 73].

Microscopic identification of sperm cells continues to be used in some forensic laboratories; its usefulness continues to be controversial due to the fact that the use of this technique in cases in which the evidence is minimal leads to the loss of such evidence besides making sperm cell identification difficult due to the lack of contrast. Fluorescent contrast techniques (FISH and immunolabeling) and LMD solve the problem of microscopic identification by allowing to separate cell mixtures from more than one contributor and producing genetic autosomal profiles free from DNA contamination [18, 25–27].

DNA extraction methods are increasingly effective in the recovery of trace evidence but are still ineffective in the analysis of mixture (separation of contributors), which is a common scenario in sexual assault. The technique used to isolate sperm cells from epithelial cells is the differential extraction, but since it is not always possible to separate both cells, it is necessary to implement other techniques [33].

Autosomal STR analysis using the PCR technique is widely used for human identification; however, DNA mixture is frequent in sex crimes, and its scope is limited. The application of next-generation sequencing in cases of mixed DNA allows the solving of the problem since the sequencing can show the construction of the bases that make up the units of alleles. Thus, even if two or three people in a mixture have the same length, next-generation sequencing (NGS) can tell them apart or, in compromised and degraded samples, regain relevance in sexual crimes [72, 76].

NGS has opened new possibilities in human identification, since it is no longer limited to one type of marker at a time. It allows analyzing a large number of individuals obtaining a significant depth of sequencing of their genomes; an analyst can sequence a multiple number of STRs, identity, ancestry, and phenotypic informative SNPs [74]. However, it is necessary to establish parameters in the admissibility of the evidence on new technologies; considering phenotypic information as a search pattern for a suspect, as well as tracking it with the information of their ancestry, is debatable from an ethical and moral point of view. There is a lot of work to be done for this area to be developed.

Conclusively, solid foundations in the development of sexual assault investigations include scrutiny, selection, and discrimination of evidence supported on the knowledge of the forensic investigator. It is the investigators who hold a crucial role in the fulfillment of the purpose of forensic sciences which is to contribute to the uphold of justice amid the threat to humanity's most fundamental rights, to life and freedom.

## Acknowledgements

The authors would like to express our gratitude to all forensic scientists of the Criminalistics and Forensic Services Institute for providing technical reference and their valuable information.

## Notes/thanks/other declarations

None.

## Author details

Benito Ramos González[1*], Miranda Córdova Mercado[1], Orlando Salas Salas[2], Juan Carlos Hernández Reyes[3], Martín Guardiola Ramos[3], Elton Solis Esquivel[4], Gerardo Castellanos Aguilar[5] and Porfirio Diaz Torres[5]

1 Scientific Research Department, Criminalistics and Forensic Services Institute, Attorney General's Office of the State of Nuevo Leon, Monterrey, Mexico

2 Forensic Genetic Laboratory, Criminalistics and Forensic Services Institute, Attorney General's Office of the State of Nuevo Leon, Monterrey, Mexico

3 Evidence Analysis Laboratory, Criminalistics and Forensic Services Institute, Attorney General's Office of the State of Nuevo Leon, Monterrey, Mexico

4 Forensic Chemistry Laboratory and Quality Assurance, Criminalistics and Forensic Services Institute, Attorney General's Office of the State of Nuevo Leon, Monterrey, Mexico

5 Crime Investigation Group, Criminalistics and Forensic Services Institute, Attorney General's Office of the State of Nuevo Leon, Monterrey, Mexico

*Address all correspondence to: benito.ramos@gmail.com

## References

[1] Magalhães T, Dinis-Oliveira RJ, Silva B, Corte-Real F, Nuno-Vieira D. Biological evidence management for DNA analysis in cases of sexual assault. The Scientific World Journal. 2015;**2015**:1-11. DOI: 10.1155/2015/365674

[2] Technical Working Group on Crime Scene Investigation United States of America. Crime Scene Investigation: A Guide for Law Enforcement [Internet]. 2000. Available from: https://www.ncjrs.gov/pdffiles1/nij/178280.pdf [Accessed: 10 July 2018]

[3] Lee W, Khoo B. Forensic light sources for detection of biological evidences in crime scene investigation: A review. Malaysian Journal of Forensic Sciences. 2010;**1**:17-28

[4] Horswell J. The practice of crime scene investigation. In: International Forensic Science and Investigation Series. 1st ed. New York: CRC Press; 2004. 421 p. ISBN: 0-748-40609-3

[5] Li R. Forensic Biology. 2nd ed. Boca Raton: CRC Press. 533 p. ISBN: 13: 978-1-4398-8972-5

[6] Buttler JM. Advanced Topics in Forensic DNA Typing: Methodology. 1st ed. San Diego: Academy Press; 2012. 652 p. ISBN: 978-0-12-374513-2

[7] Marshall PL, Stoljarova M, Larue BL, King JL, Budowle B. Evaluation of a novel material, Diomics X-Swab™, for collection of DNA. Forensic Science International: Genetics. 2014;**12**: 192-198. DOI: 10.1016/j.fsigen. 2014.05.014

[8] Chávez ML, Hernández-Cortés R, Jaramillo-Rangel G, Ortega-Martinez M. Relevance of sampling and DNA extraction techniques for the analysis of salivary evidence from bite marks: A case report. Genetics and Molecular Research. 2015;**14**:10165-10171. DOI: 10.4238/2015.August.21.23

[9] Sweet D, Lorente M, Lorente JA, Valenzuela A, Villanueva E. An improved method to recover saliva from human skin the double swab technique. Journal Forensic Science. 1997;**42**: 320-322. DOI: 10.1520/JFS14120J

[10] Sabine Hess S, Haas C. Recovery of trace dna on clothing: A comparison of mini-tape lifting and three other forensic evidence collection techniques. Journal of Forensic Science. 2017;**62**: 187-191. DOI: 10.1111/1556-4029.13246

[11] Howard S. National Best Practices for Sexual Assault Kits: A Multidisciplinary Approach [Internet]. 2017. Available from: https://www.ncjrs.gov/pdffiles1/nij/250384.pdf [Accessed: 01 July 2018]

[12] International Protocol on the Documentation and Investigation of Sexual Violence in Conflict [Internet]. 2014. Available from: https://assets.publishing.service.gov.uk/government/uploads/system/uploads/attachment_data/file/319054/PSVI_protocol_web.pdf [Accessed: 10 August 2018]

[13] Hebda LM, Doran AE, Foran DR. Collecting and analyzing DNA evidence from fingernails: a comparative study. Journal Forensic Science. 2014;**59**:1343-1350. DOI: 10.1111/1556-4029.12465

[14] Forensic Exams for the Sexual Assault Suspect [Internet]. 2013. Available from: http://www.evawintl.org/library/DocumentLibraryHandler.ashx?id=24 [Accessed: 08 August 2018]

[15] Akmal M, Aulanni'am A, Widodo MA, Sumitro BS, Purnomo BB. The important role of protamine in spermatogenesis and quality of sperm:

A mini review. Asian Pacific Journal of Reproduction. 2016;**5**:357-360. DOI: 10.1016/j.apjr.2016.07.013

[16] De Moors A, Georgalis T, Armstrong G, Modler JB, Fre'geau Chantal J. Sperm Hy-Liter$^{TM}$: An effective tool for the detection of spermatozoa in sexual assault exhibits. Forensic Science International: Genetics. 2013;**7**:367-379. DOI: 10.1016/j.fsigen.2013.02.011

[17] Westring CG, Morten W, Nielsen SJ, Fogleman JC, Old JB, Lenz C, et al. SPERM HY-LITERTM for the identification of spermatozoa from sexual assault evidence. Forensic Science International: Genetics. 2014;**12**:161-167. DOI: 10.1016/j. fsigen.2014.06.003

[18] McAlister C. The use of fluorescence in situ hybridisation and laser microdissection to identify and isolate male cell in an azoospermic sexual assault case. Forensic Science International: Genetics. 2011;**5**:69-73. DOI: 10.1016/j.fsigen.2010.04.008

[19] Fontana F, Rapone C, Bregola G, Aversa R, de Meo A, Signorini G, et al. Isolation and genetic analysis of pure cells from forensic biological mixtures: The precision of a digital approach. Forensic Science International: Genetics. 2017;**29**:225-241. DOI: 10.1016/j. fsigen.2017.04.023

[20] Han JP, Yang F, Xu C, Wei YL, Zhao XC, Hu L, et al. A new strategy for sperm isolation and STR typing from multi-donor sperm mixtures. Forensic Science International: Genetics. 2014;**13**:239-246. DOI: 10.1016/j. fsigen.2014.08.012

[21] Costa S, Ciané L, Porto MJ, Correia-de-Sá P. Laser scanning microdissection, advantages and pitfalls in forensic diagnosis. In: Antonio MV, editor. Microscopy and Imaging Science: Practical Approaches to Applied Research and Education. 1st ed. Formatex Research Center; 2017. pp. 202-207. ISBN-13: 9788494213496

[22] Elliott K, Hill DS, Lambert C, Burroughes TR, Gill P. Use of laser microdissection greatly improves the recovery of DNA from sperm on microscope slides. Forensic Science International. 2003;**137**:28-36. DOI: 10.1016/S0379-0738(03)00267-6

[23] Micke P, Ostman A, Lundeberg J, Ponten F. Laser-assisted cell microdissection using the PALM system. Methods in Molecular Biology. 2005;**293**:151-166

[24] Vandewoestyne M, Van Hoofstat D, Van Nieuwerburgh F, Deforce D. Automatic detection of spermatozoa for laser capture microdissection. International Journal Legal Medicine. 2009;**123**:169-175. DOI: 10.1007/ s00414-008-0271-1

[25] Murray C, McAlister C, Elliott K. Identification and isolation of male cells using fluorescence in situ hybridisation and laser microdissection, for use in the investigation of sexual assault. Forensic Science Internacional: Genetics. 2007;**1**:247-252. DOI: 10.1016/ j. fsigen.2007.05.003 [Epub Aug 13, 2007]

[26] Lynch L, Gamblin A, Vintiner S, Simons JL. STR profiling of epithelial cells identified by X/Y-FISH labelling and laser microdissection using standard and elevated PCR conditions. Forensic Science International: Genetics. 2015;**16**:1-7. DOI: 10.1016/j. fsigen.2014.10.017

[27] Murray C, McAlister C, Elliott K. The use of fluorescence in situ hybridisation and laser microdissection to isolate male non-sperm cells in cases of sexual assault. International Congress Series. 2006;**4**:622-624. DOI: 10.1016/j. fsigen.2010.04.008

[28] Chen J, Kobilinsky L, Wolosin D, Shaler R, Baum H. A physical method

for separating spermatozoa from epithelial cells in sexual assault evidence. Journal Forensic Science. 1998;**43**:114-118. DOI: 10.1520/JFS16097J

[29] Butler JM. Forensic DNA Typing: Biology, Technology and Genetics of STR Markers. 2nd ed. Burlington: Academy Press; 2005. 660 p. ISBN: 9781493300204

[30] Phillips K, McCallum N, Welch L. A comparison of methods for forensic DNA extraction: Chellex-100 and the QIAGEN DNA Investigator kit (manual and automated). Forensics Science International: Genetics. 2012;**6**:282-285. DOI: 10.1016/j.fsigen.2011.04.018

[31] Hu Q , Liu Y, Yi S, Huang D. A comparison of four methods for PCR inhibitor removal. Forensic Science International: Genetics. 2015;**16**:94-97. DOI: 10.1016/j.fsigen.2014.12.001 [Epub Dec 12, 2014]

[32] Singh UA, Kumari M, Iyengar S. Method for improving the quality of genomic DNA obtained from minute quantities of tissue and blood samples using Chelex 100 resin. Biological Procedures Online. 2018;**1**:12. DOI: 10.1186/s12575-018-0077-6

[33] Gill P, Jeffreys AJ, Werrett DJ. Forensic application of DNA 'fingerprints. Nature. 1985;**318**:577-579. DOI: 10.1038/318577a0

[34] Schwerdtner G, Germann U, Cossu C. The separation of male and female: A comparison of seven protocols. Forensic Science International: Genetics. 2017;**6**:e9-e11. DOI: 10.1016/j. fsigss.2017.09.021

[35] Tsukada K, Asamura H, Ota M, Kobayashi K, Fukushima H. Sperm DNA extraction from mixed stains using the Differex™ System. Elsevier: International Congress Series. 2006;**1288**:700-703. DOI: 10.1016/j.ics.2005.12.059

[36] Mudariki T, Pallikarana-Tirumala H, Ives L, Hadi S, Goodwin W. A comparative study of two extraction methods routinely used for DNA recovery from simulated post coital samples. Forensic Science International: Genetics Supplement Series. 2013;**4**:194-195. DOI: 10.1016/j.fsigss.2013.10.100

[37] Joseph W. The Blooding. 1st ed. New York: Perigord Press; 1989. 390 p. ISBN 10: 0688086179

[38] Kasai K, Nakamura Y, White R. Amplification of a variable number of tandem repeats (VNTR) Locus (pMCT118) by the polymerase chain reaction (PCR) and its application to forensic science. Journal of Forensic Sciences. 1990;**35**:1196-1200. DOI: 10.1520/JFS12944J

[39] Nicklas JA, Buel E. Quantification of DNA in forensic samples. Analytical and Bioanalytical Chemistry. 2003;**376**: 1160-1167. DOI: 10.1007/s00216-003-1924-z

[40] Gill P, Haned H, Bleka O, Hansson O, Dørum G, Egeland T. Genotyping and interpretation of STR-DNA: Low-template, mixtures and database matches: Twenty years of research and development. Forensic Science Internatinal: Genetics. 2015;**18**:100-117. DOI: 10.1016/j.fsigen.2015.03.014

[41] Butler JM. Short tandem repeat typing technologies used in human identity testing. BioTechniques. 2007;**43**:Sii-SSv. DOI: 10.2144/000112582

[42] Scientific Working Group on DNA Analysis Methods of SWGDAM. Interpretation Guidelines for Autosomal STR Typing by Forensic DNA Testing Laboratories [Internet]. 2010. Available from: http://media.wix. com ugd/4344b0_61b46a0e1a4c41ccb6 5f719a533b8e29.pdf [Accessed: 03 August 2018]

[43] Butler JM, Shen Y, McCord BR. The development of reduced size STR amplicons as tools for analysis of degraded DNA. Journal of Forensic Science. 2003;**48**:1054-1064. DOI: 10.1520/JFS2003043

[44] Martín P, García O, Albarrán C, García P, Alonso A. Application of mini-STR loci to severely degraded casework samples. International Congress Series. 2006;**1288**:522-525. DOI: 10.1016/j. ics.2005.10.044

[45] Senge T, Madea B, Junge A, Rothschild MA, Schneider PM. STRs, mini STRs and SNPs—A comparative study for typing degraded DNA. Legal Medicine. 2010;**13**:68-74. DOI: 10.1016/j. legalmed.2010.12.001

[46] Purps J, Geppert M, Nagy M, Rowwer L. Validation of a combined autosomal/Y-chromosomal STR approach for analyzing typical biological stains in sexual-assault cases. Forensic Science International: Genetics. 2015;**19**:238-242. DOI: 10.1016/j.fsigen.2015.08.002

[47] Apostolov A. Differentiation of mixed biological traces in sexual assaults using DNA fragment analysis. Biotechnology & Biotechnological Equipment. 2014;**28**:301-305. DOI: 10.1080/13102818.2014.909171

[48] Diegoli TM. Forensic typing of short tandem repeat markers on the X and Y chromosomes. Forensic Science International: Genetics. 2015;**18**: 140-151. DOI: 10.1016/j. fsigen.2015.03.013

[49] Thornton T, Zhang Q, Cai X, Ober C, McPeek MS. XM: Association testing on the X-chromosome in case-control samples with related individuals. Genetic Epidemiology. 2012;**36**: 438-450. DOI: 10.1002/gepi.21638

[50] BallantyneKaye N, Keerl V, Wollstein A, Choi Y, Zuniga SB, Ralf A, et al. A new future of forensic Y-chromosome analysis: Rapidly mutating Y-STRs for differentiating male relatives and paternal lineages. Forensic Science International: Genetics. 2012;**6**:208-218. DOI: 10.1016/j.fsigen.2011.04.017

[51] Rogalla U, Woźniak M, Swobodziński J, Derenko M, Malyarchuk BA, Dambueva I, et al. A novel multiplex assay amplifying 13 Y-STRs characterized by rapid and moderate mutation rate. Forensic Science International: Genetics. 2015;**15**:49-55. DOI: 10.1016/j. fsigen.2014.11.004

[52] Dixon LA, Dobbins AE, Pulker HK, Butler JM, Vallone PM, Coble MD, et al. Analysis of artificially degraded DNA using STRs and SNPs—Results of a collaborative European (EDNAP) exercise. Forensic Science International. 2006;**164**:33-44. DOI: 10.1016/j. forsciint.2005.11.011

[53] Ding L, Wiener H, Abebe T, Altaye M, Go RC, Kercsmar C, et al. Comparison of measures of marker informativeness for ancestry and admixture mapping. BMC Genomics. 2011;**12**:622. DOI: 10.1186/1471-2164-12-622

[54] Louhelainen J. SNP arrays. Microarrays (Basel). 2016;**5**:27. DOI: 10.3390/microarrays5040027

[55] Fondevila M, Pereira R, Gusma L, Phillips C, Lareu MV, Carracedo A, et al. Forensic performance of insertion-deletion marker system. Forensic Science International. 2011;**3**:e443-e444. DOI: 10.1016/j. fsigss.2011.09.083

[56] Vasudeva M, Lim FJ, Vijaya PS, Kumaraswamy K. Forensic identification by using insertion-deletion polymorphisms. International Journal of Human Genetics. 2015;**15**:55-59. DOI: 10.1080/09723757.2015.11886253

[57] Kidd KK, Pakstis AJ, Speed WC, Lagace R, Chang J, Wootton S, et al. Current sequencing technology makes microhaplotypes a powerful new type of genetic marker for forensics. Forensic Science International: Genetics. 2014;**12**:215-224. DOI: 10.1016/j.fsigen.2014.06.014

[58] Kidd KK, Pakstis AJ, Speed WC, Lagace R, Chang J, Wootton S, et al. Microhaplotype loci are a powerful new type of forensic marker. Forensic Science International: Genetics. 2013;**4**:e123-e124. DOI: 10.1016/j.fsigss.2013.10.063

[59] van der Gaagb Kristiaan J, de Leeuw Rick H, Laros Jeroen FJ, den Dunnen Johan T, de Knijff P. Short hypervariable microhaplotypes: A novel set of very short high discriminating power loci without stutter artefacts. Forensic Science International: Genetics. 2018;**35**:169-175. DOI: 10.1016/j.fsigen.2018.05.008

[60] Sanger F, Nicklen S, Coulson AR. DNA sequencing with chain-terminating inhibitors. Proceedings of the National Academy of Sciences of the United States of America. 1977;**74**: 5463-5467. DOI: 10.1073/pnas.74.12.5463

[61] Sanger F, Coulson AR. A rapid method for determining sequences in DNA by primed synthesis with DNA polymerase. Journal of Molecular Biology. 1975;**94**:441-448. DOI: 10.1016/0022-2836(75)90213-2

[62] Mark W, DiZinno JA, Polanskey D, Replogle J, Budowle B. Validation of mitocondrial DNA sequencing for forensic casework analysis. International Journal of Legal Medicine. 1995;**108**:68-74. DOI: 10.1007/bf01369907

[63] Melton T, Holland C, Holland M. Forensic mitochondrial DNA analysis: Current practice and future potential. Forensic Science Review. 2012;**24**:101-122

[64] Budowle B, Allard MW, Wilson MR, Chakraborty R. Forensics and Mitochondrial DNA: Applications, debates, and foundations. Annual Review of Genomics and Human Genetics. 2003;**4**:119-141. DOI: 10.1146annurev.genom.4.070802.110352

[65] Costa S, Lima G, Correla-de-Sa P, Porto MJ, Calne L. Assessment of DNA and mtDNA degradation in sperm cells collected by laser micro-dissection. Journal of Forensic Research. 2017;**8**:5. DOI: 10.4172/2157-7145.1000393

[66] Butler JM, Levin BC. Forensic applications of mitochondrial DNA. Trends in Biotechnology. 1998;**16**:158-162. DOI: 10.1016/S0167-7799(98)01173-1

[67] Ivanov PL, Wadhams MJ, Roby RK, Holland MM, Weedn VW, Parsons TJ. Mitochondrial DNA sequence heteroplasmy in the Grand Duke of Russia Georgij Romanov establishes the authenticity of the remains of Tsar Nicholas II. Nature Genetics. 1996;**12**:417-420. DOI: 10.1038/ng0496-417

[68] Yukseloglu EH, Bayhan E, Rayimoglu G, Cavus F, Dastan K, Erkan I. Identification with mitochondrial DNA typing from one sperm cell isolated by micromanipulation. Medicine Science Internatinal Medical Journal. 2018;**7**:575-579. DOI: 10.5455/medscience.2018.07.8821

[69] Zhang L, Ding M, Pang H, Xing J, Xuan J, Wang C, et al. Mitochondial DNA typing of laser-captured single sperm cells to differentiate individuals in a mixed semen stain. Electrophoresis. 2016;**37**:2273-2277. DOI: 10.1002/elps.201600009

[70] Pereira J, Neves R, Forat S, Huckenbeck W, Olek K. MtDNA typing of single-sperm cells isolated by micromanipulation. Forensic Science International. 2012;**6**:228-235. DOI: 10.1016/j.fsigen.2011.05.005

[71] Yang Y, Xie B, Yan J. Application of next-generation sequencing technology in forensic science. Genomics, Proteomics & Bioinformatics. 2014;**12**:190-197. DOI: 10.1016/j.gpb.2014.09.001

[72] Børsting C, Morling N. Next generation sequencing and its applications in forensic genetics. Forensic Science International: Genetics. 2018;**18**:78-89. DOI: 10.1016/j.fsigen.2015.02.002

[73] Buermans HP, den Dunnen JT. Next generation sequencing technology: Advances and applications. Biochimica et Biophysica Acta. 2014;**1842**:1932-1941. DOI: 10.1016/j.bbadis.2014.06.015 [Epub Jul 1, 2014]

[74] Guo F, Yu J, Zhang L, Li J. Massively parallel sequencing of forensic STRs and SNPs using the Illumina® ForenSeq™ DNA Signature Prep Kit on the MiSeq FGx™ Forensic Genomics System. Forensic Science International: Genetics. 2017;**31**:135-148. DOI: 10.1016/j.fsigen.2017.09.003 [Epub Sep 8, 2017]

[75] Finley SJ, Benbow ME, Javan GT. Potential applications of soil microbial ecology and next-generation sequencing in criminal Investigations. Applied Soil Ecology. 2015;**88**:69-78. DOI: 10.1016/j.apsoil.2015.01.001

[76] Pilli E, Agostino A, Vergani D, Salata E, Ciuna I, Berti A, et al. Human identification by lice: A next generation sequencing challenge. Forensic Science Internatinal: Genetics. 2016;**266**:e71-e78. DOI: 10.1016/j.forsciint.2016.05.006

[77] Iyengar A, HAdi S. Use of non-human DNA analysis in forensic science: A mini review. Medicine, Science and the Law. 2014;**54**:41-50. DOI: 10.1177/0025802413487522

[78] Wang Z, Zhang S, Bian Y, Li C. Differentiating between monozygotic twins in forensics through next generation mtGenome sequencing. Forensic Science international: Genetics. 2015;**5**:e58-e59. DOI: 10.1016/j. fsigss.2015.09.023

[79] Weber-Lehmann J, Schilling E, Gradl G, Richter DC, Wiehler J, Rolf B. Finding the needle in the haystack: Differentiating "Identical" twins in paternity testing and forensic by ultra-deep next generation sequencing. Forensic Science International: Genetics. 2017;**9**:42-46. DOI: 10.1016/j.fsigen.2013.10.015

[80] Santos C, Phillips C, Fondevila M, Daniel R, van Oorschot RAH, Burchard EG, et al. Pacifiplex: An ancestry-informative SNP panel centred on Australia and the Pacific region. Forensic Science International: Genetics. 2016;**20**:71-80. DOI: 10.1016/j. fsigen.2015.10.003

[81] Santos C, Phillips C, Gomez-Tato A, Alvarez-Dios J, Carracedo Á, Lareu MV. Inference of Ancestry in Forensic Analysis II: Analysis of Genetic Data. Methods in Molecular Biology. 2016;**1420**:255-285. DOI: 10.1007/978-1-4939-3597-0_19

[82] Phillips C. Forensic genetic analysis of bio-geographical ancestry. Forensic Science International: Genetic. 2015;**18**:49-65. DOI: 10.1016/j.fsigen.2015.05.012

[83] Kidd Kenneth K, Speed William C, Pakstis Andrew J, Furtado Manohar R, Rixun F, Abeer M, et al. Progress toward an efficient panel of SNPs for ancestry inference. Forensic Science International: Genetics. 2014;**10**:23-32. DOI: 10.1016/j.fsigen.2014.01.002

[84] Tvedebrink T, Eriksen PS, Mogensen HS, Morling N. Weight of the evidence of genetic investigations of ancestry informative markers. Theoretical Population Biology.

2018;**120**:1-10. DOI: 10.1016/j. tpb.2017.12.004

[85] Phillips C, Parson W, Lundsberg B, Santos C, Freire-Aradas A, Torres M, et al. Building a forensic ancestry panel from the ground up: The EUROFORGEN Global AIM-SNP set. Forensic Science International: Genetics. 2014;**11**:13-25. DOI: 10.1016/ j. fsigen.2014.02.012

[86] Butler K, Peck M, Hart J, Schanfield M, Podini D. Molecular "eyewitness": Forensic prediction of phenotype and ancestry. Forensic Science International: Genetics. 2011;**3**:e498-e499. DOI: 10.1016/j.fsigss.2011.09.109

[87] Scudder N, McNevin D, Kelty SF, Walsh SJ, Robertson J. Forensic DNA phenotyping: Developing a model privacy impact assessment. Forensic Science International: Genetics. 2018;**34**:222-230. DOI: 10.1016/ j.fsigen.2018.03.005

[88] MacLen CE, Lamparello A. Forensic DNA phenotyping in criminal investigations and criminal courts: assessing and mitigating the dilemmas inherent in the science. Recent Advances in DNA & Gene Sequences. 2014;**8**:104-112. DOI: 10.2174/23520922 09666150212001256

[89] Kayser M. Forensic DNA Phenotyping: Predicting human appearance from crime scene material for investigative purposes. Forensic Science International: Genetics. 2015;**18**:33-48. DOI: 10.1016/j.fsigen.2015.02.003

[90] Jia J, Wei YL, Qin CJ, Hu L, Wan LH, Li CX. Developing a novel panel of genome-wide ancestry informative markers for bio-geographical ancestry estimates. Forensic Science International: Genetics. 2014;**8**:187-194. DOI: 10.1016/ j.fsigen.2013.09.004

[91] The National Database of Genetic Profiles. Regulation, functioning and operating [Internet]. 2014. Available from: http://www.agmf.es/az/ LA_BASE _DE _DATOS_ NACIONAL_ DE_ PERFILES_ GENeTICOS._ HOMBREIRO_L.pdf [Accessed: 04 August 2018]

[92] Ge J, Sun H, Li H, Liu C, Yan J, Budowle B. Future directions of forensic DNA databases. Croatian Medical Journal. 2014;**55**:163-166. DOI: 10.3325/cmj.2014.55.163

# 9

# Detection of the Species Composition of Food Using Mitochondrial DNA: Challenges and Possibilities of a Modern Laboratory

*Małgorzata Natonek-Wiśniewska and Piotr Krzyścin*

**Abstract**

Monitoring food quality is an important and constant element of the food market. This need is connected with health issues, religious beliefs of consumers, and economic considerations. For analysis, mtDNA is most commonly used because it is resist to physical factors such as temperature and pressure, which very often accompany food processing. Nowadays, scientific publications present a number of methods describing species identification from both farm animals and also less common animals. The most effective methods for determining species are based on PCR, real-time PCR, and sequencing. The methods are very sensitive, limit of detection (LOD) is 0.001% for many of them. An indispensable element of performing the described research is the strict application in the laboratory of several principles, which are intended to improve the work and make it safe for the lab technician, as well as guarantee the quality and effectiveness of the experiments carried out. The high work requirements set for the crew naturally shape the quality system from which the most popular is ISO/IEC 17025. Modern methods based on mtDNA are a good tool for food analysis, creating great opportunities for the researcher, at the same time causing challenges for the contemporary laboratory.

**Keywords:** mtDNA, quality system in the laboratory, PCR, real-time PCR, sequencing, species identification

## 1. Benefits of knowledge about possibility species identification

The reliability of food products available on the market, in terms of their origin, quantitative and qualitative composition, has long been the focus of consumers. Therefore, monitoring food quality is an important and constant element of the food products market. This need arises from health issues, consumers' religious convictions, and economic reasons. According to the WHO, in Europe, 8% of children and 4% of adults are allergic to bovine milk or hen eggs. While these products can be rapidly and easily identified in pure form, their presence in complex products may be much more difficult to detect. Knowledge of the species composition of these products, although unavailable without detailed analyses, is crucial for many patients. Likewise, religious convictions of many communities provide a power-ful incentive for monitoring real composition of the food. For example, Judaism prohibits the consumption of pork, so a large part of the followers of this religion avoid the meat of pigs and replace it with beef or sheep meat, which form a considerable part of the meat market in these countries. Unfortunately, for economic reasons, food products are often intentionally adulterated by replacing declared, more expensive components with cheaper substitutes (e.g., meat of lower quality or plant fillers). There are also cases when the quantitative share of an expensive component in a complex product is lowered. By way of example, poultry meat is on average several times cheaper than pork, which, in turn, is priced lower than beef or lamb meat. Similarly, beef is cheaper and more readily available than game meat. The price differences may induce some unfair producers to adulterate and place on the market products whose components differ from manufacturer specifications.

The declaration of meat products in the EU is mandated by the Commission Directive 2002/86/EC [1] stating that meat products have to be labeled with precise information about the species and its percentage in the product. Nevertheless, as experience shows, there are numerous examples of components being misrepresented to make a product more attractive, justify a higher price, or enter new markets. Here, it suffices to mention that in products like fast food 65% of adulteration is deducted [2, 3] and in preparations of game meat, the percentage of factually inaccurate labeling is less (30%) [4], but in sausage, this percent has grown to 90% [5]. Both food products and pet foods were found to be adulterated, and Okuma found 40% of foods for animals with meat of chicken to be falsified [6]. Based on the information reported above and day-to-day practice, it could be claimed that food adulteration is becoming a global problem, which attracts consumer attention at international level and increases public concern about the quality of food products. By way of example, in 2013, the horse-meat scandal revealed gaps in the food safety system and undermined trust between producers and consumers.

It is, therefore, essential to identify the methods for (quantitative and qualitative) determination of species composition of food ingredients to monitor the conformity of a product with the description provided by the manufacturer. Research in this area can better protect consumers from illegal and undesirable adulteration, for whatever reason.

It should be also mentioned that recent years have seen increasing awareness of the importance of food safety and quality, which increases public interest in this issue and leads to changes in legislation. This necessitates continuous development and improvement of analytical methods.

## 2. The scope of the species identification tests

The analysis most often uses mtDNA, although exceptions outlined below are permitted. The advantage of mitochondrial over genomic genome results from its resistance to the action of physical factors such as temperature and pressure, which very often accompany the processing of food. These characteristics of mtDNA contribute to a very high sensitivity of the analyses. In principle, the whole mitochondrial genome can be used for the analyses, but more frequent use is made of cytochrome B and D-loop. Cytochrome B is the most conservative of the entire mitochondrial genome. Its identification and creation of a bar code were the subject of projects aimed to describe all living organisms—both the most common and the most unique. In turn, D-loop is characterized by the highest variation between species, which enables the method to be quickly determined. The mitochondrial genome is very short compared to the body's entire genome and forms a very small proportion of it. In animals, it is slightly over 16,000 bp, which means it is rela-tively easy to develop methods for identifying the panel of organisms chosen by a researcher. Current research papers present several methods from identify single farm species such as pigs [7–9], cattle [7–10], sheep [7], horses [9, 11], chickens [9, 12], turkeys [9], ducks [8, 13], fish [14] to less common animals like kangaroos [15], snails [16], and marine animals like octopuses [17], shrimp [18] and sharks [19]. This is relatively the simplest method of analysis. With proper time investment, labor inputs, and funds, a laboratory is capable of identifying a concrete species. Such methods are generally very sensitive and enable determining adulterations as low as 0.001% [20–22], although this has little practical use because determinations below 1% are generally treated as artifacts. For this reason, the laboratories that commer-cially used methods most often set the limits of determination between 0.1 and 1% [23]. In certain cases, it is more beneficial to determine a whole group of animals rather than single species. These methods are more demanding because the reaction conditions have to be adjusted as to make the method specific for several DNA fragments that differ in sequences. The primers most commonly used are compatible with DNA of several species, which necessitates finding the most homologous frag-ments. Most often, however, the primers are homologous only in a certain percentage [19, 24]. Such analysis very often yields products of similar, indeed identical, length. Sometimes, it is, therefore, more beneficial to design one primer compatible with all species and another primer specific for single species, which gives products of different length [23, 25]. The choice of method depends on needs. Increasingly often laboratories face the challenge of discriminating between animal and plant DNA in a sample. This apparently easy task is in fact more complicated than identifying smaller groups of animals and impossible to perform based on mtDNA identification. Most often, animal DNA is identified using a DNA fragment that encodes myosin, a muscle protein; that is why myosin-based methods yield a positive reaction only for samples that contain muscles. This limitation may be a problem during analysis because the method allows no identification of matrices such as bones. Another limitation is the differentiation of animals with very similar mitochondrial genomes. This problem can be seen, for example, when distinguishing between pig (*Sus scrofa scrofa*) and wild boar (*Sus scrofa*) components. The mitochondrial genome of both species is 99% homologous (according to BLAST between these species), and there are only single point mutations, so they cannot be used for species identification. Research is underway to make differentiations based on MCR 1 [26, 27], which is a color-determining gene. In the context of food, this issue is important because of differences in taste, price, and availability of meat from these two species.

All the identified DNA fragments should be short, less than 250 bp. There is the rule that the more the food product is processed, the shorter the PCR product should be.

Extreme temperature and pressure cut DNA into short segments; for example, exposure to a pressure of 3.2 Ba results in approximately 100-bp segments and only such or short DNA fragments can be identified. Of course, in raw or cooked meat, DNA is not degraded so much, but the method involving short DNA fragments is more universal and enables determinations to be made whatever the degree of processing.

Molecular methods enable determination to be made in any matrix. In practice, DNA can be identified regardless of matrix form or earlier processing. We can freely determine species composition of both raw tissues and processed tissues in the form of meat, bones, blood, eggs, dairy products such as cheese, milk and butter, drinks, gelatin, lyophilized milk products, meat preparations, and egg products [7, 12, 27–29].

It often happens that the matrices in which DNA is sought have a form that prevents its biological origin to be clearly identified, and so it may become a source

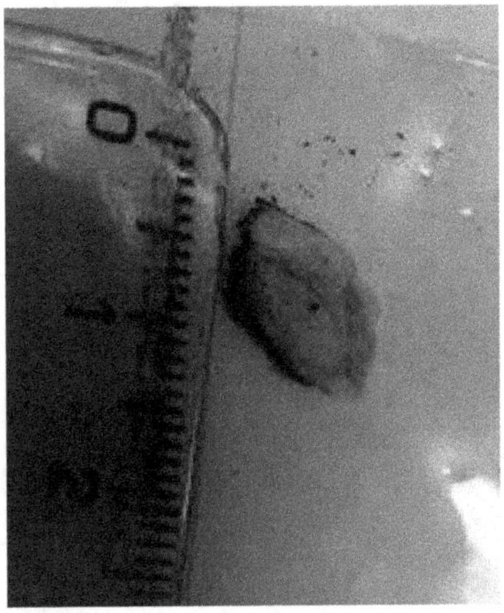

**Figure 1.**
*Biological material found by a consumer in meatballs.*

of potential problems. This is exemplified by a fragment of biological material found by a consumer in meatballs [13]. The object concerned, which was small in size and additionally resembled a human nail (**Figure 1**), was identified during the analysis as material coming from one of the breeding species, so its presence in food preparations was fully justified.

## 3. Used methods, possibility each of methods, their advantages, and disadvantages

The most effective methods of species identification are based on PCR technique. These methods use both conventional PCR and real-time PCR. Both methods can be used as monoplex or multiplex PCR. Detection in real-time PCR can use both probes and DNA-binding dyes (e.g. SYBR Green, Eva Green). A detailed schematic representation of the method is given in **Figure 2**.

Each method has its pros and cons. The simplest method, conventional monoplex PCR, is unbeatable when one concrete species is sought. These methods generally have a very high limit of determination, which is often so high that it has no practical application in commercial analyses. This figure, often below 0.001%, acquires real significance when determining undesirable trace substances or accidental artifacts.

Such methods are simplest but at the same time show the least potential, and only allow determining if a given substance contains the DNA of the species being identified.

Multiplex reactions not only offer more possibilities but also cause more problems. Since they require carrying out the reaction in one temperature, which is not necessarily optimal for all primers and as a result reactions may take place with different efficiencies, this may lead to false-negative reactions when the level of adulteration is low. Thus, although multiplex reaction unquestionably shortens the time of analysis and reduces its costs, when complex products are analyzed, the result for low content DNA can be subject to risk [30].

Another group of methods is restriction fragment length polymorphism (RFLP). This technique is based on amplification of a DNA fragment with different sequences, followed by its digestion with appropriate restriction enzymes,

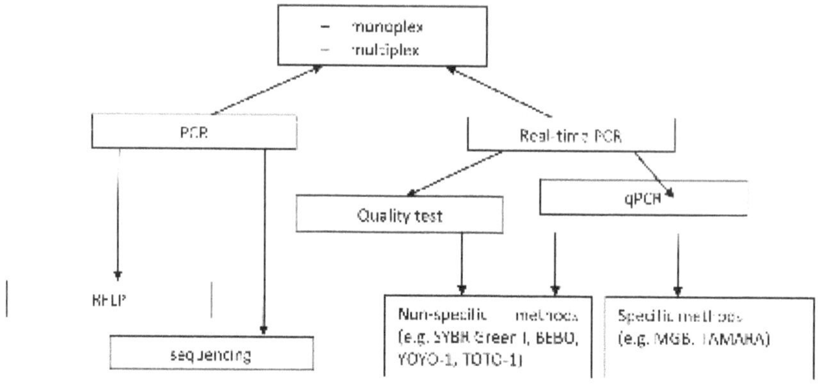


**Figure 2.**
*Detailed schematic diagram of the methods used for species identification.*

which enables even related species to be distinguished [31]. The method allows for identification of several to 25 animal species, although the latter requires the use of several restriction enzymes.

The PCR-RFLP method is simple, inexpensive, and easy to use for monitoring purposes. PCR-RFLP has been used for years and many researchers consider it outdated. However, this method works very well in the case of complex analyses, where we are interested in finding the potential presence, for example, of a group of species (e.g., birds, ruminants) and then their specific representatives. Similarly to the case of multiplex reaction, this method performs better for single-species samples, while their application for complex products may cause read errors, firstly because of similar restriction patterns for the analyzed species of animals, secondly due to the competitiveness of RFLP reaction. Another disadvantage of the PCR-RFLP method is that erroneous results may develop because of the occurrence of incomplete digestion of the restriction site or intraspecific differences, which may contribute to the removal or development of restriction sites [32].

When we analyze samples whose composition is completely unknown and has to be identified, Sanger sequencing is a very good solution. If we analyze a fragment homologous to several species, we can quickly and accurately determine its species origin. Again, this method is better applied to single-species samples and it is not a method of first choice for routine determination of specific species, if only because of higher price and the need to use more specialist equipment. However, it is an indispensable tool for analyses subject to greater uncertainty.

Another group of methods are quantitative determinations. They continue to be a major challenge for researchers because sample reactivity depends on processing method, type of matrix, and sometimes the animal. Therefore, production of the reference material that is later used to generate standard curves is subject to error of 10% or sometimes even 30%.

The production of reference material is an important issue when determining the type of meat. It should be noted that the certified reference material (CRM) is only available in the form of DNA, which in the case of quantitative tests does not work and is completely unsuitable because the mismatch of such material to the analyzed meat samples can be huge. That is why laboratories themselves produce reference materials. Usually, meat samples purchased commercially from the butcher or shop are used for this. It is important that they came from a few or several individuals. The material produced in this way is more precisely matched to the analyzed samples and has a lower risk that it will not completely match it. Before using the reference material so manufactured, it should be checked. First, the standard curve obtained from it must meet certain parameters such as slope, y-intercept, $R^2$ value, and amplification efficiency (EFF%). The appropriate numerical values for these parameters guarantee the specificity and reaction efficiency of the standard curve used. The second necessary condition is the analysis against this curve of a sample with a guaranteed concentration of the species being determined. Such samples are most often obtained as residues from proficiency tests. It should also not be forgotten that the method of isolating DNA from reference material should be the same as the test samples [33]. Many authors use methods that match the largest number of food-related matrices, e.g. CTAB [33], although this depends on the experience and preferences of each laboratory.

Standard amounts of the material needed for the analyses range from 0.1 to 0.5 g because such amounts are most often recommended by the manufacturers of DNA isolation kits, but when determining microtraces in foods, we must often settle for a fraction of this weight. Since mtDNA is most commonly used, which allows for very sensitive analyses because it is present in every cell in many million copies, often trace amounts of material are sufficient to perform the analysis.

## 4. Ensuring the quality of analyses, quality systems in the laboratory, and certificates for laboratories

The high sensitivity of mtDNA-based PCR methods is a great advantage, but at the same time, this is associated with a serious risk of cross reactions. Therefore, the tests described above must be governed by a strict application of several rules, which, by design, should make the work more efficient and safe for the laboratory technician while ensuring the quality and effectiveness of the experiments.

The overriding rule is to perform most of the procedures in a laminar flow cabinet, in which air is constantly blown out to ensure sterile conditions. Prior to the commencement of work, it is a good practice to switch on the unit for more than 10 minutes, which will allow for a complete exchange of air, and to turn on the UV lamp, which is usually part of the unit, to make the work area sterile. The working area must be wiped with a DNA-removal solution. Before starting the job, make sure all necessary equipment and materials are ready at hand. At the same time, the working space must be divided into a "clean zone" (pipettes, centrifuges, vortex mixers, reagents, pipette tips) and a "dirty zone" (used tips and basket). These zones must be separated to avoid cases where a used pipette tip is carried over the reagents, test-tube stand, etc. Laboratory technicians working in a laminar flow cabinet should be adequately prepared for work. To ensure sterility, they should wear protective aprons and disposable gloves, additionally cleaned with a DNA-removal agent.

It is also important to separate workstations at which different stages of the analysis (sample preparation, DNA isolation, PCR, electrophoresis) are performed. Any change in workstation requires that the protective apron and disposable gloves be changed. One workplace must not overlap with another. Before starting and after completing the job, working surfaces must be cleaned with a DNA-removal agent. A laboratory sample should be moved in one direction only, in accordance with each successive stage of determinations. Test equipment must be regularly verified and calibrated.

An important aspect of work at a laboratory engaged in species DNA identification is validation of methods before they are introduced. An essential requirement for every research or scientific laboratory that performs commercial testing is to use reliable methods. The methods taken from ISO/IEC or recommended by umbrella organizations (e.g., EURL-AP) have already been validated, so it is enough to check their function in the laboratory. It should be noted, however, that in the DNA research area concerned, many laboratories use their own methods. These have the advantage of being flexible and adaptable to the current needs of customers, which means that the laboratory can react quickly and optimally to the evolving market needs. Naturally, these methods have to be validated, which incurs additional charges for the laboratory:

- increased costs; before the method becomes profitable, the laboratory must usually pay high validation costs,
- time-consuming nature; there must be adequate time between the decision to introduce a method to its real application in the laboratory. The longer and more laborious the validation process, the longer the time needed,
- the need of training; it increases the costs and delays the practical implementation. However, this has a positive aspect for the laboratory in the form of better trained and more aware staff.

The high requirements placed on the personnel are naturally shaping the quality system, in which all employees are aware of their responsibilities, the work is safe, and ensures reliable results. Nowadays, most laboratories want to introduce a defined quality system. The most popular system is ISO/IEC 17025, which provides requirements for testing and calibration laboratories. Since its publication in 1999 by the International Organization for Standardization, the regulations in this document help to organize work in laboratories. Implementation of this standard certifies that all tests performed in the laboratory meet the standard and respect the chosen testing procedure. Because species identification is directly linked to food safety monitoring, introduction of the system provides measurable benefits in the form of growing prestige of the laboratory, increased efficiency, greater competencies of the managerial staff, clearly defined responsibilities and rights of the staff, increased testing accuracy, and higher number of commissioned tests.

The accreditation requirement most often results out of external pressure, from the customer or the regulatory authority [34], but sometimes it may result from the internal desire to increase the level of testing services [8] or even from institutional strategic planning [10]. However, decision to adopt ISO/IEC should consider (1) the organization's culture, (2) the actual need for pursuing accreditation—the accreditation requirement from the customer or the regulatory authority, (3) the time and the resources available, (4) the staff's knowledge and previous experience in quality, (5) the current conditions of the laboratory with reference to compliance with the standard, (6) use of standard test methods already established and known well by the laboratory staff, and (7) condition of equipment used for tests, in addition to involving appropriate costs of maintenance and calibration [34].

Modern methods based on mtDNA are a powerful tool for food analysis, creating great opportunities for the researcher, at the same time causing a number of challenges for the contemporary laboratory. The newly developed, commercially used methods are made taking into account the above-mentioned activities.

## Author details

Małgorzata Natonek-Wiśniewska* and Piotr Krzyścin
Department of Animal Molecular Biology, National Research Institute of Animal Production, Balice, Poland

*Address all correspondence to: malgorzata.natonek@izoo.krakow.pl

## References

[1] Commission Directive 2002/86/EC of 6 November 2002 amending Directive 2001/101/EC as regards the date from which trade in products not in conformity with Directive 2000/13/EC of the European Parliament and of the Council is prohibited. Available from: http://data.europa.eu/eli/dir/2002/86/oj

[2] Amjadi H, Varidi MJ, Marashi SH, Javadmanesh A, Ghovvati S. Development of rapid PCR-RFLP technique for identification of sheep, cattle and goat's species and fraud detection in Iranian commercial meat products. African Journal of Biotechnology. 2012;**11**(34):8594-8599. DOI: 10.5897/AJB11.1724

[3] Nejad FP, Tafvizi F, Ebrahimi MT, Hosseni SE. Optimization of multiplex PCR for the identification of animal species using mitochondrial genes in sausages. European Food Research and Technology. 2014;**239**(3):533-541. DOI: 10.1007/s00217-014-2249-1

[4] Amaral JS, Santos CG, Melo VS, Oliveira MBP, Mafra I. Authentication of a traditional game meat sausage (Alheira) by species-specific PCR assays to detect hare, rabbit, red deer, pork and cow meats. Food Research International. 2014;**60**:140-145. DOI: 10.1016/j. foodres.2013.11.003

[5] Prusakova OV, Glukhova XA, Afanas'eva GV, Trizna YA, Nazarova LF, Beletsky IP. A simple and sensitive two-tube multiplex PCR assay for simultaneous detection of ten meat species. Meat Science. 2018;**137**:34-40. DOI: 10.1016/j.meatsci.2017.10.017

[6] Okuma T, Hellberg R. Identification of meat species in pet foods using a real-time polymerase chain reaction (PCR) assay. Food Control. 2015;**50**:9-17. DOI: 10.1016/j.foodcont.2014.08.017

[7] Shabani H, Mehdizade M, Mousavi S, Dezfouli E, Solgi T, Khodaverdi M, et al. Halal authenticity of gelatin using species-specific PCR. Food Chemistry. 2015;**184**:203-206. DOI: 10.1016/j. foodchem.2015.02.140

[8] Hossain M, Ali M, Hamid S, Mustafa S, Desa M, Zaidul I. Targeting double genes in multiplex PCR for discriminating bovine, buffalo and porcine materials in food chain. Food Control. 2017;**73**:175-184. DOI: 10.1016/j.foodcont.2016.08.008

[9] Ha Y, Thienes C, Agapov A, Laznicka A, Han S, Nadala C, et al. Comparison of ELISA and DNA lateral flow assays for detection of pork, horse, beef, chicken, Turkey, and goat contamination in meat products. Journal of AOAC International. 2019;**102**(1):189-195. DOI: 10.5740/jaoacint.18-0128

[10] Safdar M, Yasmeen J. A multiplex-conventional PCR assay for bovine, ovine, caprine and fish species identification in feedstuffs: Highly sensitive and specific. Food Control. 2015;**50**:190-194

[11] Pegels N, García T, Martín R, González I. Market analysis of food and feed products for detection of horse DNA by a TaqMan real-time PCR. Food Analytical Methods. 2015;**8**(2):489-498. DOI: 10.1007/s12161-014-9914-7

[12] Natonek-Wiśniewska M, Krzyścin P. The use of PCR and real-time PCR for qualitative and quantitative determination of poultry and chicken meals. Annals of Animal Science. 2016;**16**(3):731-741. DOI: 10.1515/aoas-2016-0003

[13] Natonek-Wiśniewska M, Krzyścin P. Evaluation of the suitability of mitochondrial DNA for species identification of microtraces and forensic traces.

Acta Biochimica Polonica. 2017;**64**(4): 705-708. DOI: 10.18388/abp.2017_2304

[14] Rasmussen R, Morrissey M. DNA-based methods for the identification of commercial fish and seafood species. Comprehensive Reviews in Food Science and Food Safety. 2008;**7**(3):280-295. DOI: 10.1111/j.1541-4337.2008.00046.x

[15] Venkatachalapathy R, Sharma A, Sukla S, Hattacharya T. Cloning and characterization of DGAT1 gene of Riverinebuffalo. DNA Sequence. 2008;**19**(3):177-184

[16] Borgo R, Souty-Grosset C, Bouchon D, Gomot L. PCR-RFLP analysis of mitochondrial DNA for identification of snail meat species. Journal of Food Science. 1996;**61**(1):1-4. DOI: 10.1111/j.1365-2621.1996.tb14712.x

[17] Taylor A, Niall J, McKeown P, Shaw W. Molecular identification of three co-occurring and easily misidentified octopus species using PCR–RFLP techniques. Conservation Genetics Resources. 2012;**4**(4):885-887. DOI: 10.1007/s12686-012-9665-y

[18] Wilwet L, Jeyasekaran G, Shakila RJ, Sivaraman B, Padmavathy P. A single enzyme PCR-RFLP protocol targeting 16S rRNA/tRNAval region to authenticate four commercially important shrimp species in India. Food Chemistry. 2018;**239**:369-376. DOI: 10.1016/j.foodchem.2017.06.132

[19] Fotedar S, Lukehurst S, Jackson G, Snow M. Molecular tools for identification of shark species involved in depredation incidents in Western Australian fisheries. PLoS One. 2019;**14**(1):e0210500. DOI: 10.1016/S0309-1740(98)00112-0

[20] Ali M, Razzak M, Hamid S, Rahman M, Al Amin M, Rashid N. Multiplex PCR assay for the detection of five meat species forbidden in Islamic foods. Food Chemistry. 2015;**177**:214-224. DOI: 10.1016/j.foodchem.2014.12.098

[21] Cammà C, Di Domenico M, Monaco F. Development and validation of fast real-time PCR assays for species identification in raw and cooked meat mixtures. Food Control. 2012;**23**(2):400-404. DOI: 10.1016/j.foodcont.2011.08.007

[22] Karabasanavar N, Singh S, Kumar D, Shebannavar S. Detection of pork adulteration by highly-specific PCR assay of mitochondrial D-loop. Food Chemistry. 2014;**145**:530-534. DOI: 10.1016/j.foodchem.2013.08.084

[23] Rodríguez M, García M, González I, Asensio L, Hernández P, Martín R. PCR identification of beef, sheep, goat, and pork in raw and heat-treated meat mixtures. Journal of Food Protection. 2004;**67**(1):172-177. DOI: 10.4315/0362-028X-67.1.172

[24] Matsunaga T, Chikuni K, Tanabe R, Muroya S, Shibata K, Yamada J, et al. A quick and simple method for the identification of meat species and meat products by PCR assay. Meat Science. 1999;**51**(2):143-148. DOI: 10.1016/S0309-1740(98)00112-0

[25] Kesmen Z, Gulluce A, Sahin F, Yetim H. Identification of meat species by TaqMan-based real-time PCR assay. Meat Science. 2009;**82**(4):444-449. DOI: 10.1016/S0309-1740(98)00112-0

[26] Rębała K, Rabtsava A, Kotova S, Kipen V, Zhurina N, Gandzha A, et al. STR profiling for discrimination between wild and domestic swine specimens and between main breeds of domestic pigs reared in Belarus. PLoS One. 2016;**11**(11):e0166563. DOI: 10.1371/journal.pone.0166563

[27] Kaltenbrunner M, Mayer W, Kerkhoff K, Epp R, Rüggeberg H, Hochegger R, et al. Differentiation between wild boar and domestic pig

in food by targeting two gene loci by real-time PCR. Scientific Reports. 2019;**9**(1):9221. DOI: 10.1038/s41598-019-45564-7

[28] Natonek-Wiśniewska M, Krzyścin P, Bugno-Poniewierska M. Development of a sensitive and specific qPCR method to detect duck and goose DNA in meat and feathers. European Food Research and Technology. 2019;**245**(2):335-342. DOI: 10.1007/s00217-018-3165-6

[29] Lifschitz C. Cow's milk allergy: Evidence-based diagnosis and management for the practitioner. European Journal of Pediatrics. 2015;**174**:141-150. DOI: 10.1007/s00431-014-2422-3

[30] Zha D, Xing X, Yang F. A multiplex PCR assay for fraud identification of deer products. Food Control. 2010;**21**(10):1402-1407. DOI: 10.1016/j. foodcont.2010.04.013

[31] Pascoal A, Prado M, Castro J, Cepeda A, Barros-Velázquez J. Survey of authenticity of meat species in food products subjected to different technological processes, by means of PCR-RFLP analysis. European Food Research and Technology. 2004;**218**(3):306-312. DOI: 10.1007/s00217-003-0846-5

[32] Gil L. PCR-based methods for fish and fishery products authentication. Trends in Food Science and Technology. 2007;**18**:558-566. DOI: 10.1016/j.tifs.2007.04.016

[33] Boldura O, Popescu S. PCR: A powerful method in food safety field. Biochemistry, genetics and molecular biology. In: Polymerase Chain Reaction for Biomedical Applications. USA: Intech Publishers; 2016. pp. 135-158. DOI: 10.5772/65738

[34] Grochau I, Schwengber C. A process approach to ISO/IEC 17025 in the implementation of a quality management system in testing laboratories. Accreditation and Quality Assurance. 2012;**17**(5):519-527. DOI: 10.1007/s00769-012-0905-3

# 10

# Preparation, Characterization and Ion-Exchange Properties of an Organic-Inorganic Composite Cation Exchanger: Polyaniline-Bi(III) Iodovanadate and Catalytic Properties of Bi(III) Iodovanadate Inorganic Ion Exchanger

*Nainar Kohila, Kasi Sathiyaseelan and Mariyathanislas Sagaya Lourdhu Sumithra*

**Abstract**

Polyaniline based composite cation exchange materials have been used in industrial application for more than 100 years. The organic-inorganic composite cation exchanger was prepared by using sol-gel process. The organic polymer part furnishes the good chemical and mechanical properties, whereas the inorganic part improves the ion-exchange behavior and thermal stability. A new and novel polyaniline-Bi(III) iodovanadate composite cation exchanger prepared by using sol-gel process and was characterized by FT-IR, XRD, SEM-EDS studies. The ion exchange capacity, effect of size and charge of metal ion, eluent concentration, effect of time on IEC elution behavior and oxidizing properties of Bi(III) iodovanadate was also studied by column method. Polyaniline-Bi(III) iodovanadate composite exhibits ion exchange capacity for $Na^+$ ion is 1.48 meq/g.

**Keywords:** elution behavior, ion exchange capacity, thermal stability, sol-gel process

## 1. Introduction

Polyaniline has become one of the most interesting organic conducting polymer due to its high chemical and environmental stability, easy to synthesis in labora-tory, feasibility of electrical conductivity and low cost of aniline monomer [1–3]. Literature survey reported that many ion exchange composite cation exchanger used for the separation of some toxic metal ions and dyes from the environment. New organic-inorganic conducting polymer composite cation exchanger developed by incorporation of organic polymer into the inorganic heteropolyacid moiety [4]. It is extensively used in rechargeable batteries, charge storage devices, protector shield in magnetic fields, sensors, biosensors, catalytic processes, microwave absorption, image processing and infrared optic applications [5–9]. Polymer based composite cation exchanger have extended enormous applications such as ion selective electrode, photocatalyst, antimicrobial, sensors, environmental remediation, etc. Several hetropolyacids are used as a catalyst in organic synthesis and they are extensively used for reagents in qualitative and quantitative analysis [10–12]. This paper deals with preparation, characterization, ion exchange studies, chemical stability and oxidizing ability of Bi(III) iodovanadate and newly fabricated polyaniline-Bi(III) iodovanadate composite cation exchanger. The structural analysis of polyaniline-Bi(III) iodovanadate composite was done by FT-IR, XRD and SEM-EDS studies.

## 2. Experimental section

### 2.1 Reagents and instrument

The reagents used for the preparation were analytical grade and used without any further purification. FT-IR spectra were recorded on a JACSO-4100 FT-IR Spectrometer. X-ray diffraction pattern was also recorded by using analytical system Shimadzu XRD-6000 model and the spectrum was recorded 10–90° using Cu-Kα radiation. The surface morphology and elemental composition was determined by using scanning electron microscope JSM-6390Lv energy dispersive X-ray detector.

### 2.2 Preparation of polyaniline-Bi(III) iodovanadate cation exchanger

Polyaniline gel was prepared by 0.2 M solution of aniline and potassium persulfate in 1 M hydrochloric acid [13] with constant stirring. Bi(III) iodovanadate inorganic ion exchanger was prepared by mixing 1:2:3 volume ratio of 0.2 M solution of bismuth nitrate, potassium iodate and sodium meta vanadate. The mixture of solution was adjusted to pH = 1 by using 1 M $HNO_3$, and the precipitate was stirred for 1 hour [14]. The gel of polyaniline was mixed with inorganic precipitate of Bi(III) iodovanadate and the mixture was stirred thoroughly using magnetic stirrer. The green colored gel was kept for 1 day. The gel was filtered and dried it in oven at 50°C. The dried product of composite cation exchanger is crushed and the product is converted into $H^+$ form by using 1 M $HNO_3$ with occasional shaking for 1 day. The product is filtered and dried at 50°C. The $H^+$ form of polyaniline-Bi(III) iodovanadate is used for ion exchange and chemical stability studies.

### 2.3 Ion exchange capacity

The ion exchange capacity of dry $H^+$ form of polyaniline-Bi(III) iodovanadate was determined by column process using 0.1 M NaCl as eluent. The liberated $H^+$ ion was determined titrimetrically against NaOH solution by using phenolphthalein indicator and IEC was calculated by using formula,

$$\text{IEC} = \frac{(N * V)}{W}\ \text{meq/g} \tag{1}$$

where N and V are the normality and volume of NaOH respectively and W is the weight in gram.

### 2.4 Chemical stability

To find out extent of dissolution of composite cation exchange material, chemical stability was studied in different organic and inorganic solvents. Two fifty milligram of $H^+$ form of polyaniline-Bi(III) iodovanadate mixed with 25 ml of different solvents and kept for 24 hours. After 24 hours the composite material was filtered and dried. The stability of composite cation exchanger was determined by change in color and weight of composite cation exchanger.

### 2.5 Inorganic heteropolyacid as oxidizing agent

The catalytic function of heteropoly compounds has attracted much attention, particularly in the last two decades, because their acidic and redox properties can be controlled at atomic/molecular levels. As for the phase of these catalytic systems, various systems are possible: homogenous liquid, liquid/liquid (phase transfer), liquid/solid, gas/solid systems, and so on. There are actually several large-scale industrial processes that use heteropolyacid catalysts.

Polyaniline gel was prepared by 0.2 M solution of aniline and potassium persulfate in 1 M hydrochloric acid with constant stirring. About 0.5 g of Bi(III) iodovanadate inorganic ion exchanger were added to the solution of acidic aniline with constant stirring for 1 hour. Aniline was polymerized into the green color poly-aniline. The product of polyaniline was filtered and washed with DMW, ethanol, acetone and dried at 50°C.

## 3. Results and discussion

The incorporation of organic polymer polyaniline into the inorganic matrix of Bi(III) iodovanadate was confirmed by carrying out FT-IR spectral studies. The characteristic peaks of polyaniline and Bi(III) iodovanadate were observed in the FT-IR spectrum of polyaniline-Bi(III) iodovanadate composite shown in **Figure 1b**. The FT-IR peak observed at 3400 $cm^{-1}$ is due to the -OH stretching vibrations. The benzenoid and quinoid stretching frequency obtained at 1477 and 1560 $cm^{-1}$ [15]. Peaks at 1270 and 1654 $cm^{-1}$ as singed to -CN stretching and -NH bending vibration in neighboring quinoid ring [16]. The characteristic bands at 878, 784 and 671 $cm^{-1}$ may be assigned to M-O stretching [17].

The XRD pattern of polyaniline shows broad peak at two theta value of 25.43° which indicate low crystallinity of the conducting polymer XRD pattern of polyaniline-Bi(III) iodovandate composite (**Figure 2b**) exhibit high intensity peaks at two theta values 26.09, 33.6 and 32.8°. The observation in the XRD pattern of polyaniline composite shows that the composite is crystalline nature and calculated particle size is 14.96 nm.

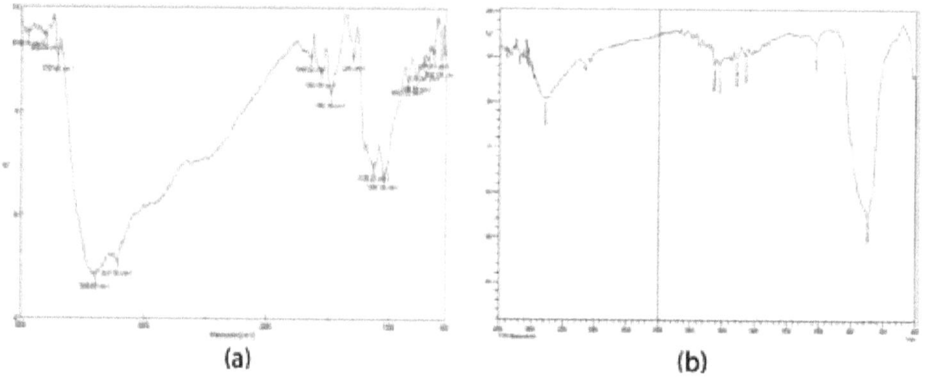


**Figure 1.**
*(a) FT-IR spectrum of polyaniline; (b) FT-IR spectrum of polyaniline-Bi(III) iodovanadate.*

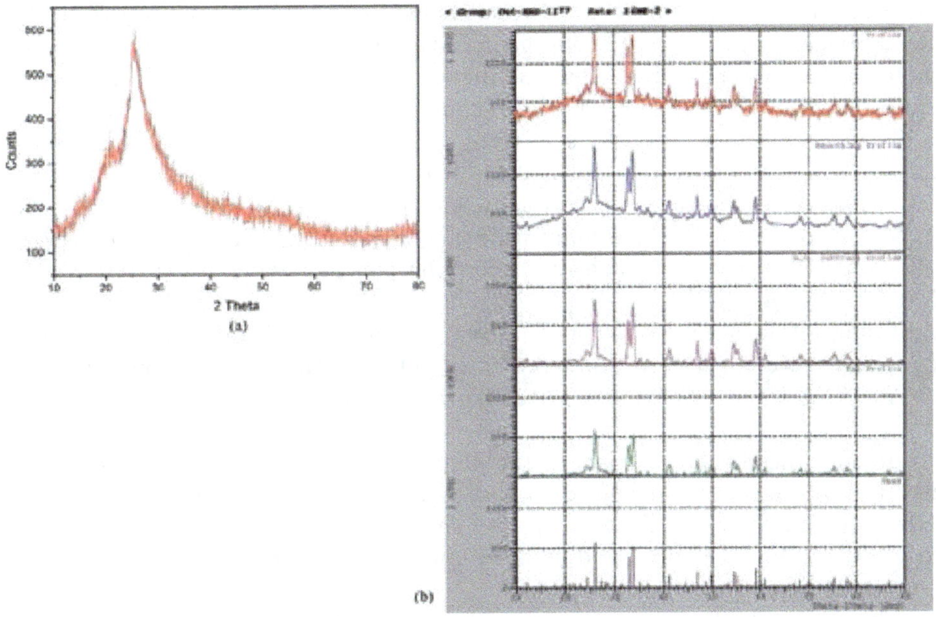


**Figure 2.**
*(a) XRD pattern for polyaniline; (b) XRD pattern for polyaniline-Bi(III) iodovanadate.*

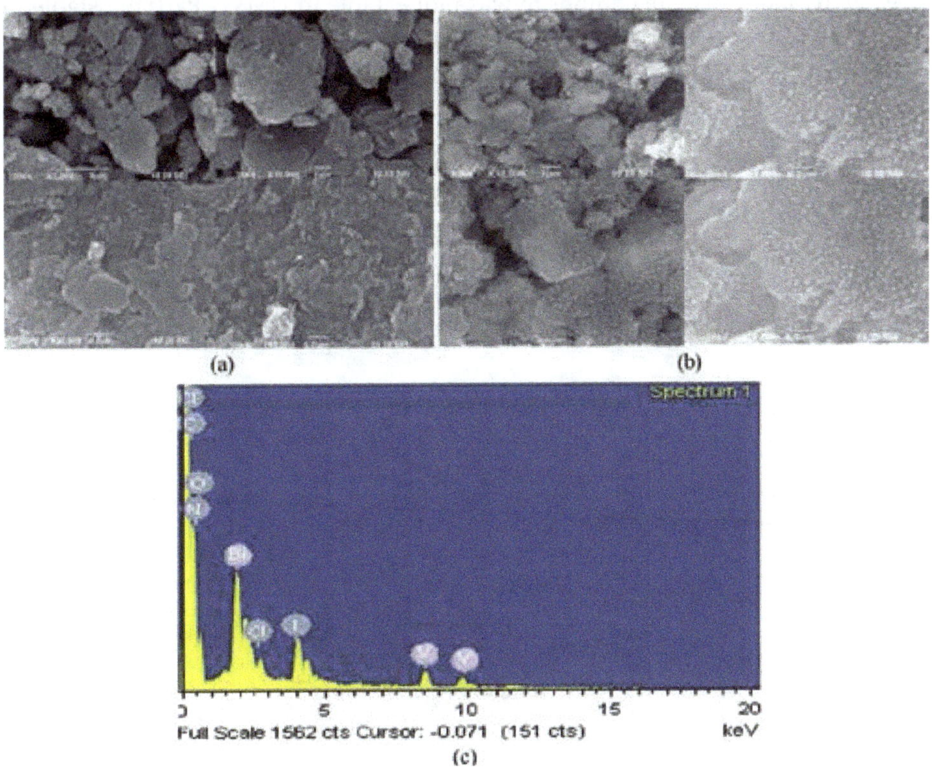


**Figure 3.**
*(a) SEM photograph for polyaniline; (b) SEM photograph for polyaniline-Bi(III) iodovandate; (c) EDS analysis for polyaniline-Bi(III) iodovanadate.*

**Figures 3a, b** represents SEM photographs of polyaniline and polyaniline-Bi(III) iodovanadate composites, respectively. SEM photograph of polyaniline salt suggests that agglomerates are randomly distributed on its surface, whereas polyaniline-Bi(III) iodovanadate composite have porous morphology with granular structure [18]. It is clearly evident from SEM study, inorganic ion exchanger homogeneously distributed on the surface of polyaniline. **Figure 3c** shows that presence of all the elements present in the material which shows purity of composite.

The ion exchange capacity is affected by the size and charge of metal ions. Ion exchange capacity of polyaniline-Bi(III) iodovanate composite for various alkali and alkaline earth metal cations follows the order $K^+ > Na^+ > Li^+$ and $Ba^{2+} > Sr^{2+} > Mg^{2+}$ respectively. The ions with smaller hydrated ionic radii easily enter the pores of cation exchanger resulting in higher adsorption. Similar observation was observed by Mesalem et al. and Nachood et al. [19, 20]. Elution behavior was carried out to find out the volume of eluent (NaCl) required for complete elution in $H^+$ ion from 1 g composite cation exchanger on $H^+$ form and represented in **Figure 4b**. It is clear from **Figure 4b** 120 ml of NaCl is enough for complete elution of $H^+$ ions.

Eluent concentration is main factor which affects the $Na^+$ ion exchange capacity. The minimum molar concentration of eluent was found to be 0.1 M. The effect of time on ion exchange capacity shows constant: ion exchange capacity after 60 minutes. The observed datas of eluent concentration and effect of time show in (**Figure 5**).

Chemical stability study was carried out to find the stability of prepared composite cation exchanger in different solvents of interest such as DMW, 2 M HCl, $H_2SO_4$, NaOH, ether, 1,2-dichloroethane, cyclohexane and benzene. The polyaniline-Bi(III) iodovanadate composite was more stable in DMW, Partially stable in mineral acid and organic solvents and unstable in 2 M NaOH because of dedoped form of ion exchanger.

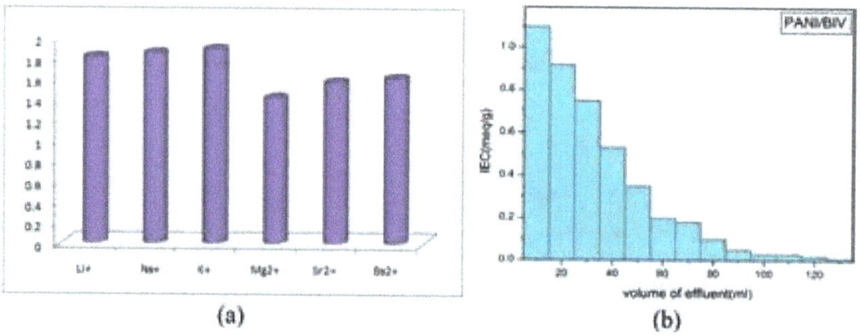


**Figure 4.**
*(a) IEC of polyaniline-Bi(III) iodovanadate composite for various metal ions; (b) elution behavior of polyaniline-Bi(III) iodovanadate composite.*

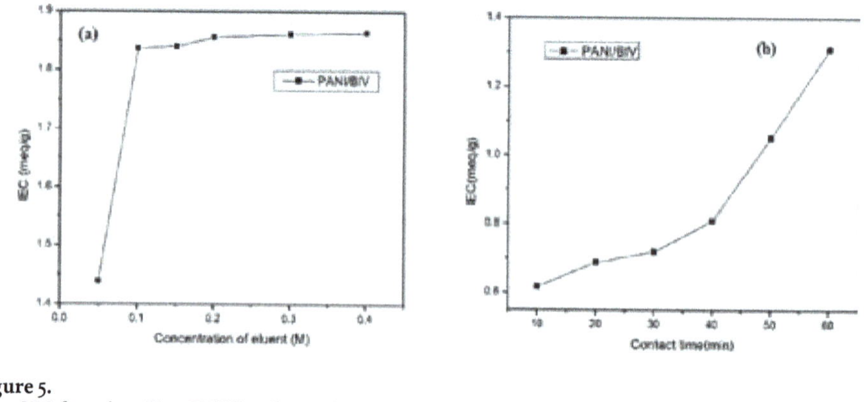


**Figure 5.**
*Na + IEC for polyaniline-Bi(III) iodovanadate as a function of (a) eluent concentration; (b) contact time.*

### 3.1 Bi(III) iodovanadate as oxidizing agent

The hetropolyacid (iodovanadate) unit of inorganic ion exchanger like Bi(III) iodovanadate is made up of oxygen and hydrogen with some metals and nonmetal (iodine and vanadium). Bi(III) iodovanadate can act as a oxidizing agent, to the polymerization of aniline monomer into green colored polyaniline gels without adding oxidizing agent [10, 21–24].

## 4. Conclusion

In this present paper, polyaniline-Bi(III) iodovanadate composite cation exchanger have enhanced $Na^+$ ion exchange capacity compared to polyaniline and Bi(III) iodovanadate. The composite cation exchanger was prepared successfully by using sol-gel method. FT-IR and SEM-EDS studies confirmed that the inorganic ion exchanger incorporated into polyaniline matrix. XRD spectral studies proved that the composite cation exchanger is in nano size range and crystalline nature. Ion exchange capacity and chemical stability studies proved that the composite material act as good ion exchanger and stable in mineral acid and organic solvents. It was concluded that polyaniline-Bi(III) iodovanadate composite act as good potential for environmental remediation.

## Author details

Nainar Kohila, Kasi Sathiyaseelan* and Mariyathanislas Sagaya Lourdhu Sumithra
Department of Chemistry, Aditanar College of Arts and Science, M.S. University, Tirunelveli, Tamil Nadu, India

*Address all correspondence to: sseelan800@gmail.com

## References

[1] Ding X, Han D, Wang Z, Xu X, Niu L, Zhang Q. Micelle assisted synthesis of polyaniline magnetic nanorods by in-situ self-assembly process. Journal of Colloid and Interface Science. 2008;**320**:341-345

[2] Kim BJ, Oh SG, Han MG, Im SS. Preparation of polyaniline nanoparticle in miceller solution as polymerization medium. Langmuir. 2000;**16**:5841-5845

[3] Gao Y, Zhou Y. Polyaniline nofibers fabricated by electrochemical polymerization: A mechanistic study. European Polymer Journal. 2007;**43**:2292-2297

[4] Khan AA, Alam MM, Mohammed IF. Electrical conductivity and ion-exchange kinetitic studies of a new crystalline type 'organic-inorganic' cation-exchange material: polyprrolr/polyantimonic acid composite system, $(Sb_2O_5)(-C_4H_4NH-).nH_2O$. Journal of Electroanalytical Chemistry. 2004;**572**:67-78

[5] Rivera-Utrilla J, Sanchez-Poloa M, Gomez-Serranob V, Alvarezc PM, Alvim MCM, Ferrazd MC, et al. Activated carbon modifications to enhance its water treatment applications. An overview. Journal of Hazardous Materials. 2011;**187**:1-23

[6] Jing G, Zhou Z, Song L, Dong M. Ultrasound enhanced adsorption and desorption of chromium (VI) on activated carbon and polymeric resin. Desalination. 2011;**279**:423-427

[7] Gupta RK, Singh RA, Dubey SS. Removel of mercury ions from aqueous solution by composite of poly aniline with red mud. Separation and Purification Technology. 2004;**38**:225-232

[8] Aia L, Jianga J, Zhang R. Synthetic Metals. 2010;**160**:762

[9] Salem MA. The role of polyaniline salts in the removal of direct blue 78 from aqueous solution: A kinetic study. Reactive and Functional Polymers. 2010;**70**:707-714

[10] Lu T, Niu M, Hou Y, Wu W, Ren S, Yang F. Catalytic oxidation of cellulose to formic acid in $H_5PV_2Mo_{10}O_{40}$ + $H_2SO_4$ aqueous solution with molecular oxygen. Green Chemistry. 2016;**18**:4725-4732

[11] Jeffery GH, Bassett J, Mendham J, Denny RC. Vogels Textbook of Quantitative Analysis. V ed. John Wiley & Sons Inc. 1991. p. 472

[12] Chalmers RA, Parker CA, Stephen WI, Mills AA, Hall RJ, Kirkbright GF. Determination of non-metals. Proceedings of the Society for Analytical Chemistry. 1966;**3**:157-164

[13] Zeng XR, Man Ko J. Structure-Conductivity relationships of iodine-doped polyaniline. Polymer. 1998;**39**:1187-1195

[14] Buhra R, Naushad M, Adnan R, Alothman ad ZA, Rafathullah M. Polyaniline supported nanocomposite cation exchanger: Synthesis, characterization and application for efficient removal of $Pb^{2+}$ ion from aqueous medium. Journal of Industrial and Engineering Chemistry. 2015;**21**:1112-1118

[15] Furukawa Y, Ueda F, Hyodo Y, Nakajima T, Kawagoe T. Vibrational Spectra and structure of polyaniline. Macromolecules. 1988;**21**:1297

[16] King ET, Neoh KG, Tan KL. Polyanilne A polymer with many interesting intrinsic redox state. Progress in Polymer Science. 1998;**23**:277

[17] Socrstes G. Infrared Characteristics Group Frequencies. NJ: Wiley; 1980. p. 145

[18] Deepti B, Patle Wasudeo B, Gurnule W, Zade AB. Synthesis, characterization and ion exchange properties of a terpolymer derived from 4-hydroxybenzophenone, biuret and formaldehyde. Der Pharmacia Lettre. 2011;**3**:341-353

[19] Mesalem MA. Sorption kinetics of cooper, zinc, cadmium and nickel ions in synthesized silico-antimonate ion exchanger. Colloids and Surfaces A: Physicochemical and Engineering Aspects. 2003;**225**:1380-1384

[20] Nachood FC, Wood W. The reaction velocity of ion exchanger. Journal of the American Chemical Society. 1944;**66**:1380-1384

[21] Mizuno N, Makoto M. Heterogenous catalysis. Chemical Reviews. 1998;**98**:199-217

[22] kozhevnikov IV. Catalysis by heteropolyacid and multicomponent polyoxometalates in liquid-phase reactions. Chemical Reviews. 1998;**98**:171-198

[23] Illies S, Kraushaar-Czarnetzki B. Processing study on the stability of heteropolyacid catalyst in the oxidation of methacrolein to methylacrylic acid. Industrial and Engineering Chemistry Research. 2016;**55**:8509-8518

[24] Silva D, Marcio J, Oliverira D, Macedo C. Catalysis by Keggin heteropolyacid salts. Current Catalysis. 2018;**7**:26-34

# 11

# Ambient Biobanking Solutions for Whole Blood Sampling, Transportation and Extraction

*Armaity Nasarabadi Fouts, Alejandro Romero, James Nelson, Mike Hogan and Shanavaz Nasarabadi*

**Abstract**

Biobanking increases the rate at which precision medicine can be used to successfully refine currently existing medical treatment methodologies. The purpose of precision medicine is to increase a patient's likelihood of defeating a chronic disease, by creating a unique and personal treatment method. However, the research necessary to develop precision medicine requires thousands of biospecimens, which is why biobanking is necessary to move precision medicine forward. Traditional biobanks are a library of preserved biological specimens, such as tissue and whole blood, that can be later accessed for further testing and analysis. Maintaining these types of biobanks is cumbersome and expensive, due to freezer care. Biobank samples are used to support therapeutic drug monitoring in clinical trials, epidemiology, public health screening, and biomarker discovery. Collecting samples for large translational studies requires making regular trips to the phlebotomist or a clinic, which is an inconvenience that is exacerbated when collecting samples in remote and/or resource-limited locations. Inconsistencies in sample collection can affect downstream clinical studies. Remedies for these procedural issues include the development of a medium that effectively preserves the samples at ambient temperature and developing a virtual biobanking system that allows for long-distance access to bioinformatic data of previously analyzed biospecimens.

**Keywords:** biobanking, precision medicine, dried blood spots (DBS), nucleic acid, ambient temperature storage, translational studies

## 1. Introduction

Biobank acts as a library for genotypic and phenotypic data for a variety of biological samples. It is the process of acquiring, storing, processing, and distributing biological materials for the purpose of clinical use, including for the development of precision medicine. The term biobanking covers a broad range of samples, including those of animal, plant, and microbial origins. For instance, animal samples can be organ tissue, marrow, and synovial fluid, while plants can be roots, leaves, bark, flowers, and lastly microbial samples. The biobanking arena has seen significant advances from collecting and cataloging samples to having detailed archives of genotypic and phenotypic information. The storing of this information is part of the newest wave in biobanking, virtual biobanking. Virtual biobanks contain full genomes of previously collected specimen that may be accessed through specialized software or portals. Virtual biobanking assists investigators in searching multiple sites for specimen worldwide, essentially, allowing for the mining of data remotely [1]. Integration of genomics, proteomics, and metabolomics, as well as introduction of highly sensitive analysis methods, has translated into a demand for high-quality specimens and the need for accurate, reliable, and standardized clinical data. However, current methods in collecting samples are strenuous, expensive, and unreliable. Samples collected in the field have to be chilled or frozen until analysis, but the shipping of large chilled containers and powering freezers are cost limitations affecting research projects. Furthermore, there is a relatively small window of time between sample collection, storage, and analysis to preserve sample integrity; reducing the reliability of data. As a result, there has been a growing demand in the market for developing ambient temperature storage methods. In the following sections, the relevance of biobanking to precision medicine will be discussed as well as advances in sample collection and ambient temperature storage methods to reduce the cost of acquiring and storing precious biospecimen.

## 2. Precision medicine

In the last decade, there has been a push to understand factors that affect an individual's health on the molecular level. These factors include an individual's unique lifestyle and environment because it is now understood that epigenetics plays a large role in a person's health as well as their development of chronic diseases, such as cancer [2]. Epigenetics and its effects on multiple "omics" (e.g., proteomics) require more than a snapshot of a single person's life. Instead, large data sets ranging from local population (i.e., a neighborhood or city block) to a statewide population, or larger, are required to truly understand the connection between lifestyle and health; but this is not the only advantage of having a large sample size. Determining treatment for a disease requires the largest possible sample size, in order to account for all the possible variables that lead to developing the illness. The marriage of genetic sequencing and external factors that affect health (i.e., lifestyle and environmental) is the foundation of precision medicine. Precision medicine is the use of multiple facets of an individual's health to develop a unique treatment plan [3] (**Figure 1**). With the cost of sequencing decreasing, it has become possible to query the whole genome in search of variants that are known to cause certain disease, and, thus, develop targeted therapies and reduce the overall strain placed on the body [4, 5]. Analysis of the human genome in the context of diagnostic medicine is one of the main facets of precision medicine. The current methodology for designing treatment plans is based on general information obtained from clinical trials; however, every person is unique and there are numerous instances where these "umbrella" treatment methods prove to be unsuccessful [3]. Thus, by being able to determine the root cause of the disease, be it lifestyle affecting gene expression, genetic inheritance of a mutation, or a random mutation itself, precision medicine provides the opportunity to have a focused approach in diagnostic medicine.

Precision medicine, and the initiative to push it forward, was strongly endorsed by the Obama administration after a young woman was able to determine the cause of her extremely unusual form of liver cancer through virtual searches of sequenced genomes of donors with the same disease [6]. The synergy between genetic markers and new therapies for cancer treatment is one powerful example of precision medicine. Typically, biopsied tissue samples or whole blood samples are used as the material for sequencing [6, 7]. For example, liquid biopsies are now routinely used

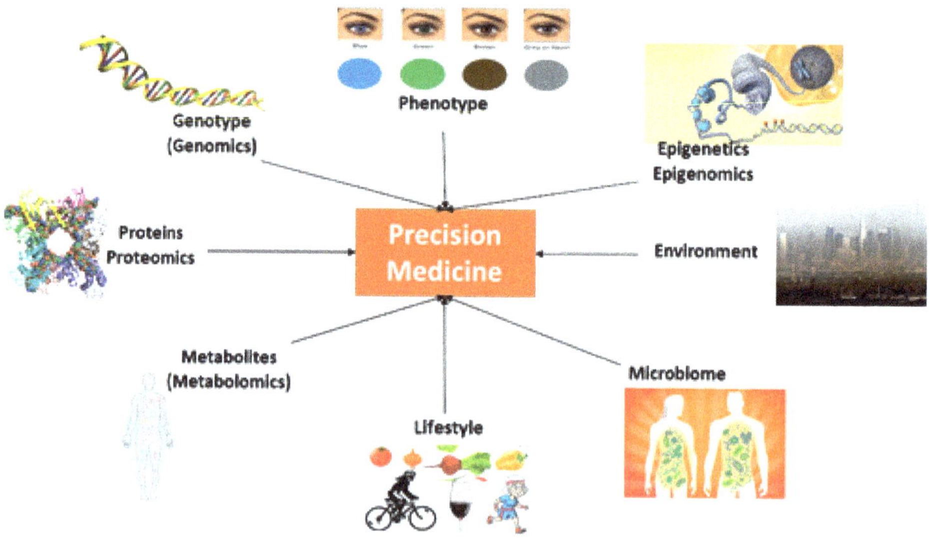


**Figure 1.**
*Precision medicine is a holistic approach to treatment where for the first time, the phenotypic, proteomic, metabolomic, and the genomic composition of the individual as well as their links to other factors such as the microbiome and even the environment will be taken into consideration. This means treating the patient as a "whole" in contrast with the current isolated symptom-treatment approach. The goal of precision medicine is to develop a personalized treatment regimen, a one-of-a-kind approach that would have a better clinical outcome for the patient.*

for the analysis of cell-free DNA to look for progression/regression of cancer as well as additional genomic mutations [8]. As technology advances, instant interdisciplinary integration has become a reality and bioinformatic biobanking makes this integration possible [4]. However, there are many instances when hospitals or research facilities will not release the sequenced data of the samples that were collected or preserved specimens, and literature sources have inadequate information of the biospecimen used [6, 8]. This increases the necessity to have both a national and global data base available so that invaluable information could be accessed seamlessly. One of the first steps in creating such a large data bank is the development of the Million Donor cohort.

Precision medicines' Million Donor cohort comes with a great responsibility for those institutions preserving these samples to answer future research questions [8]. The purpose of the cohort is to begin collecting data from one million individuals across the nation with diverse backgrounds [3, 9]. Building such a large cohort proves to be a daunting task; however, Terry [6] has shown that, when able to, patients will take the initiative to be active members in their health maintenance. Although questionnaires will be used to develop the cohort data, the success of precision medicine depends on the number of available electronic medical records (EMRs) to gain valuable insight into quantitative medical data [3]. Developing such a sample pool that can be easily accessed requires storing the data in virtual bio-banks, a topic that will be discussed in detail later in the chapter [9, 10]. Biobanking and the accessibility of data with a user-friendly network for the purpose of data mining are crucial not only to both short-term and long-term goals of the Precision Medicine Initiative but also to the future. Access to samples is necessary to fulfill the vision of combining established clinicopathological parameters with emerging molecular profiling approaches to create diagnostic, prognostic, and therapeutic solutions that are precisely tailored to an individual patient's unique requirements. Sample availability and sample preservation via biobanking are key to the future of the Precision Medicine Initiative and beyond.

## 3. Biobanking

In this section, the growth of biobanking for the purpose of current research interests will be discussed. Biobanking has been implemented in the scientific community for over 100 years by various institutions worldwide [11, 12]. Biobanks are large repertoires of biospecimen, ranging from animal samples to plant and microbes, that are used for research purposes [12]. Biobanking itself is a relatively simple concept (**Figure 2**). All types of biospecimen are stored in a biobank for long, yet finite amounts of time. These repositories of specimen in traditional biobanks remain in large freezers and other storage facilities until needed [12, 13]. Thus, biobanks are extremely valuable for translational research studies since generations of specimen may be stored and received.

Standardization of samples is key to successful biobanking. Reliability of samples collected in an ethical and legal manner with the oversight of the Institutional Review Board (IRB) or equivalent ethics committee for the biobanking institutions in their respective countries is crucial to ensuring reproducibility of results. International standards are being established by both the European Union [14] and International Society of Biological and Environmental Repositories (ISBER) in the United States [15] to establish standardization metrics for biobanked samples. ISBER coordinated the launch of the International Repository Locator (IRL) website in early 2015. This centralized locator, analogous to a "repository directory," was created to increase the profile of individual repositories including ISBER, researchers, funding bodies, governments, and private industry. However,

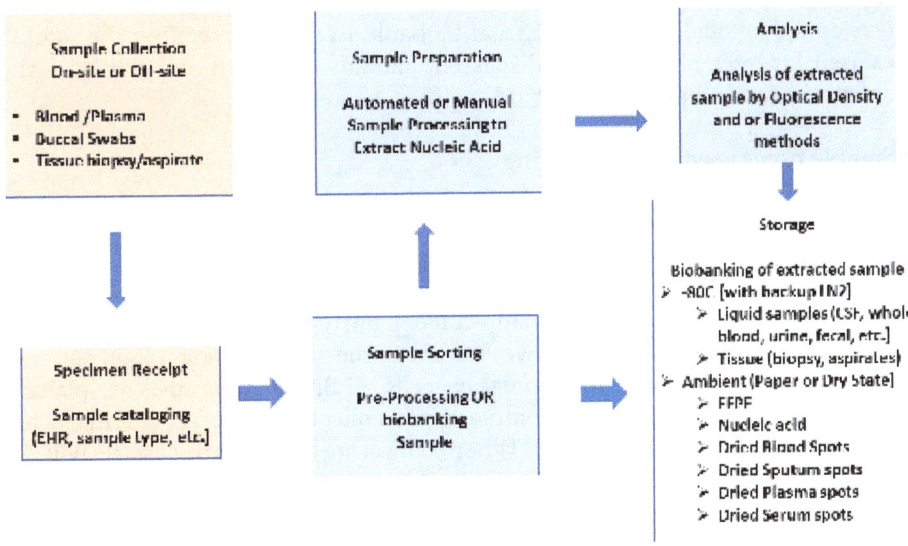


**Figure 2.**
*Infrastructure needed to biobank samples in a laboratory (left to right). Blood, buccal swab, or tissue biopsy/ aspirate samples collected from donors are transported to the research facility if the collection site is remote to the sample processing laboratory. Maintaining traceability and transparency should be mandatory to all human samples, the samples are cataloged electronically when received and connected with the donor's electronic health record (EHR) if available, along with other relevant information before either biobanking or processing the sample for nucleic acid or other biomolecule (DNA, RNA, buffy coat, proteins) extraction. Depending on the size of the donor pool, the most efficient means of processing the large cohort of samples is by automation. The biomolecule extracted is analyzed for quantity and quality before storage in a biobank at −80, or −196°C (liquid nitrogen). Consequently, for an economical means of storage of a large cohort of samples, the extracted biomolecules can be stored in a chemistry matrix for dry state or "glassy state" such as RNA/DNA stable (Biomatrica), RNAsecure (ThermoFisher) or GenTegra RNA/DNA (GenTegra LLC) or on treated paper such as Whatman FTA or GenSaver or untreated paper such as Whatman 903 or GenCollect. The choice of media for storage for biobanking is institute dependent.*

not all facilities with biobanks are willing or able to share these invaluable biospecimens [3, 12, 13]. With such limitations on the use of biobanks, the importance to develop new methods for biobanking continues to grow as technology and research methods have advanced and become more refined to solve previously, seemingly impossible medical mysteries. The Million Donor cohort acts as one solution to the problem of free-data sharing since the project's purpose is to create a comprehen-sive virtual biobank for the purpose of precision medicine and bettering healthcare [3, 16]. Making biobanking a realistic tool and more accessible requires the blending of specimen collection and analysis.

To streamline biobanking and research analysis, a new concept in biobanking is bioinformatic biobanking. This refers to the querying of sequenced genomes that have been stored in virtual biobanks. Bioinformatic biobanks are large databases of information pertaining to the sequenced and analyzed specimen [12, 17]. In this concept of a biobank, an individual's immutable genetic markers form a library to be queried over a lifetime for continuing patient management. Once a sample is analyzed and the data stored, it is then a simple matter of querying the data when and as required for specific gene regions, biomarkers, variations, etc. This approach is currently being implemented by Helix personal genomics with whole genome sequencing as the first step. Using this method negates the need for long-term sample storage because the whole genome can be virtually analyzed for specific biomarkers that may correlate with a disease. Once the genomic analysis is completed, any future test queries involve only a bioinformatic search, as opposed to additional sample collection and repeated analysis. This reduces the cost of biomedical research significantly; however, the collection, transportation, and storage of samples until analysis occurs still pose a significant cost and slow efforts in developing a globally available virtual biobank. In the next section, the sample types used for biobanking will be discussed, and advances that will eliminate the cost of storing and shipping such liquid samples.

## 3.1 Sample types used for biobanking

The most common sample types collected for precision medicine and biobanking of human specimen are tissue samples and whole blood. Tissue samples can be further subdivided into liquid biopsy samples for circulating tumor cells (CTC), tissue biopsy samples such as formalin-fixed paraffin-embedded (FFPE) tissue, and fresh frozen tissue samples or wet mount tissue slides. Whole blood samples can be: peripheral blood mononuclear cells (PBMCs), serum, or plasma. Additional, albeit less common, sample types collected are cerebrospinal fluid (CSF), urine, and fecal material. When collecting these samples, it will be imperative to have "True Control" samples from surrounding disease-free tissue and corresponding known disease-state samples for comparisons; but, it is not always practical for tissue biopsy samples or CSF. In such instances "external" matched controls must serve as acceptable substitutes for "True Control" samples.

### *3.1.1 Tissue*

Tissue samples such as formalin-fixed paraffin-embedded (FFPE) blocks have been stored since the early twentieth century. FFPE tissue samples are a common sample type collected from biopsies. Although core biopsy samples yield a healthy amount of tissue, tissue biopsy procedure is a painful process for the patient and can potentially cause considerable trauma to the surrounding tissue. Fine needle aspirate (FNA) biopsy with a 21-gauge needle to remove tissue samples for pathology is less traumatic to the patient and to the surrounding tissue. Compared to core biopsy samples that are typically about 17 mg or more, the FNA samples are just 2–10 mg and the amount of sample that is donated to research is often less than 1 mm as priority for testing of the biopsy sample is to perform cytopathology. The best outcome for nucleic acid-based testing from tissue samples is to isolate nucleic acid from fresh or flash-frozen at −196°C tissue samples. There is no ambient temperature method available to preserve tissue samples for extracting good-quality nucleic acid.

FFPE is the most common method of preserving tissue samples at ambient temperature. FFPE tissue storage has been used for three decades [18] as a means of keeping tissue samples at ambient temperature for future research [19, 20]. This has created a large resource of pathologically interesting human and animal samples. Fixing tissue samples with formalin and embedding in paraffin preserves the pathology of the tissue. But formalin fixation can cause both inter and intra protein cross-linking [21–23] as well as cross-linking of histones to DNA [24]. Other factors affecting the quality of nucleic acid from FFPE samples include buffering formalin, time and temperature of fixation and penetration of formalin into the tissue by stasis, or by ultrasound, or microwave irradiation. The nucleic acid and protein quality are additionally dependent on the time of collection of tissue following postmortem interval and cold ischemia. Acceptable time for collection of tissue samples is between 4 h postmortem and 12 h after cold ischemia has set in. Acceptable time for formalin fixation of tissue postmortem is <48 h for RNA [25, 26], <24 h for proteins [27–32], and <72 h for DNA [33–36]. It would be best to isolate the nucleic acids from FFPE samples within the acceptable time to ensure the best outcome for the quality of the nucleic acid isolated. The isolated nucleic acid can be further stored at ambient temperature by removing the aqueous media from the nucleic acid sample or by adding some commercially available stabilizers for ambient temperature storage of nucleic acid. Although cross-linking of nucleic acid is of concern with aged FFPE samples [18, 33], nucleic acid extracted from FFPE samples have been successfully used for amplification, single cell analysis, and methylation studies. Decalcification of the FFPE sample using EDTA allows for longer PCR product [37], stronger fluorescence in situ hybridization (FISH) signals, lower background staining [38], and superior comparative genomic hybridization [1] results as compared to other methods.

### *3.1.2 Blood and blood components*

Perhaps the most economical samples are blood samples collected in EDTA tubes [35]. A host of specialized blood collection tubes are commercially available for stabilization of transcripts such as Tempus Blood RNA tubes and PAXgene Blood RNA tubes [39]. The strategy for storing the samples for short-term usage and long-term biobanking needs will determine the quality of the sample. Acceptable short-term storage of weeks to months of blood and blood components such as serum, plasma, peripheral mononuclear cells (PBMCs) etc. is at 4 to −20°C and long-term storage is at −80 to −196°C. Liquid samples such as whole blood, saliva, plasma, and serum samples as dried spots can be stored for decades at ambient temperature if sampled on chemically treated substrate such as FTA or GenSaver paper cards [40, 41].

Serum and plasma samples can be stored at ambient temperature for extended periods of time on chemically treated bead matrix such as GenTegra LLC's Matrix Chaperone (MC) (**Figure 3**). Up to 250 μL of serum or plasma sample can be applied to the MC for storage and for biobanking. Downstream analysis of the MC can be performed simply by adding back equivalent volume of water to the MC. The full complement of analytes, proteins, enzymes, and nucleic acid in serum and clotting factors in plasma samples (data not shown) have been successfully stored for 25 days at ambient temperature on MC consisting of a randomly packed chemically

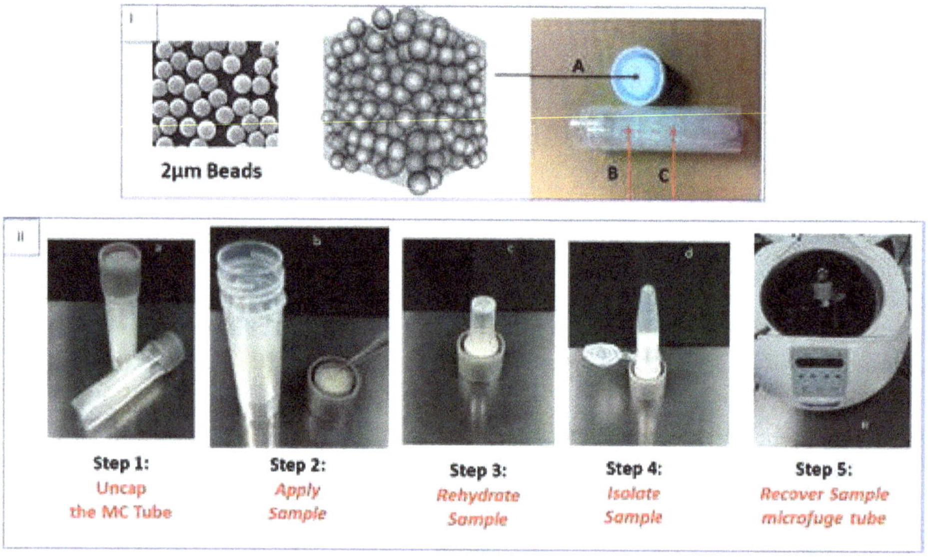


**Figure 3.**
*Ambient storage of serum samples. It is possible to store the entire complement of biological molecules in serum (or any other biological fluids) on a simply made storage device consisting of 2-µm polystyrene beads coated with stabilization chemistry collectively called matrix chaperone (MC) for ambient storage and transportation. The collection device was a simple three-component device (i), containing a cap (A) with the resuspended stabilization chemistry in a matrix of polystyrene beads and the holding chamber (B) with silica gel (C) to facilitate drying. A volume of 250 µL of serum sample was added to the cap containing the chemical matrix of polystyrene beads (iib) and capped. The assembly was placed in an upside-down position for at least 12 h to facilitate drying. To initiate analysis, the sample was reconstituted with 250 µL of water (iic). The sample is recovered by transferring the sample to a 1.5-mL tube (iid) and centrifuging the sample at maximum speed for a minute (iie). A complete metabolic panel and a lipid panel test were performed on this reconstituted serum sample (**Table 1**).*

treated microsphere wafer when compared to pristine always frozen at −20°C serum samples (**Table 1**).

### *3.1.3 Dried blood spots microsample biobanking*

Dried blood spots (DBSs) can be used for both real-time microsampling and subsequent ambient temperature biobanking for epidemiology and biomarker discovery (**Figure 4**). DBS samples can be particularly effective as a means of sample collection from participants in clinical trials. A survey by Tasso Inc. determined that a trial candidate may be more compliant to sample collection when given a less painful option for sample collection such as the OnDemand automated blood collection device for DBS collection and when done in the comfort of their own home (data unpublished) (**Figure 5**). Blood stabilized on the DBS can then be mailed by local postal services at the patient's own convenience. Although storage of whole blood as DBS is an old technology, historically poor stability outside the lab environment, as well as low recovery levels and generally low quality of extracted nucleic acids and numerous blood proteins, has hindered its acceptance. In recent years, there has been development for a completely new, "smart health care," paper-based sampling technology, which overcomes many of these known drawbacks. Deployed as a simple, painless skin prick onto a chemically treated collection card, the dried blood may then be recovered by ordinary magnetic bead or column-based DNA purification. With the resurgence of interest in the use of DBS for sample collection, research is being done to develop novel chemistries to yield RNA, DNA,

**Table 1.a: Enzyme Panel**

| Enzymes | Normalized to Frozen Ambient | Always Frozen Values | Initial Raw Values Ambient | Normal Range (IU/L) |
|---|---|---|---|---|
| ALT | 75% | 25 | 19 | 7-40 |
| AST | 70% | 34 | 24 | Oct-38 |
| CK | 74% | 107 | 120 | 30-[illegible] |
| ALP | 45% | 68 | 25 | 44-147 |
| Amylase | 84% | 61 | 51 | 25-102 |
| GGT | [illegible] | 47 | 47 | [illegible] |

**Table 1.b: Lipid Panel**

| Lipids | Normalized to Frozen Ambient | Always Frozen | Initial Raw Values Ambient | Normal Range |
|---|---|---|---|---|
| Cholesterol | 75% | 254 | 191 | ≤200 |
| Triglycerides | 75% | 237 | 177 | [illegible] |
| HDL | [illegible] | [illegible] | [illegible] | [illegible] |

**Table 1.c: Protein Panel**

| Proteins | Normalized to Frozen Ambient | Always Frozen Values | Initial Raw Values Ambient | Normal Range |
|---|---|---|---|---|
| Total Protein | 83% | 7 | 6 | 6-8.3 g/dL |
| Albumin | [illegible] | 5 | 4 | [illegible] |
| Ferritin | 75% | 255 | 190 | [illegible] |
| TSH | 79% | 11 | 9 | 0.4-4.0 mIU/L |
| T4 | [illegible] | 5.3 | 5.2 | 4.5-11.2 mcg/dL |
| SHBG | 77% | 40 | 31 | [illegible] nmol/L |
| IgA | 75% | 205 | 153 | 80-350 mg/dL |
| IgG | 77% | 927 | 716 | 620-1400 mg/dL |
| IgM | 68% | 95 | 65 | 45-250 mg/dL |

**Table 1.d: Metabolite Panel**

| Metabolites | Normalized to Frozen Ambient | Always Frozen | Initial Raw Values Ambient | Normal Range |
|---|---|---|---|---|
| Creatine | 54% | 1 | 0.5 | 0.6-1.2 mg/dL |
| Cortisol | 70% | 17 | 13 | 7-25 mcg/dL |
| Testosterone | 121% | 169 | 159 | 250-827 ng/dL |

**Table 1.**

*Stability of serum enzymes, proteins, lipids, and metabolites at ambient when stored on the polystyrene bead matrix, MC, containing ambient stabilization chemistry for all biomolecules. A volume of 250 μL of CAP-certified serum samples spotted on the polystyrene matrix (MC) tubes and dried before storing for 25 days at ambient. Corresponding control serum samples were stored at −20°C. After 25 days of storage at ambient, the experimental and control samples were hydrated with 250 μL of water. All the analytes from the rehydrated MC serum samples and the fresh always frozen −20°C control samples were quantified for the complete metabolic panel and the lipid panel with the cobas® 6000 analyzer. The percent recovery of the analytes was calculated for the ambient stored samples compared to the control serum samples from the initial raw values of the test biomolecules. International units per liter (IU/L) of the enzyme panel for alanine amino transferase (ALT), aspartate aminotransferase (AST), enzyme marker creatine phosphokinase (CK), amylase, and gamma glutamyl transferase (GGT) are within the normal range for the ambient stored serum samples when compared to −20°C control samples. The level of the alkaline phosphatase (ALP) was low compared to the control serum samples indicating that the stabilizer is not able to protect the labile ALP enzyme. All the molecules tested in the protein panel and the lipid panel were in the normal range and maintained at between 68% and 83% of control indicating stability at ambient of all biomolecules in this panel on the MC. Of the three metabolites tested in the metabolic panel, the stabilizing matrix of the serum sample stored at ambient could not stabilize the metabolite creatine but cortisol and testosterone were stabilized.*

and proteins with quality and quantity enough to support advanced analytical methods such as next generation sequencing and multiplex proteomics.

DBS is also associated with a 100-fold lower carbon footprint, being 100 times more compact (in terms of sample size) and it is readily suited for automated recovery from such solid-state blood specimens [42]. Long-term storage for multiple decades requires storage at −196°C under liquid nitrogen or on treated paper such as Whatman® FTA, Ahlstrom-Munksjö GenSaver™ cards. For short-term storage (weeks to a month), untreated paper such as Whatman 903 paper, Ahlstrom-Munksjö GenCollect™ paper, etc. may be used. The paper products work by drawing the water out of the sample causing localized dehydration of the sample. Specifically treated papers such as Whatman FTA and Ahlstrom-Munksjö GenSaver cards further stabilize the sample by either lysis of the cells and/or by prevention of various oxidative damage to the sample. Ribosomal RNA (18S and 28S rRNA) is more labile when stored in DBS, as demonstrated with a less than ideal RNA integrity number (RIN) below 6.5 (RIN values will be explained later on). Storage of blood on treated paper is superior to untreated paper for decades-long storage of DNA but to date there is no product available for storage of total RNA in whole blood for decades other than storing at −196°C.

Advances in non-invasive diagnostics for cancer where routine blood sample collection can be used for tracking progression of the disease is much more affordable and less painful than a solid tissue biopsy alternative. DBS microsample is a good alternative to collecting liquid whole blood in EDTA tubes by phlebotomy for individuals where tracking of progression of disease is crucial for prescribing treatment options. Advances in the quality and availability of highly sensitive

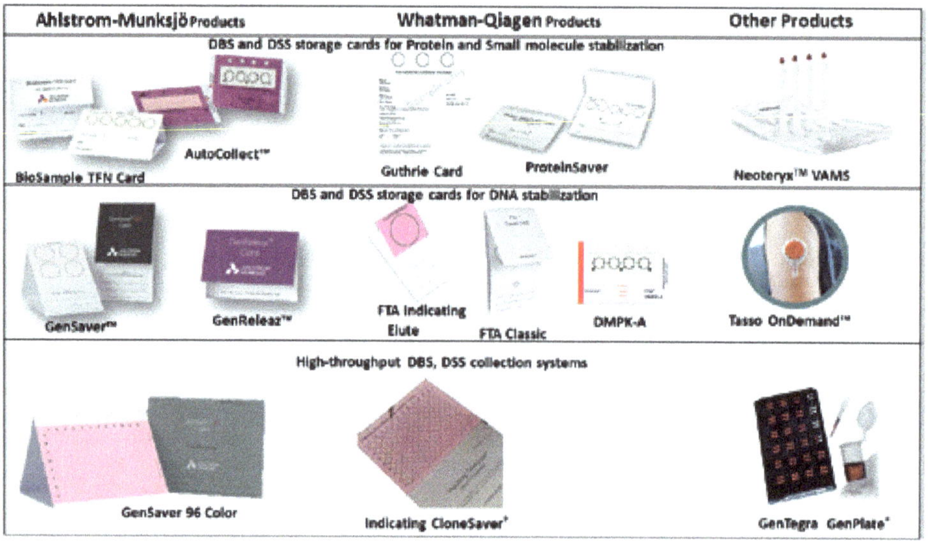


**Figure 4.**
*Commercially available DBS collection devices for ambient storage and transportation. Biological samples collected remotely or stored at designated biobanks can utilize any one of the various products available for ambient storage of liquid samples. A number of formats of high-quality fiber-based media are available from Ahlstrom-Munksjö, Whatman-Qiagen, and others. The colorless format of cards is ideal for storage of colored biological samples such as fecal matter, plants, and whole blood. The colored cards are for storage of colorless biological samples such as serum, saliva, and other organics, at ambient. Biosample TFN card, AutoCollect card from Ahlstrom-Munksjö and Guthrie card, protein saver card from Whatman-Qiagen are ideal for collection of DBS needed for protein and small molecule analysis. The VAMS storage device from Neoteryx™ is convenient for patient-centric remote collection of microvolume samples. Some of the collection cards such as FTA, GenSaver, GenSaver Color cards and GenPlates are chemically treated for long-term preservation of DNA at ambient. These cards are ideal for biobanking and forensics application. AutoCollect card with perforated DBS circles are designed for automated sample preparation. GenPlates allow for high-throughput automated spotting of biological samples. The Tasso OnDemand collection device with integrated VAMS or paper cards is a painless alternative for volumetric collection of DBS. GenSaver, GenPlates, and GenReleaz cards allow for the convenience of direct downstream analysis (PCR, NGS, STR, etc.), from a 1-mm punch of DBS without any need for sample extraction. The 96-well format is ideal for storage of biobank samples and for screening and health monitoring applications. GenSaver 96 color, Indicating CloneSaver and GenPlates are designed for high-throughput biobanking needs.*

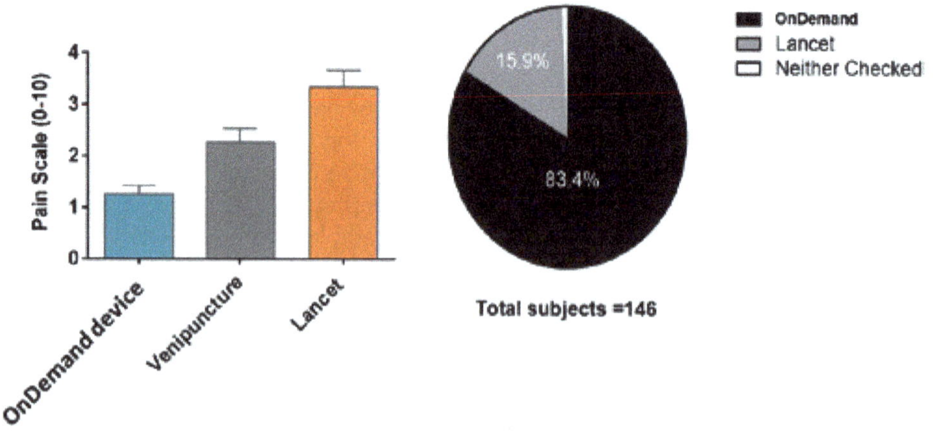


**Figure 5.**
*Patient survey for acceptance of blood collection method. Convenience, lack of pain, and simplicity of sample collection from donor will ensure compliance. As determined from a survey of 146 subjects, by Tasso Inc., on a pain scale of 1–10, the surveyed subjects graded the OnDemand device at 1.25, venipuncture at 2.25, and the lancet method of blood collection at >3.0. Of the 146 subjects, 83.4% preferred the least painful OnDemand method of blood collection compared to 15.9% by the lancet method (data courtesy of Tasso Inc.).*

instruments coupled with the development of software and methodological platforms for improved qualitative and quantitative analyses have made adoption of microsampling mainstream [43]. At home, blood collection of finger stick blood redundancy on microsampling devices such as GenSaver or FTA paper or on polymer compound, for example, Volumetric analytical MicroSampling (VAMS), is a convenient and a less painful alternative to phlebotomy. Advances in automation of almost painfree microsampling of blood with the OnDemand by Tasso Inc. or Tap by Seventh Sense Biosystems are great alternatives for finger prick collection. Although opponents of precision medicine [34, 44] argue that matching the genotypic and phenotypic makeup of the individual to the treatment will not work, thus far, an individual-refined approach to selecting treatments has yielded demonstrable if still limited success.

### *3.1.4 Postmortem samples biobanking*

Living donors contribute tissue samples only if it is a medical necessity. A possible viable source of large quantities of tissue samples is through postmortem collection of whole organs and tissues from consenting families. This avenue incurs a whole host of new challenges such as the donation consent process, recovery of organs and tissues in a limited time frame, postmortem, impact of donation on the donor families and steps necessary for creating a postmortem biobank such as IRB and registries. Carithers et al. [45] describe development of eligibility requirements aligned with scientific needs of the project and implementation of a successful infrastructure for biospecimen procurement to support the prospective collection, annotation, and distribution of blood, tissue, and cell lines and associated clinical data from postmortem samples. The development of donor eligibility criteria is crucial since limited donor history is available within the time frame needed for the collection of potential donor samples as degradation of biomolecules starts immediately with death. This proposition incurs a whole host of new challenges such as the donation consent process, recovery of organs and tissues in a limited time frame, postmortem, impact of donation on the donor families, and steps necessary for creating a postmortem biobank such as IRB and registries.

Sample collection is an intricate process that involves having many facets of the process coming together such as participant willingness, maintenance of anonymity of the sample source, and collecting samples in an ethically appropriate manner. Primarily, sample collection depends on the individual's willingness to participate in the clinical study and their trust in the collecting institution. Higher participation rates may be anticipated if the need for the study results is focused on the greater good of the community [46]. Often the most problematic aspect of sample collection when collecting large number of donor samples is maintaining donor anonymity. The Kaiser foundation admits that donor personal information is vulnerable and has placed controls to mitigate the issue such as educating the donor as well as assigning an alternate identification to donor samples that are for the Precision Medicine Initiative [47].

Both private and federally funded institutions have set up repositories to collect and archive biological samples to be then made available to researchers globally through for-profit, paid services or for free through not-for-profit organizations. One such globally recognized institute, the Kaiser Foundation, launched their initiative in October of 2015 and has thus far accumulated over 220,000 samples through volunteers with an end goal of collecting a total of 500,000 samples [47, 48]. The Kaiser Foundation has the added advantage of possession of the patient's lifestyle and EMRs that can be integrated along with the sample.

### *3.1.5 Nucleic acid extraction and storage*

The quality and quantity of nucleic acid (DNA and RNA) depends on the quality of the nucleic acid in the starting material and the extraction method. There are many commercially available kits for extracting nucleic acids from varied sample types such as blood, PBMCs, serum, DBS, and fresh or frozen tissue samples. Decalcified FFPE samples are treated in the same manner as tissue samples. The common mechanism by which most nucleic acid extraction kits work is through lysis of the cells to release the nucleic acid followed by capture of the nucleic acid in chaotropic agents such as guanidinium salt on paramagnetic silica beads or on glass fiber filters. The silica beads or glass fiber filters are then washed to remove the proteins and cellular debris leaving relatively clean nucleic acid on the beads/filter. The nucleic acid is then released with a buffered elution solution most commonly Tris-EDTA at pH 8.3.

Extraction methods or kits must be chosen so they are well suited to the sample type [49–52]. Nonetheless, there is varying opinion in the literature regarding the choice of nucleic acid extraction kit to use for different sample types. Molteni et al. [50] determined that the efficiency of extracting DNA from DBS on plain paper and Whatman FTA paper was better with Masterpure kit than with Qiagen's QIAamp Blood mini kit and GenSolve DNA extraction kit (GenTegra LLC) being next best. In contrast, Daniels et al. of Broad Institute [52] determined that the efficiency of extracting DNA from DBS on Whatman FTA paper was superior with GenSolve DNA extraction kit than Qiagen's QIAamp Blood mini kit. McClure et al. [53] report comparable quality of DNA extracted from DBS on Whatman FTA paper with the GenSolve DNA extraction kit to that extracted from whole blood samples. These DNA samples were compared on the Illumina BovineSNP50 iSelect BeadChip, which requires unbound, relatively intact (fragment sizes ≥2 kb), and high-quality DNA. Superior-quality total RNA can be extracted with the time-tested phenol extraction using the commercially available Tri Reagent, although good-quality total RNA can also be obtained by using commercially available RNA extraction kits (Zymo Research, Qiagen, and ThermoFisher). Agitation of the DBS sample in lysis solution at 40°C ensures more efficient extraction of RNA.

Clearly, the choice of method of nucleic acid extraction is dependent on prior sample expertise and analysis methods to be used for the study. A distinction between the quantity of nucleic acid extracted vs. the quality of nucleic acid is crucial, since having a large quantity of compromised nucleic acid will still result in an unsatisfactory outcome (**Figure 6**). A good check for the quality of DNA and RNA is by calculating the DNA Integrity Number (DIN) [54] or the RNA integrity number (RIN) [15, 55] of the sample by electrophoresis in the Agilent TapeStation or similar devices. Alternatively, the quality of DNA can also be assessed by amplification of a 3-kb to 7-kb fragment of a low copy housekeeping gene such as glyceraldehyde 3-phosphate dehydrogenase (GAPDH) [56] and for RNA fragment >0.9 kb of a low copy gene such as RNase P [51].

Nucleic acid (both DNA and RNA) extracted from samples can be stored either at very low temperatures of −20, −80, or − 196°C or in a dry state by spray drying; lyophilization; air drying in the presence of commercially available protective chemistries such as RNAstable, DNAstable, (Biomatrica Inc., San Diego, CA), GenTegra-DNA [57] or GenTegra-RNA (GenTegra LLC, Pleasanton, CA); or by spotting on paper. The ribose-phosphate backbone in RNA molecules makes them susceptible to degradation. RNA consequently needs to be stored short term at −80°C or long term at ultra-low temperature of −196°C or in a precipitated form under ethanol. It is also possible to store RNA vitrified in the dry state at ambient temperature in the presence of protectants (GenTegra-RNA) that form the "glassy

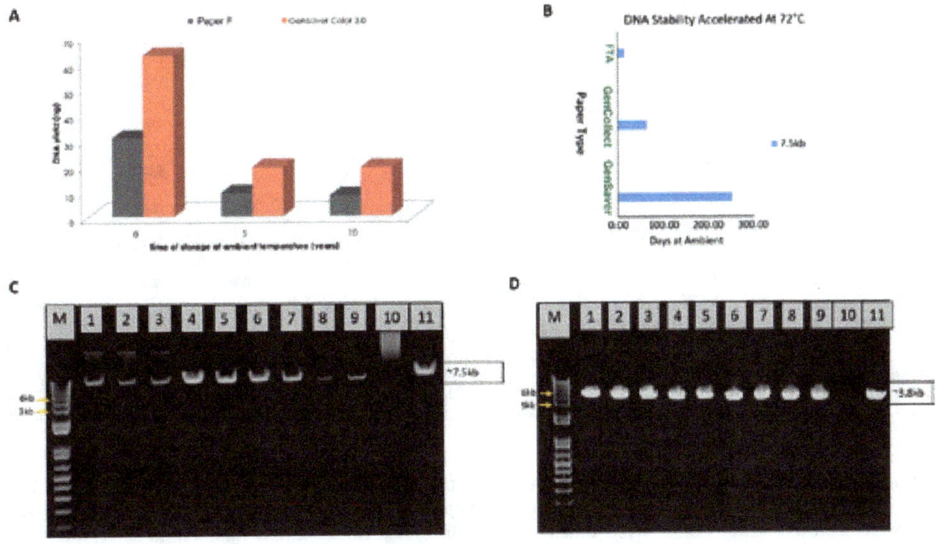


**Figure 6.**
*Quality and quantity of DNA from DBS. A volume of 125 µL of whole blood spotted on GenSaver, GenCollect, and Paper F (FTA) paper cards, these were stored for up to 10 years at ambient temperature (A). DNA extraction yields obtained from three 6-mm DBS punches of GenSaver and FTA paper cards, the amount of DNA obtained from the sample reduces with aging of the DBS. The quality of the DNA is influenced by the chemical protective agents added to the card (B). A 7.5-kb fragment of a single copy gene (GAPDH) was amplified from 5 ng of DNA from DBS on GenSaver, GenCollect, and FTA cards. DNA in DBS on GenSaver cards is 18× more stable than FTA and 4× more stable than GenCollect cards when a 7.5-kb fragment of a single copy gene (GAPDH) was amplified by polymerase chain reaction (PCR). A volume of 20 µL of the PCR product was subjected to electrophoresis on a 1.2% agarose gel (C). Gel electrophoresis of GAPDH 7.5-kb PCR product. Bands showing the PCR product from DBS in lanes 1–3 (GenCollect) and lanes 7–9 (FTA) are less intense than those of lanes 4–7 (GenSaver) for the 7.5-kb GAPDH product (D). Gel electrophoresis of GAPDH 3.8-kb PCR product. Product on lanes 1–3 comes from GenCollect DBS, product on lanes 4–6 comes from GenSaver DBS, products on lanes 7–9 comes from FTA DBS, lane 10 is the negative control and 11 is the positive amplification control, the intensities of the PCR product is not distinguishable among paper types. The differences in the intensities of the 7.5-kb and the 3.8-kb product indicate that the DNA is better protected from environmental effects in GenSaver cards. Samples were run on an E-gel.*

state" to prevent oxidative, hydrolytic, or RNase damage to the ribonucleic acid. The protective chemistries also allow the dry nucleic acid to redissolve easily because the chemistries prevent the formation of the gel-state that pure nucleic acids often form at high concentration. The gel-state makes solubilization very difficult without using mechanical forces that will also break the nucleic acid strands.

Oxygen and water are essential components in the generation of reactive molecules with the degradation process accelerating with increased temperature, reduced ionic strength of storage solution, increased concentration of divalent cations (greater than 5 ppb) or nucleolytic enzymes. In aqueous solutions (a convenient format for storage), nucleic acids are sensitive to depurination, depyrimidination, deamination, and hydrolytic cleavage. To inhibit this acid-catalyzed degradation of DNA, sample storage solutions for DNA need to be slightly alkaline buffered solutions such as tri-buffered to pH of 8.3. Nucleic acid extracted from clinical samples likely contain up to 30–40 ppb of iron (from heme or haem). Presence of even trace amounts of divalent cations (greater than 5 ppb) increases the oxidative degradation of nucleic acid due to the formation of highly reactive free radicals via Fenton reaction [58]. Adding chelating agents such as EDTA and EGTA to a concentration of 500 mM to the nucleic acid storage solution would ensure that the intrinsic divalent cations present in the clinical samples are chelated.

Dry storage of nucleic acids in the presence of protective chemistries causes the molecules to lose the ability to diffuse as the sample undergoes a non-crystalline

| 4 Years at Ambient (26°C) | | | | | | | | | |
|---|---|---|---|---|---|---|---|---|---|
| | HeLa RNA | | | WBC RNA | | | | Rat liver RNA | |
| Buffer | Recovery | Bioanalyzer | Strand Breaks | Recovery | Bioanalyzer | Strand Breaks | Recovery | Bioanalyzer | Strand Breaks |
| | % | RIN | Average per kb | % | RIN | Average per kb | % | RIN | Average per kb |
| | 115 | 8.2 | | 116 | 5.3 | | 110 | 4.9 | |
| Water | 121 | 8.7 | 0.08 | 118 | 5.4 | 0.37 | 109 | 4.8 | 0.50 |
| | 116 | 7.7 | | 115 | 4.8 | | 113 | 5.0 | |
| | 115 | 8.4 | | 112 | 5.0 | | 109 | 4.3 | |
| Citrate | 116 | 7.7 | 0.08 | 111 | 4.1 | 0.38 | 110 | 3.8 | 0.67 |
| | 112 | 8.5 | | 111 | 4.8 | | 111 | 4.3 | |
| | 115 | 8.2 | | 115 | 4.7 | | 115 | 4.8 | |
| EDTA | 118 | 7.9 | 0.08 | 112 | 3.6 | 0.38 | 114 | 4.3 | 0.46 |
| | 115 | 8.3 | | 121 | 3.8 | | 114 | 4.7 | |

**Table 2.**
*HeLa RNA, WBC RNA, and rat liver RNA in water, citrate, or EDTA buffer were stored in GenTegra-RNA for 4 years at ambient temperature (25°C). All of the samples were hydrated with water at the end of 4 years. RIN scores were analyzed by Bioanalyzer and average strand breaks calculated per kilobase (as determined by the negative natural log of ratio of peak heights of 28S-18S) at time equals 4 months to time zero ($R_n$) for the sample groups. The source of RNA and degree of RNase carryover was the key factor in determining the maintenance of a stable RIN score and development of number of strand breaks per kb and is independent of the type of buffer used for storage of the RNA.*

amorphous phase or a "glassy state." In this dry "glassy state," the movement of protons is expected to be approximately one atomic diameter in 200 years, thus preventing both oxidative and nucleolytic degradation of the nucleic acid. Storage at ultra-low temperatures of −196°C also vitrified as the water becomes solid ice and the molecules lose their ability to move. If moisture is added to the dry sample or the temperature is raised in ultra-cooled samples above the glass transition temperature of water, DNA/RNA damage can occur as the proton movement and reactivity resume [59]. Trace amounts of RNase would also become active upon rehydration causing RNA damage.

Successful storage of biomolecules including nucleic acid is ultimately dictated by the purity of the extracted material. Highly pure total RNA samples from HeLa cells with a RIN of 10 can be stored dry for up to 6 years and 2 months at room temperature in GenTegra-RNA (**Table 2**) without appreciable loss of RNA integrity or strand breakage, but rat liver RNA that carries along cellular impurities in the extracted total RNA shows degradation of up to 0.2 strand breaks per kilobase, deterioration in RIN from 9.0 to 4.0 and a short storage life of 1 year and 8 months. Human blood lymphocyte RNA, like rat liver RNA (at an intermediate level of residual purity), displays more damage after 4 years as assessed by RIN analysis, suggestive of 0.4–0.5 RNA breaks per kilobase after 4 years of ambient temperature dry-state storage in GenTegra-RNA. WBC RNA, like rat liver RNA samples stored with additional 1 mm of EDTA, incurred much less damage upon 7 months of storage at 56°C (only about 0.1break/kb) (data not shown). RNA strand breakage (X) is determined from the calculated RIN value of the aged RNA to the RIN value of the unaged RNA stored at −20°C [60].

## 4. Conclusion

Located in hospitals, universities, non-profit organizations, and pharmaceutical companies, biobanks are key infrastructures for research and development; however, these vaults for biospecimen are expensive to maintain and are precious samples that are not willingly shared. Nonetheless, biobanking provides invaluable insight into biomedical mysteries. The long-term translational studies allowed by maintaining archives of samples could provide valuable insight for future generations to treat chronic diseases. Currently, the research community hopes to use biobanking to push forward precision medicine, an initiative set forth by the Obama administration to form unique, targeted treatments for each individual [3, 6]. There is a two-fold challenge for sample collection, ensuring privacy and getting volunteers. Transparency and traceability of samples are key to governance of all human biospecimens. Living donors contribute tissue samples only if it does not directly affect the quality of their life. The development of donor eligibility criteria is crucial since limited donor history is available within the time frame needed for the collection of potential donor samples as degradation of biomolecules starts immediately with death. Thus, it is incredibly important to minimize the amount of time samples spend out of storage, in biobanks.

Biobanking biospecimens is an expensive endeavor both in terms of manpower and natural resources used. For example, a single −80°C freezer consumes as much energy as a small studio apartment. Most biobanks install a bank of −80°C freezers to store the biospecimen samples as each sample needs to be stored for 10 years as per CLIA and CAP guidelines. Many institutions store samples for longer than a decade for research, test development, and validation purposes. FFPE blocks are cataloged at ambient temperature room temperature making them the most efficient way of storing biospecimen samples. Most other sample types are presently stored in −80°C freezers for short-term storage or −196°C under liquid nitrogen for longer term storage. Although not yet mainstream, advances in microsampling analytical technologies has popularized ambient temperature biobanking of whole blood, whole blood components, fecal, urine, plasma, and serum biospecimens on paper such as treated GenSaver or FTA papers or on VAMS tips or Matrix chaperone. Nucleic acid can be stored at ambient temperature in the presence of protective stabilizers in a dry state with a choice of commercially available time-tested products such as GenTegra-DNA, GenTegra-RNA, DNAstable, RNAstable, RNAlater, etc. Ambient temperature storage is the most economical, environmentally friendly, low-carbon footprint, and practical way of storage when long-term storage for decades is needed. In addition to reducing molecular mobility, drying the samples removes water that can participate in hydrolytic reactions. Furthermore, storing samples in a dry state at ambient temperature is independent of environmental factors such as electrical supply, temperature, and humidity.

The Precision Medicine Initiative aimed at precisely and rapidly analyzing many more cancer genomes will bring about a deeper understanding of cancers fueled by discoveries of molecular diagnostic methods. The first fruits of precision medicine are already apparent as a wide range of nucleic acid and antibody/protein-based drugs have been optimized for individuals with favorable genetic makeup. With a goal of collecting a million samples for the Precision Medicine Initiative, storing the samples such as blood at −80 or −196°C for prolonged period (decades) is going to become impractical at some time. Consideration needs to be given to space and energy requirements for such an undertaking. A more practical approach is to consider dry ambient temperature storage of biomolecules that have a commercially available solution for storage. Although dry storage of nucleic acid and DBS at ambient temperature is an economical alternative, adoption of this concept by the research community would be a paradigm shift from the time-tested method of preservation by cryogenics. This could be due simply to availability of freezers for storing other sample types that yet do not have an ambient temperature storage method. A new technology introduced to the marketplace has a 30-year adoption cycle and dry storage is a couple decades into that cycle with increasing number of research facilities converting to ambient temperature storage where applicable. Biobanking of human samples has many ramifications that go beyond the science and technology of their storage. There are national, state, and even local regulations that must be met to ensure the protection of individual rights and individual privacy. Educating donors on the purpose of biospecimen collection and assurance of maintaining the privacy of the donor has favorable outcome. Perhaps it is reasonable to consider in a future review these legal and privacy issues.

## Author details

Armaity Nasarabadi Fouts, Alejandro Romero, James Nelson, Mike Hogan and Shanavaz Nasarabadi*
GenTegra, LLC, Pleasanton, California, United States

*Address all correspondence to: shanavazn@gentegra.com

## References

[1] Alers JC, Krijtenburg P-J, Vissers KJ, van Dekken H. Effect of bone decalcification procedures on DNA In situ hybridization and comparative genomic hybridization: EDTA is highly preferable to a routinely used acid decalcifier. The Journal of Histochemistry and Cytochemistry. 1999;**47**:703-709

[2] Alegría-Torres JA, Baccarelli A, Bollati V. Epigenetics and lifestyle. Epigenomics. 2011;**3**:267-277

[3] Jaffe S. Planning for US precision medicine initiative underway. Lancet. 2015;**385**:2448-2449

[4] Ashley EA et al. Clinical assessment incorporating a personal genome. Lancet (London, England). 2010;**375**:1525-1535

[5] Side Effects of Cancer Treatment—National Cancer Institute. Available from: https://www.cancer.gov/about-cancer/treatment/side-effects

[6] Terry SF. Obama's precision medicine initiative. Genetic Testing and Molecular Biomarkers. 2015;**19**:113-114

[7] Alix-Panabieres C, Pantel K. The circulating tumor cells: Liquid biopsy of cancer. Kliniceskaja Laboratornaja Diagnostika. 2014; **118**:60-64

[8] Simeon-Dubach D, Perren A. Better provenance for biobank samples. Nature. 2011;**475**:454-455

[9] NIH awards $55 million to build million-person precision medicine study | National Institutes of Health (NIH). Available from: https://www.nih.gov/news-events/news-releases/nih-awards-55-million-build-million-person-precision-medicine-study

[10] Gaziano JM et al. Million veteran program: A mega-biobank to study genetic influences on health and disease. Journal of Clinical Epidemiology. 2016;**70**:214-223

[11] Eiseman E, Haga SB. Handbook of Human Tissue Sources. 1999. pp. 11-77

[12] De Souza YG, Greenspan JS. Biobanking past, present and future: Responsibilities and benefits. AIDS. 2013;**27**:303-312

[13] Ntai A, Baronchelli S, Pellegrino T, De Blasio P. Biobanking shifts to "precision medicine". Journal of Biorepository Science for Applied Medicine. 2014;**2**:11-15

[14] Alizadeh AA et al. Toward understanding and exploiting tumor heterogeneity. Nature Medicine. 2015;**21**:846-853

[15] Schroeder A et al. The RIN: An RNA integrity number for assigning integrity values to RNA measurements. BMC Molecular Biology. 2006;**7**:1-14

[16] Riegman PHJ, Morente MM, Betsou F, de Blasio P, Geary P. Biobanking for better healthcare. Molecular Oncology. 2008;**2**:213-222

[17] Kroth PJ, Schaffner V, Lipscomb M. Technological and administrative factors implementing a virtual human biospecimen repository. In: AMIA ... Annual Symposium proceedings. AMIA Symposium. 2005. p. 1011

[18] Fox CH, Johnson FB, Whiting J, Roller PP. Formaldehyde fixation. The Journal of Histochemistry and Cytochemistry. 1985;**33**:845-853

[19] Bass BP, Engel KB, Greytak SR, Moore HM. A review of preanalytical factors affecting molecular, protein, and morphological analysis of formalin-fixed, paraffin-embedded (FFPE) tissue: How well do you know

your FFPE specimen? Archives of Pathology and Laboratory Medicine. 2014;**138**:1520-1530

[20] Hood BL et al. Proteomic analysis of formalin-fixed prostate cancer tissue. Molecular & Cellular Proteomics. 2005;**4**:1741-1753

[21] Fraenkel-Conrat H, Olcott HS. The reaction of formaldehyde with proteins. V. Cross-linking between amino and primary amide or guanidyl groups. Journal of the American Chemical Society. 1948;**70**:2673-2684

[22] Metz B et al. Identification of formaldehyde-induced modifications in proteins: Reactions with model peptides. The Journal of Biological Chemistry. 2004;**279**:6235-6243

[23] Middlebrook WR, Phillips H. The action of formaldehyde on the cystine disulphide linkage of wool; the conversion of subfraction A of the combined cystine into combined lanthionine and djenkolic acid and subfraction B into combined thiazolidine-4-carboxylic acid. The Biochemical Journal. 1947;**41**:218-223

[24] Jackson V. Studies on histone organization in the nucleosome using formaldehyde as a reversible cross-linking agent. Cell. 1978;**15**:945-954

[25] Beaulieu M et al. Analytical performance of a qRT-PCR assay to detect guanylyl cyclase C in FFPE lymph nodes of patients with colon Cancer. Diagnostic Molecular Pathology. 2010;**19**:20-27

[26] Chu WS et al. Ultrasound-accelerated formalin fixation of tissue improves morphology, antigen and mRNA preservation. Modern Pathology. 2005;**18**:850-863

[27] Boenisch T. Effect of heat-induced antigen retrieval following inconsistent formalin fixation. Applied Immunohistochemistry & Molecular Morphology. 2005;**13**:283-286

[28] Ibarra JA, Rogers LW, Kyshtoobayeva A, Bloom K. Fixation time does not affect the expression of estrogen receptor. American Journal of Clinical Pathology. 2010;**133**:747-755

[29] Leong AS-Y, Milios J, Duncis CG. Antigen preservation in microwave-irradiated tissues: A comparison with formaldehyde fixation. The Journal of Pathology. 1988;**156**:275-282

[30] O'Rourke MB, Padula MP. Analysis of formalin-fixed, paraffin-embedded (FFPE) tissue via proteomic techniques and misconceptions of antigen retrieval. BioTechniques. 2016;**60**

[31] Wasielewski R, Mengel M, Nolte M, Werner M. Influence of fixation, antibody clones, and signal amplification on steroid receptor analysis. The Breast Journal. 1998;**4**:33-40

[32] Williams JH, Mepham BL, Wright DH. Tissue preparation for immunocytochemistry. Journal of Clinical Pathology. 1997;**50**:422-428

[33] Ferrer I et al. Effects of formalin fixation, paraffin embedding, and time of storage on DNA preservation in brain tissue: A BrainNet Europe study. Brain Pathology. 2007;**17**:297-303

[34] Prasad V. Perspective: The precision-oncology illusion. Nature. 2016;**537**:S63

[35] Reineke T et al. Ultrasonic decalcification offers new perspectives for rapid FISH, DNA, and RT-PCR analysis in bone marrow trephines. The American Journal of Surgical Pathology. 2006;**30**:892-896

[36] Zsikla V, Baumann M, Cathomas G. Effect of buffered formalin on amplification of DNA from paraffin wax

embedded small biopsies using real-time PCR. Journal of Clinical Pathology. 2004;**57**:654-656

[37] Sarsfield P et al. Formic acid decalcification of bone marrow trephines degrades DNA: Alternative use of EDTA allows the amplification and sequencing of relatively long PCR products. Journal of Clinical Pathology—Molecular Pathology. 2000;**53**:336

[38] Babic A et al. The impact of pre-analytical processing on staining quality for H&E, dual hapten, dual color in situ hybridization and fluorescent in situ hybridization assays. Methods. 2010;**52**:287-300

[39] Mirnezami R, Nicholson J, Darzi A. Preparing for precision medicine. New England Journal of Medicine. 2012;**366**:489-491

[40] Rahikainen AL, Palo JU, de Leeuw W, Budowle B, Sajantila A. DNA quality and quantity from up to 16 years old post-mortem blood stored on FTA cards. Forensic Science International. 2016;**261**:148-153

[41] Saieg MA et al. The use of FTA cards for preserving unfixed cytological material for high-throughput molecular analysis. Cancer Cytopathology. 2012;**120**:206-214

[42] Li F, Ploch S. Will 'green aspects of dried blood spot sampling accelerate its implementation and acceptance in the pharmaceutical industry? Bioanalysis. 2012;**4**:1259-1261

[43] Freeman JD et al. Review state of the science in dried blood spots. Clinical Chemistry.2018;**64**(4):656-679

[44] Tannock IF, Hickman JA. Limits to personalized cancer medicine. The New England Journal of Medicine. 2016;**375**:1289-1294

[45] Carithers LJ et al. A novel approach to high-quality postmortem tissue procurement: The GTEx project. Biopreservation and Biobanking. 2015;**13**:311-317

[46] Gillespie K et al. Patient views on the use of personal health information and biological samples for biobank research. Journal of Patient-Centered Research and Reviews. 2017;**4**:171

[47] Kaiser Permanente Expands Precision Medicine Biobanking Effort. Available from: https://healthitanalytics.com/news/kaiser-permanente-expands-precision-medicine-biobanking-effort

[48] Permanente, K. Kaiser Permanente Research Bank Consent Form. Available from: https://researchbank-econsent.kaiserpermanente.org/(X(1)S(p1ehme24jg1janrg12rgdzfm))/ConsentForm/ConsentForLocalPrint/?AspxAutoDetectCookieSupport=1

[49] Choi EH, Lee SK, Ihm C, Sohn YH. Rapid DNA extraction from dried blood spots on filter paper: Potential applications in biobanking. Osong Public Health and Research Perspectives. 2014;**5**:351-357

[50] Molteni CG et al. Comparison of manual methods of extracting genomic DNA from dried blood spots collected on different cards: Implications for clinical practice. International Journal of Immunopathology and Pharmacology. 2013;**26**:779-783

[51] Olsvik PA, Lie KK, Jordal AEO, Nilsen TO, Hordvik I. Evaluation of potential reference genes in real-time RT-PCR studies of Atlantic salmon. BMC Molecular Biology. 2005;**6**

[52] Daniels R, Volkman SK, Milner DA, Mahesh N, Neafsey DE, Park DJ, et al. A general SNP-based molecular barcode for *Plasmodium falciparum* identification and tracking.

Malaria Journal. 2008;**7**:223. DOI: 10.1186/1475-2875-7-223

[53] McClure MC, McKay SD, Schnabel RD, Taylor JF. Assessment of DNA extracted from FTA® cards for use on the Illumina iSelect BeadChip. BMC Research Notes. 2009;**2**:107

[54] Padmanaban A. DNA Integrity Number (DIN) For the Assessment of Genomic DNA Samples in Real-Time Quantitative PCR (qPCR) Experiments Application Note Author

[55] Mueller O, Lightfoot S, Schroeder A. RNA Integrity Number (RIN)-Standardization of RNA Quality Control Application

[56] Kozera B, Rapacz M. Reference genes in real-time PCR. Journal of Applied Genetics. 2013;**54**:391-406

[57] McDevitt SL, Hogan ME, Pappas DJ, Wong LY, Noble JA. DNA storage under high temperature conditions does not affect performance in human leukocyte antigen genotyping via next-generation sequencing (DNA integrity maintained in extreme conditions). Biopreservation and Biobanking. 2014;**12**:402-408

[58] Graf E, Mahoneys JR, Bryant RG, Eaton JW. Iron-catalyzed hydroxyl radical formation. The Journal of Biological Chemistry. 1984;**259**:3620-3624

[59] Williams RJ, Leopold AC. The glassy state in corn embryos. Plant Physiology. 1989;**89**:977-981

[60] Nasarabadi S, Hogan M, Nelson J. Biobanking in precision medicine. Current Pharmacology Reports. 2018;**4**:91-101

# 12

# A Review of the Analytical Methods Based on Chromatography for Analyzing Glyphosate in Foods

*Pasquale Avino, Ivan Notardonato and Mario Vincenzo Russo*

**Abstract**

Glyphosate is a pesticide widely used in agriculture, horticulture, and silviculture as well as around homes and gardens. It was introduced by Monsanto in the early 1970s, and it is a broad spectrum, nonselective, post-emergence herbicide that inhibits plants' shikimic acid pathway. Glyphosate is considered as "difficult herbicide" in terms of trace analysis. It has low molecular weight, low volatility, thermal lability, and good water solubility. These properties cause problems in its extraction, purification, and detection. The determination often requires additional processes that may allow quantification by chromatographic methods. Several analytical procedures have been developed based on solid-phase extraction, ion-exchange chromatography, or matrix solid phase dispersion. Most published methods involve liquid extraction followed by clean-up. This review would like to revise the literature on this issue discussing the relevant chromatographic methods reported in the literature in terms of analytical parameters for analyzing such compound in food chain.

**Keywords:** glyphosate, pesticide, herbicide, chromatography, GC, LC, MS, LOD/LOQ, food, recovery, human health

## 1. Introduction

Glyphosate (GLYP) (or, less commonly, but still used, glyphosphate), a broad-spectrum herbicide, is one of the most used pesticides in the world [1], nearly $5 billion in sales and an annual global production about 825,800,000 kg [2]. Glyphosate is a nonselective herbicide; therefore, it is a molecule that eliminates all weeds without distinction.

Glyphosate [IUPAC N-(phosphonomethyl)glycine; CAS registry number 1071-83-6] is an aminophosphoric analogue of glycine and an important amino acid. It was discovered in the early 1950s by Henri Martin and was patented by Monsanto and sold as a Roundup® product for about 20 years; after 2001 (patent expiration date), free production of glyphosate was legally permitted [3, 4]. As of 2010, more than 750 glyphosate products have been on the market [5, 6]. The first important worldwide warning about the GLYP occurred in 2017: the Canadian Food Inspection Agency (CFIA) confirmed that 36.6% of the Canadian wheat samples had a high presence of GLYP (3.9% above the legal limits, which in Canada is 5 ppm) [7]. In Canada, GLYP-based products are widely used for improving the wheat ripening and drying. Such occurrence has created a big supply problem in Europe where this practice is prohibited: for instance, Italy imported large amounts of wheat to make flour for pasta from Canada (and from the United States as well).

GLYP inhibits the 5-enolpyruvylshikimate-3-phosphate (EPSP) enzyme produced by plants, which is involved in the synthesis of three essential amino acids such as tyrosine, tryptophan, and phenylalanine. The mechanism of action is absorption through the foliage, and to a small extent through the roots, and transport to growth points. Since this enzyme is present only in the plant kingdom, glyphosate acts only on plant organisms.

GLYP is a leaf herbicide (it is absorbed by the leaves of the plant), systemic (once absorbed, it passes toward the growth points, causing the death of the plant), nonselective (in fact, it is active on all plants, if not genetically modified). Glyphosate-based products are activated by the addition of a surfactant, polyoxy-ethylene amine (POEA), which promotes penetration through the leaf surface of plants; other additives used are sulfuric acid and phosphoric acid. Its main metabolite is aminomethylphosphonic acid (AMPA). It should be noted that a fraction of AMPA could be due to degradation processes of the detergents/surfactants rather than from glyphosate. GLYP does not penetrate deeply into the soil (maximum 20 cm) and is easily degraded by bacteria. This means that the probability that it reaches the aquifers is very low and that its presence is certainly lower than that of other dangerous pollutants.

The half-life of GLYP in the soil is between 2 and 197 days, a typical half-life of 47 days has been suggested. The soil and climate conditions on the persistence of glyphosate in the soil are very important. The average half-life of GLYP in water varies from few to 91 days. The AMPA metabolite of glyphosate has been found in Swedish forest soils for up to 2 years after a glyphosate application. In this case, the persistence of AMPA has been attributed to frozen soil for most of the year. The adsorption of glyphosate into the soil, and then its release from the soil, varies according to the type of soil. GLYP is generally less persistent in water than in land, with 12–60 days persistence observed in Canadian ponds, although persistence of more than a year has been recorded in American lake sediments.

GLYP (**Figure 1**) is a weak acid commonly used in the form of salt, distributed as a powder or as a water-soluble concentrate. At room temperature, it appears as a colorless crystalline solid, is completely soluble in water, and is highly insoluble in common organic solvents such as benzene and dichloromethane. GLYP is a nonvolatile and photo-resistant molecule, and its dissolution in water generates four chemical equilibria represented by the respective acid dissociation constants ($K_a$). In logarithmic form, pKa acquires the following values: 2.0, 2.6, 5.6, and 10.6. This aspect makes the molecule highly polar and amphoteric [8].

During the reactions involving the enzymes glyphosate oxidase and glypho-sate N-acetyl transferase, glyphosate can form different metabolites: the main is

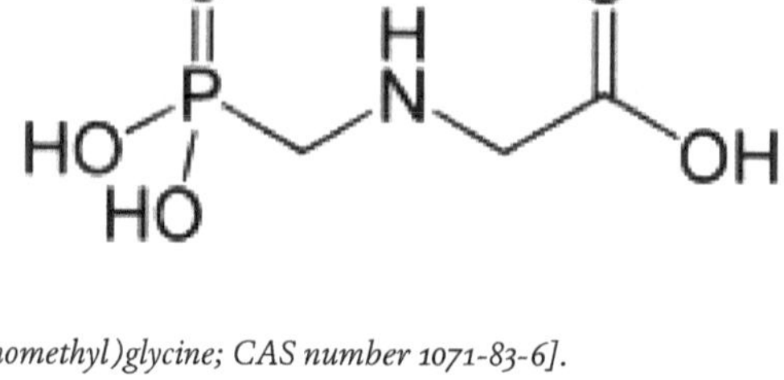


**Figure 1.**
*Glyphosate [N-(phosphonomethyl)glycine; CAS number 1071-83-6].*

considered the amino-methylphosphonic acid (AMPA), whereas the others are glyoxylate, N-acetyl glyphosate, N-acetyl-AMPA, methylphosphonic acid, sarcosine, N-methyl-aminomethylphosphonic acid (MAMPA), hydroxymethylphosphonic acid, and phosphonoformic acid [9]. This behavior is important: these compounds should be considered when toxicity and environmental studies are performed for the risk assessment. Similarly, compounds used as adjuvants in commercial glyphosate formulations should be considered: for instance, polyoxyethylene amine (POEA), used as a surfactant in Roundup [10] or isopropylamine, ammonium and trimesium salts, or formulation impurities such as N-(phosphonomethyl)iminodiacetic acid and bis(phosphonomethyl)amine. This occurrence is really important because the adjuvants can modify the toxicity of pesticides based on glyphosate as active ingredient; so, the result is the need of a novel toxicological evaluation [11].

All these considerations play an important role in the GLYP toxicity. The toxicity of a substance is assessed according to its median lethal dose (lethal dose, 50% – $LD_{50}$), that is, the dose that causes the death of 50% of the individuals taking the test substance: Class 1, high acute toxicity, $LD_{50}$ less than 50 mg per kg of live weight; Class 2, moderate toxicity, $LD_{50}$ between 50 and 500; Class 3, mild toxicity, $LD_{50}$ between 500 and 5000; and Class 4, harmless, $LD_{50}$ of over 5000 mg. The GLYP is in Class 3, while in Class 2, we find, for example, caffeine, aspirin, and boiling chloride, and in Class 1, the vitamin $D_3$. In **Table 1**, acute toxicity assessment is reported.

Also, important is the concept of daily limit dose (expressed in milligrams per kilogram of body weight considered) definable as the maximum amount of herbicide that can be consumed daily without causing damage. Based on this concept, the glyphosate content of a food or a drink should be correctly evaluated using the milligrams of glyphosate per kilogram of body weight that can be taken per day as a unit of measurement. In this way, the European Food Safety Authority (EFSA) has set a daily limit dose of 0.5 mg $kg^{-1}$ of weight per day [12].

A tumor associated with glyphosate would be the non-Hodgkin lymphoma (NHL). In 2013, the German Federal Institute for Risk Assessment (BfR) found that "the available data are contradictory and far from convincing" in terms of the relationship between exposure to glyphosate formulations and the risk of various cancers, including the NHL [13–18]. A meta-analysis published in 2014 identified an increased risk of NHL in workers exposed to glyphosate formulations [19, 20].

| | **High toxicity** | **Moderate toxicity** | **Low toxicity** | **Very low toxicity** |
|---|---|---|---|---|
| Acute oral[a] | ≤50 mg $kg^{-1}$ | >50–500 mg $kg^{-1}$ | >500–5000 mg $kg^{-1}$ | >5000 mg $kg^{-1}$ |
| Inhalation[b] | ≤0.05 mg $L^{-1}$ | >0.05–0.5 mg $L^{-1}$ | >0.5–2.0 mg $L^{-1}$ | >2.0 mg $L^{-1}$ |
| Dermal[a] | ≤200 mg $kg^{-1}$ | >200–2000 mg $kg^{-1}$ | >2000–5000 mg $kg^{-1}$ | >5000 mg $kg^{-1}$ |
| Primary eye irritation | Corrosive or corneal involvement | Corneal involvement (8–21 days) | Corneal involvement (7 days) | Minimal effects clearing in 24 hours |
| Primary skin irritation | Corrosive | Severe irritation at 72 hours | Moderate irritation at 72 hours | Mild or slight irritation at 72 hours |

[a]*As $LD_{50}$.*
[b]*As lethal concentration, 50% ($LC_{50}$).*

**Table 1.**
*Relationship between GLYP levels and toxicity.*

In March 2015, the International Agency for Research on Cancer (IARC) classified glyphosate "probably carcinogenic to humans" (Group 2a) based on epidemiological studies, animal studies, and in vitro studies: in particular, GLYP has been defined genotoxic through at least two mechanisms known to be associated with human carcinogens [21–23]. In contrast, EFSA concluded in November 2015 that "the substance is unlikely to be genotoxic (i.e., harmful to DNA), or pose a threat to humans." Subsequently, EFSA itself states that while there may be formulations containing glyphosate that are carcinogenic, studies relating only to glyphosate as an active ingredient do not show this effect [24, 25]. The European Chemicals Agency (ECHA), on the basis of "the scientific evidence available at the moment," classified GLYP, according to the CLP Regulation, as a chemical causing eye damage (H318) and being toxic to aquatic life with long-lasting effects (H411), but "the available scientific evidence did not meet the criteria in the CLP Regulation to classify glyphosate for specific target organ toxicity, or as a carcinogen, as a mutagen or for reproductive toxicity" [26]. The United States Environmental Protection Agency (US EPA) has classified glyphosate as a Group E chemical, meaning the agency has determined that there is "evidence of noncarcinogenicity to humans" [27, 28]. In any case, US EPA has established tolerances for GLYP residues in different commodities [29]. The difference of point of views depends on the fact that IARC and US EPA have analyzed different studies and applied different statistics. Further, EFSA analyses concern only the glyphosate molecule, whereas the studies considered by IARC also concern glyphosate-based products placed on the market [30].

This brief analysis shows that, in any case, international pesticide regulatory agencies and scientific organizations agree that there is no evidence that GLYP as an active substance is carcinogenic to humans, only IARC has classified glyphosate as "probably carcinogenic."

Finally, it should be considered an interesting hypothesis by Samsel and Seneff [31]: they propose a relationship between celiac disease and imbalances in gut bacteria generated by the known GLYP effects on them.

The EFSA has renewed the authorization for GLYP, establishing the acute reference dose (ARfD) at 0.5 mg $kg^{-1}$ of body weight, while the acceptable operator exposure level (AOEL) was set at 0.1 mg $kg^{-1}$ body weight per day and the acceptable daily intakes (ADIs) for consumers are in line with the ARfD threshold, 0.5 mg $kg^{-1}$ body weight per day.

There are several exposure sources of humans to GLYP in the environment, for example, air, water, application to crops and target weeds, and food [32–34]. Solomon deeply reviewed the exposure data from the literature (PubMed and Google Scholar) and unpublished reports in different papers [35, 36]: in both papers, he reaches a similar conclusion: "In all cases, measured and estimated systemic exposures to glyphosate in humans and animals were less than the ADIs and the RfD. Based on this large dataset, these exposures represent a *de minimis* risk." The conclusion reached by Gillezeau et al. [33] is instead intermediate by reviewing the same literature (PubMed and Google Scholar): they state that "additional studies are urgently needed to evaluate levels of glyphosate and related metabolites in the general population and in workers." Further, they observe the great differences in the analyzed papers: they detected some bias such as the few studies on potential occupational GLYP exposure, or no study designed to address the hypothesis of seasonality in exposure, or the use of a few populations of farmers and relative collection of one-time spot urine. They rise serious doubts about the data generalizability, which they consider rather limited.

This paper would like to critically revise the literature on chromatographic methods developed for analyzing GLYP and AMPA in food matrices, specifically grains (e.g., rice, wheat, soybean, and maize), honey, olive and oil, vegetables, fruit, beverages (e.g., drinking water, milk, tea, and coffee), cheese, and meat/fish products. In literature (source: Scopus database), there are 2666 papers using keywords "glyphosate" and "analysis" by the end of April 2020 and 361 using "chromatography" as third keyword.

## 2. Glyphosate determination in different food matrices

Starting from the Canadian study performed in 2017, the scientific attention on GLYP has become stronger, and several papers are annually published dealing the determination of such compound, along with its main metabolite AMPA, on different agricultural and food matrices. For avoiding dispersive information due to the big amount of studies aimed to this determination, the authors have focused their attention on the main innovative analytical methods based on chromatographic methods for determining both compounds in such matrices. It is also necessary to advise the reader that different matrices could be determined with same analytical protocols, at least showing different analytical parameters (multiresidue analyses), as well as in literature are present papers dealing important toxicological studies with no analytical information.

### 2.1 Approaching the determination

Before approaching the discussion on the different analytical methodologies developed for analyzing GLYP and AMPA in agricultural and food matrices, it should be necessary to resume some toxicological information on it along with some chemical characteristics to be taken into account for evaluating the analytical process.

First, a maximum residue level (MRL) is defined as the highest level of a pesticide residue legally tolerated in or on food or feed when pesticides are applied correctly [37]. For each product, an MRL of GLYP has been determined [38]. An example of this database is reported in **Table 2**.

A preliminary important information comes from the EU Reference Laboratories for Residues of Pesticides (EURL-SRM): for all the analytical steps, it is highly recommended the use of plastic vials because there is an interaction between the pesticide and the glass surface, especially when aprotic solvents are used. These interactions greatly affect the precision and accuracy, especially at low GLYP concentration. This statement is important because it influences its stability and degradation as well. Among the different solvents, water with 10% acetonitrile is considered a good storage solvent, whereas the compound is not stable in water and methanol. At room temperature, the degradation is very low within 14 days, whereas if extract is stored in the refrigerator, it is stable over 7 months [39].

Finally, the authors would like to remember some definitions regarding the parameter of an analytical method. Recovery is the term used in analytical and preparative chemistry to denote the fraction of the total quantity of a substance recoverable following a chemical procedure [40]. Accuracy is the difference between the mean of some measurements and the value considered as the true or correct value for the quantity measured, whereas precision is the measurement reproducibility, that is, the dispersion around a central value. In regard to the chromatographic separation, a signal-to-noise (S/N) ratio of 3 is acceptable for determining the limit of detection (LOD), that is, the lowest amount of analyte in a sample, which can be detected, whereas a ratio of 10 for the limit of quantification (LOQ), that is, the lowest amount of analyte in a sample, which can be quantitatively determined with precision and accuracy [41–44]. The S/N definition for chromatography is the ratio

| Product | MRL |
|---|---|
| Tangerines, clementines, oranges, and grapes | 0.5 |
| Lemons, grapefruits, cedars, kumquats, apples, pears, peaches, apricots, cherries, plums, almonds, hazelnuts, strawberries, and table olives | 0.1 |
| Oil olives | 1 |
| Potatoes | 0.5 |
| Wild mushrooms | 50 |
| Other vegetables | 0.1 |
| Baked beans | 2 |
| Grain peas, lupines, and lentils | 10 |
| Other leguminous vegetables | 0.1 |
| Flax seeds, rapeseed, mustard, and cotton | 10 |
| Sunflower and soybeans | 20 |
| Other oil seeds | 0.1 |
| Wheat and rye | 10 |
| Barley, oats, and sorghum | 20 |
| Corn | 1 |
| Other cereals | 0.1 |
| Sugar beets (roots) | 15 |
| Forage from meadows and pastures, and alfalfa | 0.1 |

**Table 2.**
*Maximum residue levels for some food products.*

of the peak height relative to the middle of the noise range (S) to the difference between the maximum and minimum baseline signal values for the noise (N) [45].

## 2.2 Cereal grains

Grain is the largest and well-studied matrix in this field. Many papers deal the glyphosate determination in cereals and legumes due to the worldwide use of such herbicide in the relative cereal crops. We must remember that, as said at the beginning, the first warning came precisely by analyzing several Canadian wheat samples and finding almost 37% of them with high presence of the pesticide. So, after this occurrence, scientific and health attention has been very high and focused on cereals in general, for example, maize corn, millet, barley, oats, rice, wheat wild rice, amaranth, and quinoa.

The literature analysis for the GLYP determination in such matrix is very large; for this reason, the authors focused their attention on the main publications starting from the last deep review, that is, by Tadeo et al. [46]. The same method will be applied to the revision of the analytical methods for GLYP determination in vegetables and fruit matrices.

A routine control method based on extraction with water by ultrasonication was developed by Granby et al. [47] for analyzing several Danish mill products. It was one of the first studies based on green chemistry, that is, the authors used no organic solvents or chemicals except diluted solutions of $NaHCO_3$ (as eluent) and, in some cases, $H_2SO_4$. The samples (rye or wheat in grain and flour) were subjected to online clean-up and separation by in-series system of ion chromatography

(IC) and high performance liquid chromatography (HPLC) with detection by electrospray ionization mass spectrometry in the negative-ion mode. The method, investigated in the range of 0.03–0.33 mg $kg^{-1}$, shows a GLYP recovery of 85%, a repeatability between 1 and 14%, a reproducibility from 4 to 16%, and a LOD of 0.02 mg $kg^{-1}$ (LOQ was not reported).

A very interesting paper was published by Tseng et al.: they used the gas chromatography coupled with a pulsed flame photometric detector (PFPD) for a simultaneous determination of GLYP and glufosinate (DL-homoalanin-4-yl-(methyl)phosphinic acid, GLUF) along with their main metabolites including AMPA [48] after a single-step derivatization with trimethyl orthoacetate (TMOA). In particular, the authors studied the influence of the heating temperature (70–90°C) and time (90–120 min) on the AMPA and 3(methylphosphinico) propionic acid (3-MPPA, a GLUF metabolite) derivatization. They optimized the method on soybean sprouts and rice samples and determined the different analytical parameters (recoveries 72–81, 71–86, 101–119, and 83–90%; LOD of 0.02, 0.03, 0.02, and 0.01 μg $g^{-1}$; and LOQ of 0.06, 0.10, 0.06, and 0.04 μg $g^{-1}$ for glyphosate, AMPA, GLUF, and 3-MPPA, respectively; RSD < 10%). On the other hand, Li et al. used fluorenylmethyl chloroformate (FMOC-Cl) as derivatizat-ing agent followed by HPLC-MS/MS for analyzing GLYP and AMPA residues in different matrices such as rice, wheat, vegetables, fruits and tea, pig and chicken muscles, aquatic products, chestnut, and honey [49]. Further, they also used an isotope-labeled 1,2-$1^{3}C^{15}N$ GLYP for increasing the accuracy and the precision of the measurements at low GLYP concentration. In this way, they obtained recoveries between 80.0 and 104% and RSDs from 6.7 and 18.2% with a LOQ of 0.05 mg $kg^{-1}$ for both compounds and a correlation of 0.998 in the linear range of 0.20–10 μg $L^{-1}$.

In 2007, Granby's group published a paper on the (six) laboratory intercomparison for determining GLYP, chlormequat, and mepiquat (these two are plant growth regulators, also used for the growth reduction of the lowest straw part) residues in cereals [50]. GLYP was analyzed by treating the samples twice with MilliQ water by ultrasonication followed by centrifugation, filtration, clean-up on polystyrene-based reverse phase column, and separation by IC-HPLC-MS/MS, whereas the other two compounds were extracted by UltraTurrax and cleaned-up by SPE-$C_{18}$. The results showed very different LOQs and recoveries reached by the six laboratories (0.01 and 0.3 mg $kg^{-1}$ and 29 and 109% for GLY) with a good within-laboratory precision and a poor between-laboratory precision [51]. For glyphosate, the authors stated the presence of a systematic component between laboratories to be the reason of such large data variability.

Simple sample preparation and fast chromatographic analysis are the main features of the paper by Martins-Júnior et al. [52]. They analyzed GLYP and AMPA in soybean samples by means of liquid-liquid partition with dichloromethane and protein precipitation followed by HPLC-MS/MS determination (in positive and negative electrospray ionization, ESI, mode). This paper highlights the choice of the liquid-liquid partition and protein precipitation. Particularly, the paper evidences the importance of the second step, that is, the protein precipitation for eliminating the matrix interference: different solvents, that is, acetone, acetonitrile, and methanol, were tested, and methanol was found the best for reducing it (but it does not eliminate it). The authors took advantage of the great performance of the tan-dem mass spectrometry (MS/MS) and reached LODs of 0.09 and 0.1 mg $kg^{-1}$ and LOQs of 0.30 and 0.34 mg $kg^{-1}$ for GLYP and AMPA, respectively, with recoveries between 79.6 and 109.1% and RSD below 12.2%. Further, the authors suggested to apply this analytical protocol to other crop matrices, where GLYP is largely used, for instance, corn and cotton.

As just noted above and especially using the LC-MS/MS as GLYP detection, the matrix effect is not negligible. In literature, different possibilities have been studied for reducing this artifact: for instance, sample dilution [53], injection of smaller volumes [54], the optimization of sample preparation and/or chromatographic parameters [55], or the use of expensive internal standard (IS). Ding et al. [56] developed a combination of C18 and SAX cartridge for reducing the matrix effect. After to have optimized the analytical conditions, the authors used hydrophilic interaction chromatography (HILIC)/WAX mixed-mode stationary phases for glyphosate retention and LC-MS/MS in negative ion mode for the detection. They used this methodology for analyzing soybean, corn, spicy cabbage, apple, and carrot samples. GLYP is investigated in a linear range between 0.02 and 10 mg $kg^{-1}$, with a $R^2 > 0.999$, and the intra- and inter-day errors are 2.7 and 1.8%, respectively, whereas the precision as RSD is below 7%. Using the developed analytical procedure, the authors reached good LOD and LOQ , 0.02 and 0.005 mg $kg^{-1}$, respectively, and recoveries ranging between 83.1 and 100.8% according to the different matrices analyzed, specifically 89–96% for soybean, 84–101% for corn, 86–94% for carrot, 85–93% for spicy cabbage, and 83–100% for apple. Quite interesting in this paper are both the possibilities to quantify such herbicides in different plant-derived or processed foods (this is not so common in the literature) and to use solution calibration curves instead of matrix-match calibration curve for the analysis.

Botero-Coy et al. explored for first the possibility to analyze GLYP in rice, maize, and soybean without derivatization step but just direct LC–MS/MS with a triple quadrupole instrument after water extraction and SPE using Oasis HLB cartridge [57]. The method has allowed to reach high correlation coefficients (<0.99) in the range of 1–250 μg $L^{-1}$, recoveries between 77 and 100% with RSDs below 17%, and good LODs and LOQs (0.007–0.12 mg $kg^{-1}$ and 0.1 and 2 m $kg^{-1}$, respectively) for all matrices.

A Chinese-French scientific paper in 2018 dealt the determination of GLYP and GLUF in 136 food samples, of which 34% of samples with high (banana, apple, orange, potato, carrot, and juice) and low (biscuits or bread) water contents and 66% of animal origin samples (milk-based foods included, e.g., milk, cheese, and butter) [58]. After a solvent extraction (acidified water, methanol, and dichloromethane), the authors performed a derivatization by means of FMOC and a solid phase extraction (SPE) $C_{18}$ for purifying and concentrating the extract and a HPLC-MS/MS analysis for determining the two compounds. Using these conditions, recoveries between 82 and 112%, LODs and LOQs of 1.7 and 5 μg $kg^{-1}$, respectively, and RSDs below 20% for both compounds were achieved.

An ion chromatography-tandem mass spectrometry-based method was developed by Adams et al. for analyzing 14 polar pesticides including GLYP in cereal and grape samples [59]. The extraction is based on quick polar pesticide (QuPPe). Although the method is interesting, not all the analytical parameters are reported except the recoveries for cereals (specifically, oat flour) ranging between 85 and 104%.

A simple method based on acidified methanol solution extraction followed by centrifugation and filtration and LC-MS/MS analysis was developed by Santilio et al. for analyzing GLYP in rice and maize [60]. The authors highlighted the importance of using GLYP isotope labeled in the matrix effect reduction. LODs of 2 μg $kg^{-1}$ for rice and 4 μg $kg^{-1}$ for maize and a LOQ of 10 μg $kg^{-1}$ for both matrices were reached in a linearity range of 0.01–1.5 mg $kg^{-1}$ ($R^2$ 0.9982) with recoveries ranging between 74 and 98% and RSD < 20%. Finally, it should be reported that the authors' principal aim was to develop a method to be routinely used for analyzing rice and maize, taking into account the relative MRLs established, and to extend it to other matrices.

Finally, Herrera López et al. set up a multiresidue analysis for determining 14 highly polar pesticides (parents and metabolites) in 352 samples including oat and soya beans, lettuce, grapes, and oranges [61]. After a solvent extraction step, a LC system coupled with a hybrid quadrupole/linear ion trap mass spectrometer system (with ESI source) (LC-ESI-QTRAP-MS) was used for reaching high analytical performances: linearity range between 0.01 and 10 mg kg$^{-1}$ with $r^2 > 0.99$, recoveries between 70 and 120% with RSD < 22% (specifically, GLYP between 83 and 118% with RSD < 22%), LOQs between 20 and 500 μg kg$^{-1}$ (particularly, for GLYP 500 and 20 μg kg$^{-1}$) for all the investigated matrices. The clean-up procedure was not involved because no appropriate sorbent was found to increase the protocol, and the derivatization step was not necessary, whereas these authors also stated that the use of an isotopically labeled internal standard helps in the matrix effect correction.

## 2.3 Meat, fish, and cheese

The scientific attention on GLYP contamination in this food class is on the rise recently. Only few papers are available on such matrices. In fact, if the GLYP behavior in the aquatic environment is studied since many years [62–64], poor information is presented on its presence in foods.

Starting from the paper by Botero-Coy et al. [57], Chiesa et al. developed a method based on IC coupled to high resolution mass spectrometry (IC-HRMS) for determining GLYP, GLUF, and AMPA in different foods of animal origin without a derivatization step [65]. The authors focused their attention on the matrix, particularly on the lipid composition, which is the major interfering group because coextracted with the analytes. The main contribution of this study was to identify the best extraction solvent: among different assays, the best solution is 30% of metha-nol and 70% of acidified water (1% formic acid). Thirty samples among fish (bass), bovine muscle, and organic honey were analyzed. The detector, an orbitrap quipped with heated electrospray ionization (HESI) source, allowed to reach very low LOQs (4.26–5.38 ng g$^{-1}$, 6.25–6.47 ng g$^{-1}$, and 4.30–9.26 ng g$^{-1}$ for fish, bovine, and honey, respectively), good recoveries (96.9, 76.1, and 97.0%, respectively), RSDs <13.1%, and good correlation coefficients ($R^2 > 0.992$).

Actually, in literature, there are other few papers showing the determination of GLYP and AMPA in muscle meat (bovine, cow, pig, and chicken), but the LODs are higher (50 ng g$^{-1}$) [49, 66, 67], whereas the only paper on fish does not report any information on LOQ [68].

A communication dealing with the determination of GLYP and GLUF in animal feeds shows linearity more than 0.999, instrumental detection limits (IDLs) of 8.3 μg kg$^{-1}$ and 1.1 μg kg$^{-1}$, respectively, accuracy between 102 and 112%, and precision below 6% in both matrices [69].

Finally, about the GLYP determination in cheese or, basically, in milk-based foods, the authors just discussed above the only paper present in the literature [58]. Please note that the milk as beverage will be discussed in other section.

## 2.4 Vegetables

Some papers dealing with the GLYP determination in such food matrices are just discussed previously [46, 49, 56, 58, 61]: here the attention is focused on papers showing novelty or improvements in the analytical methodology or large studies on the herbicide content. The first interesting paper is dated in 1992: Tanaka and coauthors developed a very easy method employing routinely available instrumentation, that is, HPL C with a flu orescence d etection [ 70 ]. The anal ytical parameters are quite weak (recoveries >68% and >88% for GLYP and AMPA, respectively, and LOD 0.05 ppm for both), but it is to be appreciated the use of common equipment. In 1996, two papers investigated the GLYP presence in green lentils, fresh beans [71], and carrot [72]. The first paper introduced a post-column reaction, a denitrozation, for obtaining a N-nitroso-GP (NGP) derivate to be analyzed by HPLC coupled with thermal energy analyzer (TEA), that is, a chemiluminescence detector. Over vegetables, the authors also analyzed beverages (water and beer) and cereals (rice flour, corn, barley, and rye). They obtained recoveries between 83 and 97% for vegetables, 70–100% for beverages, and 67–100% for cereals with LODs ranging between 0.005 and 1 μg $g^{-1}$. On the other hand, the second paper presents a GC analysis coupled with flame photometric detection (FPD) for analyzing GLY, AMPA, and GLU. The use of instrumentation commonly present in each laboratory is to be appreciated also in this case. The three compounds were derivatized with N-isopropoxycarbonyl (isoPOC) for obtaining the relative isoPOC methyl ester derivatives: 0.5–1 μL of this solution wax injected in the GC-FPD. The authors determined the LODs (12, 8, and 20 pg injected for GLY, AMPA, and GLU, respectively), the recoveries (91–104, 94–104, and 91–100%, respectively), and the correlation coefficients ($R^2$ 0.9992, 0.9982, and 0.9992, respectively) in a linearity range of 5–200 ng.

Hooijschuur and coauthors explored the possibility to use the microcolumn liquid chromatography with FPD detection (μLC-FPD) and compared these results with those obtained by capillary electrophoresis (CE) with FPD (CE-FPD) [73]. They used a silica column (25 cm × 320 μm ID, 450 μm OD) with 5 μm LiChrosorb RP-1 bonded silica. Although CE-FPD was faster than μLC-FPD, this is more sensitive for the GLYP and AMPA analysis: LODs are 15 and 7.5 ng $mL^{-1}$, respectively, versus LOD of 1.0 μg $mL^{-1}$ for both compounds by CE-FPD. Grey et al. applied the LC-ESI/MS analysis after the derivatization with FMOC-Cl of GLYP and AMPA [74]. They evaluated the use of isotope-labeled compounds: their conclusions were positive in the GLYP determination (LODs 0.11 μg $g^{-1}$ and 0.06 μg $L^{-1}$ for lettuce and water samples, respectively), whereas they did not find any contribution for the accurate AMPA analysis (LODs 0.53 μg $g^{-1}$ and 0.3 μg $L^{-1}$, respectively). Finally, the recoveries increased from 23.2 to 98.4% for GLYP and from 33.8 to 99.4% using the isotope dilution mass spectrometry (IDMS)-based glyphosate analytical method. Finally, Takahashi et al. determined GLYP and GLUF in cabbage, Chinese cabbage, carrot, onion, strawberry, lemon, kiwi fruit, over soybean, corn, and brown rice after derivation with FMOC-Cl and analysis by HPLC with fluorescence detection [75]. Another interesting paper came from Japan in 2004: Watanabe set up a rapid method for determining GLYP, GLUF, and 3-MPPA in vegetables (cucumber and spinach) and fruits (apple, mandarin, and orange) using an anion exchange resin and elution with acetic acid, followed by derivatization with trimethyl orthoacetate and clean-up on SPE Florisil cartridge and GC-FPD analysis [76]. The method allows to reach LODs of 0.01, 0.01, and 0.005 μg $g^{-1}$ and recoveries of 83.5–89.8, 77.9–92.2, and 75.0–87.2% for GLYP, GLUF, and 3-MPPA, respectively.

A Chinese group proposed an original method for determining GLYP in apple samples [77]: after clean-up with SPE-C18, a derivatization step was performed using 4-chloro-3,5-dinitrobenzotrifluoride (CNBF). The quantification occurred by reverse ion-pair liquid chromatography using cetyltrimethylammonium bromide (CTAB) as ion-pair reagent. The strengths of the method are the formation of a stable derivative (5% degradation after 7 storage days at room temperature) and the easy pretreatment procedure. LOD of 0.01 μg $g^{-1}$, recoveries from 86.0 to 99.5%, and RSDs from 1.43 to 6.32 were achieved applying this method to apple samples.

Rembisz and coauthors started from a different idea: GLYP (as well GLU) is an aminophosphonic acid, analogous of the amino acid. So, they proposed a derivatization with phenyl isothiocyanate (PITC) for obtaining phenylthiocarbamyl derivatives (PTC derivatives): a thin-layer chromatography (TLC) with iodine-azide detection allowed to detect such compounds in parsley and lettuce samples [78]. The method was sensitive, accurate, and inexpensive showing recoveries between 95 and 104%, LODs 0.99–4 μg per spot, LOQs 1.78–8.45 μg per spot, and RSDs <7.7 for both compounds.

A fast routine analysis was developed by Boušová et al. for routinely determining the polar pesticides, including GLYP, AMPA, GLUF, and 3-MPPA, in lettuce, orange, and flour samples [79]. The coupling of ion chromatography to a triple quadrupole mass spectrometer allowed the authors to reach very good LODs and LOQs (1–10 μg $kg^{-1}$ and 10–20 μg $kg^{-1}$), recoveries ranging between 71 and 116% according to the matrix, and RSD < 18%. Rajski et al. implemented this procedure using an orbitrap detector and validating the method for aubergine, zucchini, cabbage, orange, and watermelon samples [80], achieving good recoveries (70–120%) and LOQ (0.01 mg $kg^{-1}$) for GLY, AMPA, and GLU. Melton et al. still used the ion chromatography but coupled with the tandem mass spectrometry (IC-MS/MS) for determining highly polar pesticides (including GLY, AMPA, and GLU) in 288 samples of melon, peas without pods, and pineapple [81]. Finally, a paper by Savini et al. worth to be mentioned: the authors used the UHPLC coupled with a orbitrap detector for analyzing GLPY, AMPA, GLUF, and other polar pesticides in 98 samples (83 processed fruits and vegetables and 15 infant foods) [82]. Using the developed method, the authors obtained LOQ of 0.003 mg $kg^{-1}$ for all three compounds, recoveries 75–113% in all matrices, RSDs below 18.5%, and a $R^2$ between 0.9954 and 0.9998 in the linear range of 0.001–0.1 mg $L^{-1}$. Another important advantage of this method is the simultaneous determination of six polar pesticides (i.e., AMPA, glyphosate, phosphonic acid, chlorate, fosetyl-Al, and perchlorate) in 25 min.

## 2.5 Olives and olive oil

Two papers dealt with the determination of GLYP in olives and olive oil [83, 84]. Both papers deal the difficulty of analyzing such matrices, and there is strong matrix effect. In the first paper, two different methods were developed, that is, UHPLC-TOFMS and UHPLC-MS/MS using HILIC separation: in this way, the authors reached LOQ of 0.3 μg $kg^{-1}$ and 0.1 μg $kg^{-1}$, respectively, and recoveries between 57.2 and 117.6% with a linearity >0.99 and an RSD < 3.9%. The two different LOQs were calculated using time-of-flight mass spectrometry (TOFMS) and triple quadrupole instruments: as expected, the MS/MS shows lower quantifiable levels. The second paper presented a green fast-analytical method based on vortexing (1 min with acidified water) and centrifugation (10 min at 3700 rpm) and extract injection in UHPLC–MS/MS for determining GLYP, AMPA, and GLUF in different olive oils, that is, extra virgin olive oil, virgin olive oil, olive pomace oil, and soy oil. Particularly, the paper reported the determination with no internal standards nor matrix-matched calibration. The authors tested the linearity in the concentration range of 5–250 μg $L^{-1}$: they fixed LOQs at 5 μg $kg^{-1}$ for AMPA and at 10 μg $kg^{-1}$ for GLYP and GLUF and determined recoveries between 81.4 and 119.4% with intra and inter-day precision lower than 19%.

## 2.6 Honey

During the past few years, the important question has emerged about GLYP contamination in natural honey samples. Different papers have been published dealing this issue. Some of them have already been discussed previously [49, 65].

A first interesting paper dealing with such of matrix was this of Karise and coauthors[85].They set up a multiresidue method for analyzing GLYP along with other

47 pesticides in 33 honey samples collected from beehives of Estonia. The paper was focused on the detection of the pesticide concentration and the relative maximum residue levels and the possible impact of the agriculture on the product. In any case, the authors largely used the analytical methodology based on using QuEChERS (acronym of Quick, Easy, Cheap, Effective, Rugged and Safe) extraction meth-odology followed by detection using GC-MS and ultra-high-performance liquid chromatography-MS/MS (UHPLC-MS/MS): the method shows recovery between 78 and 115%, repeatability from 3.0 to 16%, LOQ for GLYP of 0.050 mg $kg^{-1}$ (and 0.010 mg $kg^{-1}$ for the other pesticides), and correlation coefficients >0.990 for all compounds.

In 2018, Zoller et al. found GLYP at very low levels in 15 of 16 honey samples analyzed; in addition, they also analyzed pulses (tofu and soy sauce), breakfast cereals (corn flakes and pops), durum wheat, pastry and snacks (crisps, etc.), bread, flour and baking mixtures, and beverages (beer, wine, milk, fruit juices, and mineral water) for a total of 243 samples [86]. The authors applied a well-tested analytical method based on solvent extraction with methanol and LC-MS/MS analysis for determining GLYP and AMPA (LODs 0.2–0.4 and 0.5–1 μg $kg^{-1}$, respectively; LOQs 0.5–1 and 1–2.5 μg $kg^{-1}$; recoveries 92–103 and 92–115%; RSDs <9.5 and <13.9%). Further, in this paper, the authors assessed a dietary risk of each food for a child of 15 kg body weight and for an adult of 60 kg body weight. The first findings of this work were that the GLYP maximum residue levels did not exceed more than the legally tolerated ones (0.1 mg $kg^{-1}$ for plant products and 0.05 mg $kg^{-1}$ for animal products). So, the scores reported by authors for the risk assessment highlighted a low exposure only for the pulses (5% of the acceptable daily intake, ADI, and acute reference dose, ARfD), whereas in all the other cases, honey samples included, the exposure to GLYP is less than 1% of the ADI/ARfD, meaning there is no any human health issue in all samples. Further, the authors, simulating a daily ingestion of the different investigated foods, estimated the probable GLYP content in urine. They found levels in agreement with those found by other authors in German [3, 17] and Swiss [87] populations, whereas some differences could be expected in AMPA concentration comparison [17].

A pilot study for monitoring GLYP and AMPA in 32 honey samples was set up by Pareja et al. based on IC coupled to a Q-Orbitrap accurate high-resolution mass spectrometry [88]. It is still confirmed that the use of IC simplifies the polar pesticide determination, whereas the use of an orbitrap detector allows to reach a GLYP LOQ of 5 μg $kg^{-1}$ (20 μg $kg^{-1}$ for AMPA), less than the allowed EU MRL (50 μg $kg^{-1}$) and recoveries ranging between 80 and 110% with RSDs <20% in the linearity range of 5–500 μg $kg^{-1}$.

Still in 2019, a Canadian group developed an easy method for analyzing GLYP, AMPA, and GLUF at low μg $kg^{-1}$ levels based on both the derivatization with FMOC-Cl in acetonitrile solution and online SPE(C18)-LC-MS/MS analysis [89] and the use of isotopically labeled internal standards (as just evidenced previously). In particular, for all the investigated compounds, the authors obtained accuracies ranging between 95.2 and 105.3% (intraday precision 1.6–7.2%) and LOQ 1 μg $kg^{-1}$. By this method, 200 honey samples were analyzed: GLYP was found in 196 samples at maximum level of 49.8 μg $kg^{-1}$ with a 95th percentile of 14.2 μg $kg^{-1}$, evidencing no risks for the consumers. Further, the authors performed a survey between their data with others from worldwide studies (the United States, Estonia, Switzerland, some just cited in this review) [85, 86, 90–92].

A 2020 paper evaluated the exposure risk of bees and humans to GLYP and AMPA residues in three different bee matrices, that is, beebread, wax, and paired samples of wax/honey collected from 379 Belgian apiaries using an analytical method based on clean-up on SPE-C18 followed by derivatization step with FMOC-Cl and

HPLC-ESI-MS/MS analysis [93]. LOD and LOQ of 1 ng $g^{-1}$ and 10 ng $g^{-1}$, respectively, were achieved for both compounds in all matrices with recoveries ranging between 72.2 and 112.9% and RSDs from 0.1 to 4.5%. The authors stated that the GLYP levels were below the EU regulation in all samples. In any case, they suggest particular attention because recent studies deal the effects of GLYP [94] and AMPA [95] below the allowed concentrations.

## 2.7 Beverages

This last matrix is really important considering the large use of beverages in the daily dietary intake. Beverages such as water, beer, milk, and fruit juices are under strict attention by the different national authorities. For instance, in 2019, a study by Cook of the CalPIRG Education Fund (available at https://uspirg.org/sites/pirg/files/reports/WEB_CAP_Glyphosate-pesticide-beer-and-wine_REPORT_022619.pdf?_ga=2.33097086.1581849178.1551185850-857148262.1551185850) reported that 19 of wine (5) and beer (14) brands contained GLYP at levels ranging between 4.8 and 51.4 ppb. Several papers have been published in recent years, some of which have already been mentioned in this review [49, 58, 71, 73, 86].

The first interesting paper by Hao et al. describes a method for analyzing GLYP, AMPA, and GLUF in drinking water, surface water, and groundwater samples [96], that is, a LC-MS/MS method with reversed-phase and weak anion-exchange mixedmode Acclaim® WAX-1 column. Good analytical parameters were obtained: LODs of 1.5, 3.9, and 1.7 μg $L^{-1}$ for GLYP, AMPA, and GLUF, respectively; LOQs of 4.5, 11.6, and 5.3 μg $L^{-1}$; and recoveries between 62 and 102%. The main aspect is the analysis by direct injection of aqueous samples without derivatization or clean-up procedures with the risk of artifacts.

In 2015, a Chinese group developed a procedure for analyzing GLYP and GLUF in tea samples by means of FMOC-Cl derivatization and UPLC–MS/MS analysis [97]. The method shows good linearity ($r > 0.990$) in the range of 0.003–0.1 mg $L^{-1}$, LODs of 0.03 mg $kg^{-1}$ for both compounds, and recoveries between 81.4 and 99.1% with RSDs <2.3%.

Two papers published in 2015 reported the GLYP, AMPA, and GLUF determination in milk and milk-based products. Ehling and Reddy carried out a derivatization with FMOC-Cl followed by means of LC-MS/MS in different nutrition3Yal milk matrices such as cow's milk, human breast milk, soy milk, and whole milk powder [98]. This study is important because the reported analytical method does not require any analytical treatment such as clean-up, evaporation, or concentration; so, the possible artifact formation is drastically reduced. Further, the importance of the use of a triple-quadrupole mass spectrometry is still confirmed in terms of selectivity and fragment analysis. This occurrence gives good analytical parameters: $R^2 > 0.99$ in the entire investigated linearity range (5–500 ng $mL^{-1}$); recoveries between 91.1 and 115.2%; LODs of 0.012 and 0.01 μg $g^{-1}$ for GLY and AMPA, respectively; LOQ of 0.05 μg $g^{-1}$ for both; high intra-day (<4.0 and <7.7% for GLYP and GLUF, respectively) and inter-day (<8.4 and <3.8, respectively) precision.

The second paper investigates the direct injection of milk extract after deproteination and SPE on Oasis cartridge [99]: the LC–MS/MS analysis under the negative ion-spray ionization mode allowed to reach low method detection limits (MDLs), that is, 0.3, 1.4, and 0.4 ng $mL^{-1}$ for GLYP, AMPA, and GLUF, respectively, and low method quantification limits (MQLs), 1, 4, and 1 ng $mL^{-1}$, respectively, with recoveries ranging between 81 and 107% and RSDs 2.04–8.36%. A LC-MS/MS method (6 min chromatographic run) was successfully applied to a sample of fortified milk with a very low herbicides concentration (0.025 μg $mL^{-1}$). Further, the use of negative mode ion spray offers high sensitivity and selectivity. According to the study's authors (and these authors agree), this methodology could be competitive with the enzyme-linked immunosorbent assay (ELISA) method.

Steinborn et al. reported of a survey on the GLYP content in 114 breast milk samples collected in Bavaria and Lower Saxony, Germany, by comparing the data obtained by LC-MS/MS and GC-MS/MS analyses [100]. The two analyses required (a) an ultrafiltration and chromatography on an anion exchange column for LC–MS/MS and (b) a clean-up step on a cation exchange column and derivatization with trifluoroacetic acid anhydride (TFAA) and heptafluorobutanol (HFB) for the GC–MS/MS. The authors deeply investigated the difference between the chromatograms obtained with the two methods, especially for evaluating parameters such as precision, accuracy, LOD, and LOQ . Basically, GC–MS/MS allowed to reach instrumental detection limit (IDL) lower than that found in LC–MS/MS (0.02 vs. 0.5 ng $mL^{-1}$), but they detected an interference on a GLYP peak, which they did not manage to identify (all reagents, ultrapure water, all components were tested). Therefore, they fixed the LOQ at 1 ng $mL^{-1}$, the same concentration determined by LC–MS/MS (whose LOD is 0.5 ng $mL^{-1}$). The recoveries ranged between 83 and 128% with RSD < 17% for LC–MS/MS and between 71 and 102% with RSD < 13% for GC–MS/MS. Resuming, the GC–MS/MS is powerful at lower concentrations, but it simultaneously gives more bias than LC–MS/MS; both methods manage to investigate concentration above 1 ng $mL^{-1}$ with high precision and accuracy.

Two papers investigated the GLYP and AMPA content in human milk and urine samples. In the first, a high-throughput LC–MS/MS method using stable isotope labeled internal standard and clean-up with methylene chloride allowed to reach very low LODs (0.92 and 1.2 for GLYP and AMPA in human milk samples and 0.023 and 0.033 μg $mL^{-1}$ in human urine samples) and LOQs (10 μg $mL^{-1}$ for both in breast human milk samples and 0.1 μg $mL^{-1}$ in human urine samples), high recoveries (GLYP ranging between 92 and 107% in both matrices, AMPA between 89 and 107%) with low RSDs (<7.4 and <11.6% in human milk and urine samples, respectively) [101]. The authors also studied the matrix stability over a storage in 5°C (refrigerator) and at –20°C (freezer): in the first case, the recoveries were acceptable also after 24 hours, whereas in the second case, they were good also after 3 months. On the other hand, the second paper investigated the presence of GLYP and AMPA in milk (41 samples) and urine (40 samples) from healthy lactating women from Russia and the United States [102]. The authors used the same analytical procedure as reported above (i.e., LC-MS/MS, the use of stable isotope labeled internal standard and two fragments, such as precursor and product ion transitions, for the quantification) for the analysis, that is, the same analytical parameters. The results showed GLYP and AMPA in milk samples at levels below the LODs, whereas at low concentrations (<LOD and 1.93 μg $mL^{-1}$ and <LOD and 1.33 μg $mL^{-1}$, respectively, in urine samples). The authors extrapolated the maximum intake of milk containing 1 μg $mL^{-1}$ of GLYP for a 5-kg infant: their conclusions were that the expected levels should be 12,000 times lower than the health concern.

The presence of MRLs for GLYP in barley, wheat, rye, and hops is regulated by EU Regulation (EC) No. 396/2005 (i.e., 20, 10, and 0.1 mg $kg^{-1}$) [37, 38]. These are the raw agricultural commodities for beer beverage. Jansons et al. (2018) analyzed 100 beer samples from 24 different producers and distributors in Latvia with LC–MS/MS method ($R^2 > 0.999$ in the range of 0.2–25 μg $kg^{-1}$; LOD 0.2 μg $kg^{-1}$; LOQ 0.5 μg $kg^{-1}$; RSD < 4.1%) [103]. Among the numerous samples analyzed, 8 samples showed levels below the LOD and 9 samples below the LOQ , whereas 80 samples reported a GLYP concentration below 15 μg $kg^{-1}$ and 1 sample reached a GLYP content of 150 μg $kg^{-1}$. The authors pointed out the attention on beer brands of "undisclosed" origin, that is, no country production reported on the labeling (it sounds strange to the authors of this review considering the restrictions on food labeling in the EU, but we reported the authors' considerations), which could have higher GLYP content than the locally produced beer. Further, they also compared beers by malt type (barley or combined/other), color (light or dark), packaging (canned or bottled), the presence of precipitate (precipitate or no precipitate), filtration (filtered or not filtered), and pasteurization (pasteurized or not pasteurized), finding no significant differences in these cases.

Over these papers, it should be underlined two other paper dealing the GLYP determination in river water and soil samples. This particular occurrence regards the analytical protocol used by authors. In the first paper, Kudzin et al. developed a procedure based on derivatization with TEA-trifluoroacetic anhydride (TFAA)-trimethyl orthoacetate reagent and analyses by GC-CI(or EI)/MS (LOD 2.5–5.0 pmol) and GC-flame ionization detection (GC-FID; LOD 30–80 pmol, recovery 97%) [104]. In the second paper, Hu et al. investigated the performance of a method based on GC with nitrogen-phosphorus detector (GC-NPD): they estimated a LOD of $9 \times 10^{-12}$ g and a LOQ of 0.01 mg $kg^{-1}$ in samples, recoveries between 84.4 and 94.0%, and RSDs between 8.1 and 13.7% [105]. These two papers deserve to be mentioned for having introduced the possibility to analyze GLYP by two very easy, cheap, and worldwide available detectors such as FID and NPD.

## 3. Conclusion

This long *excursus* wanted to cover the novel or advanced methodologies based on chromatographic analysis reported in the literature. The GLYP determination in foods is a really important issue, even if the different international agencies still do not totally agree on the human health concern. The importance of a continuous monitoring of such compound (and its main metabolite, AMPA), and GLUF as well, is well known by scientists and politics worldwide due to its large use in agriculture. The suggestion is to continuously develop new methods, more accurate and sensitive, based on GC-MS/MS or LC-MS/MS analysis but also routine method based on inexpensive or use-friendly detectors (FID, FPD, or NPD).

In any case, the fear for the future is that the refinement of analytical methods increasingly leads to alarmist attitudes based on the discovery of very low quantities of GLYP, which is possible for a very wide range of products, even extremely toxic, without forgetting that in nature the zero residue does not exist.

## Acknowledgements

The authors would like to thank Giuseppe Ianiri for his help in the database revision and data analysis.

## Author details

Pasquale Avino*, Ivan Notardonato and Mario Vincenzo Russo
Department of Agricultural, Environmental and Food Sciences, University of Campobasso, Campobasso, Italy

*Address all correspondence to: avino@unimol.it

## References

[1] Duke SO. Glyphosate: The world's most successful herbicide under intense scientific scrutiny. Pest Management Science. 2018;**74**:1025-1026

[2] Benbrook CM. Trends in glyphosate herbicide use in the United States and globally. Environmental Sciences Europe. 2016;**28**:1-15

[3] Conrad A, Schroter-Kermani C, Hoppe HW, Ruther M, Pieper S, Kolossa-Gehrig M. Glyphosate in German adults—Time trend (2001 to 2015) of human exposure to a widely use herbicide. International Journal of Hygiene and Environmental Health. 2017;**200**:8-16

[4] Bai SH, Ogbourne SM. Glyphosate: Environmental contamination, toxicity and potential risks to human health via food contamination. Environmental Science and Pollution Research. 2017;**23**:18988-19001

[5] Duke SO, Powles SB. Glyphosate: A once-in-a-century herbicide. Pest Management Science. 2008;**64**:319-325

[6] Duke SO. The history and current status of glyphosate. Pest Management Science. 2018;**74**:1027-1034

[7] Canadian Food Inspection Agency (CFIA). Safeguarding with science: glyphosate testing in 2015-2016. Science Branch Survey Report; Ottawa, Canada, 2017. Available from: https://www.inspection.gc.ca/food-safety-for-industry/chemical-residues-microbiology/food-safety-testing bulletins/2017-04-13/executive- summary/glyphosate-testing/eng/1491846907641/1491846907985 [Accessed: 19 April 2020]

[8] Stalikas CD, Konidari CN. Analytical methods to determine phosphonic and amino acid group-containing pesticides. Journal of Chromatography A. 2001;**907**:1-19

[9] Van Bruggen AHC, He MM, Shin K, Mai V, Jeong KC, Finckh MR, et al. Environmental and health effects of the herbicide glyphosate. The Science of Total Environment. 2018;**616**:255-268

[10] Cuhra M, Bohn T, Petr C. Too much of a good thing? Frontiers in Environmental Science. 2016;**4**:28

[11] Mesnage R, Bernay B, Séralini GE. Ethoxylated adjuvants of glyphosate-based herbicides are active principles of human cell toxicity. Toxicology. 2013;**313**:122-128

[12] EFSA (European Food Safety Authority). Conclusion on the peer review of the pesticide risk assessment of the active substance glyphosate. EFSA Journal. 2015:13, 4302, 107 pp. DOI: 10.2903/j.efsa.2015.4302

[13] European Commission. Review report for the active substance glyphosate. Health & Consumer Protection Directorate-General, 6511/VI/99-final; 2002

[14] Acquavella JF, Alexander BH, Mandel JS, Gustin C, Baker B, Chapman P, et al. Glyphosate biomonitoring for farmers and their families: Results from the farm family exposure study. Environmental Health Perspective. 2004;**112**:321-326

[15] BfR. The BfR has finalised its draft report for the re-evaluation of glyphosate. In: Press Release. Berlin: Bundesinstitut fu€r Risikobewertung (BfR); 2014. Available from: http://www.bfr.bund.de/en/the_bfr_has_finalised_its_draft_report_for_the_re_evaluation_of_glyphosate-188632.html [Accessed: 08 April 2020]

[16] BfR. Does glyphosate cause cancer? In: BfR Communication No. 007/2015, 23 March 2015. Berlin: Bundesinstitut fu€r Risikobewertung (BfR); 2015.

Available from: http://www.bfr.bund.de/cm/349/does-glyphosate-cause-cancer.pdf [Accessed: 08 April 2020]

[17] Niemann L, Sieke C, Pfeil R, Solecki R. A critical review of glyphosate findings in human urine samples and comparison with the exposure of operators and consumers. Journal of Consumer Protection and Food Safety. 2015;**10**:3-12

[18] Glyphosate [Internet]. 2016. Available from: https://www.bfr.bund.de/en/a-z_index/glyphosate-193962.html [Accessed: 21 April 2020]

[19] Schinasi L, Leon ME. Non-Hodgkin lymphoma and occupational exposure to agricultural pesticide chemical groups and active ingredients: A systematic review and meta-analysis. International Journal of Environmental Research and Public Health. 2014;**11**:4449-4527

[20] Alavanja MC, Hofmann JN, Lynch CF, Hines CJ, Barry KH, Barker J, et al. Non-Hodgkin lymphoma risk and insecticide, fungicide and fumigant use in the Agricultural Health Study. PLoS One. 2014;**9**:e109332

[21] IARC. Glyphosate. Leon, France: International Agency for Research on Cancer Report; 2015. pp. 1-92

[22] Williams GM, Aardema M, Acquavella J, Berry SC, Brusick D, Burns MM, et al. A review of the carcinogenic potential of glyphosate by four independent expert panels and comparison to the IARC assessment. Critical Reviews in Toxicology. 2016;**46**:3-20

[23] Davoren MJ, Schiestl RH. Glyphosate-based herbicides and cancer risk: A post-IARC decision review of potential mechanisms, policy and avenues of research. Carcinogenesis. 2018;**39**:1207-1215

[24] Tarazona JV, Court-Marques D, Tiramani M, Reich H, Pfeil R, Istace F, et al. Glyphosate toxicity and carcinogenicity: A review of the scientific basis of the European Union assessment and its differences with IARC. Archives of Toxicology. 2017;**91**:2723-2743

[25] European Food Safety Authority (EFSA). Review of the existing maximum residue levels for glyphosate according to Article 12 of Regulation (EC) No, 396/2005-revised version to take into account omitted data. EFSA Journal. 2019;**17**:e05862

[26] ECHA. Committee for Risk Assessment. 2016. Available from: https://echa.europa.eu/it/-/glyphosate-not-classified-as-a-carcinogen-by-echa [Accessed: 02 April 2020]

[27] Greim H, Saltmiras D, Mostert V, Strupp C. Evaluation of carcinogenic potential of the herbicide glyphosate, drawing on tumor incidence data from fourteen chronic/carcinogenicity rodent studies. Critical Reviews in Toxicology. 2015;**45**:185-208

[28] US EPA. Revised glyphosate issue paper: Evaluation of carcinogenic potential. 2017. Available from: https://www.regulations.gov/document?D=EPA-HQ-OPP-2009-0361-0073 [Accessed: 07 April 2020]

[29] Environmental Protection Agency. Tolerances and exemptions for pesticide chemical residues in food. §180.364 Glyphosate; tolerances for residues. 45 FR 64911. 1980. Available from: https://www.ecfr.gov/cgi-bin/text-idx?SID=2c85909360c7c5aff63ddd1447545d6a&mc=true&node=se40.24.180_1364&rgn=div8 [Accessed: 27 March 2020]

[30] Benbrook CM. How did the US EPA and IARC reach diametrically opposed conclusions on the genotoxicity of glyphosate-based herbicides? Environmental Sciences Europe. 2019;**31**:2

[31] Samsel A, Seneff S. Glyphosate, pathways to modern diseases. II: Celiac sprue and gluten intolerance. Interdisciplinary Toxicology. 2013;**6**:159-184

[32] Li Z, Jennings A. Worldwide regulations of standard values of pesticides for human health risk control: A review. International Journal of Environmental Research and Public Health. 2017;**14**:826

[33] Gillezeau C, Van Gerwen M, Shaffer RM, Rana I, Zhang L, Sheppard L, et al. The evidence of human exposure to glyphosate: A review. Environmental Health: A Global Access Science Source. 2019;**18**:2

[34] Ledoux ML, Hettiarachchy N, Yu X, Howard L, Sun-Ok L. Penetration of glyphosate into the food supply and the incidental impact on T the honey supply and bees. Food Control. 2020;**109**:106859

[35] Solomon KR. Glyphosate in the general population and in applicators: A critical review of studies on exposures. Critical Reviews in Toxicology. 2016;**46**:21-27

[36] Solomon KR. Estimated exposure to glyphosate in humans via environmental, occupational, and dietary pathways: An updated review of the scientific literature. Pest Management Science. 2019. DOI: 10.1002/ps.5717

[37] Regulation (EC) NO 396/2005 of the European Parliament and of the Council of 23 February 2005. Official Journal of the European Union. 2005;**70**:1-16

[38] EU Database on current MRL values. Part A of Annex I to Reg. 396/2005, 2005. Available from: https:// ec.europa. eu/food/ plant/ pesticides/ eu-pesticides- database/ public/? event=pesticide. residue. CurrentMRL&language=EN [Accessed: 07 April 2020]

[39] Anastassiades M, Kolberg DI, Eichhorn E, Benkenstein A, Wachtler A-K, Zechmann S, et al. Quick method for the analysis of numerous highly polar pesticides in foods of plant origin via LC-MS/MS involving simultaneous extraction with methanol (QuPPe-Method). I. Food of Plant Origin. Version 10. 2019. Available from: https:// www.eurl-pesticides.eu/userfiles/file/ EurlSRM/meth_QuPPe-PO_EurlSRM. pdf [Accessed: 28 March 2020]

[40] Duffus JH. Glossary for chemists of terms used in toxicology. Pure & Applied Chemistry. 1993;**65**:2003-2122

[41] Knoll JK. Estimation of the limit of detection in chromatography. Journal of Chromatographic Science. 1985;**23**:422-425

[42] Armbruster DA, Tillman MD, Hubbs LM. Limit of detection (LQD)/ limit of quantitation (LOQ ): Comparison of the empirical and the statistical methods exemplified with GC-MS assays of abused drugs. Clinical Chemistry. 1994;**40**:1233-1238

[43] Armbruster DA, Pry T. Limit of blank, limit of detection and limit of quantitation. The Clinical Biochemist Reviews. 2008;**29**:S49-S52

[44] Boqué R, Heyden YV. The limit of detection. LC-GC Europe. 2009;**22**:82-85

[45] Dolan JW. Method reproducibility —A case study. LC-GC North America. 2006;**24**:1274-1278

[46] Tadeo JL, Sánchez-Brunete C, Pérez RA, Fernández MD. Analysis of herbicide residues in cereals, fruits and vegetables. Journal of Chromatography A. 2000;**882**:175-191

[47] Granby K, Johannesen S, Vahl M. Analysis of glyphosate residues in

cereals using liquid chromatography-mass spectrometry (LC-MS/MS). Food Additives and Contaminants. 2003;**20**:692-698

[48] Tseng SH, Lo YW, Chang PC, Chou SS, Chang HM. Simultaneous quantification of glyphosate, glufosinate, and their major metabolites in rice and soybean sprouts by gas chromatography with pulsed flame photometric detector. Journal of Agricultural and Food Chemistry. 2004;**52**:4057-4063

[49] Li B, Deng X, Guo D, Jin S. Determination of glyphosate and aminomethylphosphonic acid residues in foods using high performance liquid chromatography-mass spectrometry/mass spectrometry. Chinese Journal of Chromatography. 2007;**25**:486-490

[50] Andersen JH, Bille RL, Granby K. An intercomparison study of the determination of glyphosate, chlormequat and mepiquat residues in wheat. Food Additives and Contaminants. 2007;**24**:140-148

[51] Horwitz W. Evaluation of analytical methods used for regulation of foods and drugs. Analytical Chemistry. 1982;**54**:67A-76A

[52] Martins-Júnior HA, Lebre DT, Wang AY, Pires MAF, Bustillos OV. An alternative and fast method for determination of glyphosate and aminomethylphosphonic acid (AMPA) residues in soybean using liquid chromatography coupled with tandem mass spectrometry. Rapid Communications in Mass Spectrometry. 2009;**23**:1029-1034

[53] Van Eeckhaut A, Lanckmans K, Sarre S, Smolders I, Michotte Y. Validation of bioanalytical LC-MS/MS assays: Evaluation of matrix effects. Journal of Chromatography B. 2009;**877**:2198-2207

[54] Heller DN. Ruggedness testing of quantitative atmospheric pressure ionization mass spectrometry methods: The effect of coinjected matrix on matrix effects. Rapid Communications in Mass Spectrometry. 2007;**21**:644-652

[55] Hernandez F, Sancho JV, Pozo OJ. Critical review of the application of liquid chromatography/mass spectrometry to the determination of pesticide residues in biological samples. Analytical Bioanalytical Chemistry. 2005;**382**:934-946

[56] Ding J, Jin G, Jin G, Shen A, Guo Z, Yu B, et al. Determination of underivatized glyphosate residues in plant-derived food with low matrix effect by solid phase extraction-liquid chromatography-tandem mass spectrometry. Food Analytical Methods. 2016;**9**:2856-2863

[57] Botero-Coy AM, Ibáñez M, Sancho JV, Hernández F. Direct liquid chromatography–tandem mass spectrometry determination of underivatized glyphosate in rice, maize and soybean. Journal of Chromatography A. 2013;**1313**:157-165

[58] Liao Y, Berthion J-M, Colet I, Merlo M, Nougadère A, Hu R. Validation and application of analytical method for glyphosate and glufosinate in foods by liquid chromatography-tandem mass spectrometry. Journal of Chromatography A. 2018;**1549**:31-38

[59] Adams S, Guest J, Dickinson M, Fussell RF, Beck J, Schoutsen F. Development and validation of ion chromatography–tandem mass spectrometry-based method for the multiresidue determination of polar ionic pesticides in food. Journal of Agricultural and Food Chemistry. 2017;**65**:7294-7304

[60] Santilio A, Pompili C, Giambenedetti A. Determination of

glyphosate residue in maize and rice using a fast and easy method involving liquid chromatography–mass spectrometry (LC/MS/MS). Journal of Environmental Science and Health Part B. 2019;**54**:1-6

[61] Herrera López S, Scholten J, Kiedrowska B, de Kok A. Method validation and application of a selective multiresidue analysis of highly polar pesticides in food matrices using hydrophilic interaction liquid chromatography and mass spectrometry. Journal of Chromatography A. 2019;**1594**:93-104

[62] Grunewald K, Schmidt W, Unger C, Hanschmann G. Behavior of glyphosate and aminomethylphosphonic acid (AMPA) in soils and water of reservoir Radeburg II catchment (Saxony/Germany). Journal of Plant Nutrition and Soil Science. 2001;**164**:65-70

[63] Laitinen P, Rämö S, Siimes K. Glyphosate translocation from plants to soil—Does this constitute a significant proportion of residues in soil? Plant and Soil. 2007;**300**:51-60

[64] Mercurio P, Flores F, Mueller JF, Carter S, Negri AP. Glyphosate persistence in seawater. Marine Pollution Bulletin. 2014;**85**:385-390

[65] Chiesa LM, Nobile M, Panseri S, Arioli F. Detection of glyphosate and its metabolites in food of animal origin based on ion-chromatography-high resolution mass spectrometry (IC-HRMS). Food Additives & Contaminants: Part A. 2019;**36**:592-600

[66] Alferness PL, Iwata Y. Determination of glyphosate and (aminomethyl) phosphonic acid in soil, plant and animal matrixes, and water by capillary gas chromatography with mass-selective detection. Journal of Agricultural Food Chemistry. 1994;**42**:2751-2759

[67] Picó Y, Fernández M, Ruiz MJ, Font G. Current trends in solid-phase-based extraction techniques for the determination of pesticides in food and environment. Journal of Biochemical and Biophysical Methods. 2007;**70**:117-131

[68] Wang YS, Jaw CG, Chen YL. Accumulation of 2,4-D and glyphosate in fish and water hyacinth. Water Air Soil Pollution. 1994;**74**:397-403

[69] Lee J, Kim W, Lee J, Park S, Yang H, Kho Y, et al. Development of analytical method for glyphosate and glufosinate in animal feeds using LC-MS/MS. Environmental Epidemiology. 2019;**3**:225

[70] Tanaka T, Imou M, Tamase K, Umoto F, Okayama A, Oobayashi H, et al. Simple determination of glyphosate and its metabolite (aminomethylphosphonic acid) in fruits and vegetables by high performance liquid chromatography with fluorescence detection. Shokuhin Eiseigaku Zasshi. 1992;**33**:125-132 (in Japanese)

[71] Sen NP, Baddoo PA. Determination of glyphosate as N-nitroso derivative by high performance liquid chromatography with chemiluminescence detection. International Journal of Environmental Analytical Chemistry. 1996;**63**:107-117

[72] Kataoka H, Ryu S, Sakiyama N, Makita M. Simple and rapid determination of the herbicides glyphosate and glufosinate in river water, soil and carrot samples by gas chromatography with flame photometric detection. Journal of Chromatography A. 1996;**726**:253-258

[73] Hooijschuur EWJ, Kientz CE, Dijksman J, Brinkman UAT. Potential of microcolumn liquid chromatography and capillary electrophoresis with flame photometric detection for determination

of polar phosphorus-containing pesticides. Chromatographia. 2001;**54**:295-301

[74] Grey L, Nguyen B, Yang P. Liquid chromatography/electrospray ionization/isotopic dilution mass spectrometry analysis of n-(phosphonomethyl) glycine and mass spectrometry analysis of aminomethyl phosphonic acid in environmental water and vegetation matrixes. Journal of AOAC International. 2001;**84**:1770-1780

[75] Takahashi K, Horie M, Aoba N. Analysis of glyphosate and its metabolite, aminomethylphosphonic acid, in agricultural products by HPLC. Shokuhin Eiseigaku Zasshi. 2001;**42**:304-308 (in Japanese)

[76] Watanabe S. Simultaneous analysis of glyphosate and glufosinate in vegetables and fruits by GC-FPD. Shokuhin Eiseigaku Zasshi. 2004;**45**:38-43 (in Japanese)

[77] Qian K, Tang T, Shi T, Li P, Li J, Cao Y. Solid-phase extraction and residue determination of glyphosate in apple by ion-pairing reverse-phase liquid chromatography with pre-column derivatization. Journal of Separation Science. 2009;**32**:2394-2400

[78] Rembisz Z, Zakrzewski R, Skowron M, Ciesielski W. Image analysis of phenylisothiocyanate derivatised and charge-couple device-detected glyphosate and glufosinate in food samples separated by thin-layer chromatography. International Journal of Environmental Analytical Chemistry. 2016;**96**:320-331

[79] Boušová K, Bruggink C, Godula M. Fast routine analysis of polar pesticides in foods by suppressed ion chromatography and mass spectrometry. ThermoScientific Application Note. 2016;**661**

[80] Rajski L, Díaz Galiano FJ, Cutillas V, Fernádez-Alba AR. Coupling ion chromatography to Q-orbitrap for the fast and robust analysis of anionic pesticides in fruits and vegetables. Journal of AOAC International. 2018;**101**:352-359

[81] Melton LM, Taylor MJ, Flynn EE. The utilisation of ion chromatography and tandem mass spectrometry (IC-MS/MS) for the multi-residue simultaneous determination of highly polar anionic pesticides in fruit and vegetables. Food Chemistry. 2019;**298**:125028

[82] Savini S, Bandini M, Sannino A. An improved, rapid, and sensitive ultra-high-performance liquid chromatography-high-resolution orbitrap mass spectrometry analysis for the determination of highly polar pesticides and contaminants in processed fruits and vegetables. Journal of Agricultural and Food Chemistry. 2019;**67**:2716-2722

[83] Nortes-Méndez R, Robles-Molina J, López-Blanco R, Vass A, Molina-Díaz A, Garcia-Reyes JF. Determination of polar pesticides in olive oil and olives by hydrophilic interaction liquid chromatography coupled to tandem mass spectrometry and high resolution mass spectrometry. Talanta. 2016;**158**:222-228

[84] Chiarello M, Jiménez-Medina ML, Marín Saéz J, Moura S, Garrido Frenich A, Romero-González R. Fast analysis of glufosinate, glyphosate and its main metabolite, aminomethylphosphonic acid, in edible oils, by liquid chromatographycoupled with electrospray tandem mass spectrometry. Food Additives & Contaminants: Part A. 2019;**36**:1376-1384

[85] Karise R, Raimets R, Bartkevics V, Pugajeva I, Pihlik P, Keres I, et al. Are pesticide residues in honey related to oilseed rape treatments? Chemosphere. 2017;**188**:389-396

[86] Zoller O, Rhyn P, Rupp H, Zarn JA, Geiser C. Glyphosate residues in Swiss market foods: Monitoring and risk evaluation. Food Additives & Contaminants: Part B. 2018;**11**:83-91

[87] RTS. Du glyphosate, herbicide contesté, découvert dans l'urine des Romands. 2015. Available from: https://www.rts.ch/info/suisse/7125072-du-glyphosate-herbicide-conteste-decouvert-dans-l-urine-des-romands.html [Accessed: 12 April 2020]

[88] Pareja L, Jesús F, Heinzen H, Hernando MD, Rajski L, Fernández-Alba AR. Evaluation of glyphosate and AMPA in honey by water extraction followed by ion chromatography mass spectrometry. A pilot monitoring study. Analytical Methods. 2019;**11**:2123-2128

[89] Thompson TS, van den Heever JP, Limanowka RE. Determination of glyphosate, AMPA, and glufosinate in honey by online solid-phase extraction-liquid chromatography-tandem mass spectrometry. Food Additives & Contaminants: Part A. 2019;**36**:434-446

[90] Rubio F, Guo E, Kamp L. Survey of glyphosate residues in honey, corn and soy products. Journal of Environmental & Analytical Toxicology. 2014;**4**:249

[91] Chamkasem N, Vargo JD. Development and independent laboratory validation of an analytical method for the direct determination of glyphosate, glufosinate, and aminomethylphosphonic acid in honey by liquid chromatography/tandem mass spectrometry. Journal of Regulatory Science. 2017;**5**:1-9

[92] Berg CJ, King HP, Delenstarr G, Kumar R, Rubio F, Glaze T. Glyphosate residue concentrations in honey attributed through geospatial analysis to proximity of large-scale agriculture and transfer off-site by bees. PLoS One. 2018;**13**:e0198876

[93] El Agrebi N, Tosi S, Wilmart O, Scippo M-L, de Graaf DC, Saegerman C. Honeybee and consumer's exposure and risk characterisation to glyphosate-based herbicide (GBH) and its degradation product (AMPA): Residues in beebread, wax, and honey. The Science of the Total Environment. 2020;**704**:135312

[94] Mesnage R, Defarge N, Spiroux de Vendômois J, Séralini GE. Potential toxic effects of glyphosate and its commercial formulations below regulatory limits. Food and Chemical Toxicology. 2015;**84**:133-153

[95] Mañas F, Peralta L, Raviolo J, García Ovando H, Weyers A, Ugnia L, et al. Genotoxicity of AMPA, the environmental metabolite of glyphosate, assessed by the Comet assay and cytogenetic tests. Ecotoxicology and Environmental Safety. 2009;**72**:834-837

[96] Hao C, Morse D, Morra F, Zhao X, Yang P, Nunn B. Direct aqueous determination of glyphosate and related compounds by liquid chromatography/tandem mass spectrometry using reversed-phase and weak anion-exchange mixed-mode column. Journal of Chromatography A. 2011;**1218**:5638-5643

[97] Wu X, Chen X, Xiao H, Liu B. Simultaneous determination of glyphosate and glufosinate-ammonium residues in tea by ultra performance liquid chromatography-tandem mass spectrometry coupled with pre-column derivatization. Se Pu. 2015;**33**:1090-1096 (in Chinese)

[98] Ehling S, Reddy TM. Analysis of glyphosate and aminomethylphosphonic acid in nutritional ingredients and milk by derivatization with fluorenylmethyloxycarbonyl chloride and liquid chromatography-mass spectrometry. Journal of Agricultural and Food Chemistry. 2015;**63**:10562-10568

[99] Chamkasema N, Morrisa C, Harmon T. Direct determination of glyphosate, glufosinate, and AMPA in milk by liquid chromatography/ tandem mass spectrometry. Journal of Regulatory Science. 2015;**2**:20-26

[100] Steinborn A, Alder L, Michalski B, Zomer P, Bendig P, Martinez SA, et al. Determination of glyphosate levels in breast milk samples from Germany by LC-MS/MS and GC-MS/ MS. Journal of Agricultural and Food Chemistry. 2016;**64**:1414-1421

[101] Jensen PK, Wujcik CE, McGuire MK, McGuire MA. Validation of reliable and selective methods for direct determination of glyphosate and aminomethylphosphonic acid in milk and urine using LC-MS/MS. Journal of Environmental Science and Health, Part B. 2016;**51**:254-259

[102] McGuire MK, McGuire MA, Price WJ, Shafii B, Carrothers JM, Lackey KA, et al. Glyphosate and aminomethylphosphonic acid are not detectable in human milk. The American Journal of Clinical Nutrition. 2016;**103**:1285-1290

[103] Jansons M, Pugajeva I, Bartkevičs V. Occurrence of glyphosate in beer from the Latvian market. Food Additives & Contaminants: Part A. 2018;**35**:1767-1775

[104] Kudzin ZH, Gralak DK, Drabowicz J, Luczak J. Novel approach for the simultaneous analysis of glyphosate and its metabolites. Journal of Chromatography A. 2002;**947**:129-141

[105] Hu JY, Chen C-L, Li J-Z. A simple method for the determination of glyphosate residues in soil by capillary gas chromatography with nitrogen phosphorus. Journal of Analytical Chemistry. 2008;**63**:371-375

# 13

# Applications of Ionic Liquids in Gas Chromatography

*Umaima Gazal*

## Abstract

The environment offers an enormous innovative panorama of prospects intended for the research of novel biodegradable diluents. Regular composites have been lately recycled to formulate the anionic and cationic fraction of RTIL. Numerous applications of ionic liquids have been explored in segregation discipline. Attributable to the extraordinary polarization as well as exceptional current steadiness, IL-centered immobile segments have been applied to resolution of varied series of critically stimulating complexes frequently extremely polar composites using great boiling points plus physical resemblances comprising elongated sequence fatty acids, essential oils, polycyclic aromatic sulfur heterocycles (PASHs) and PCBs. IL-centered immobile segments facilitated the gas chromatography study for effective as well as precise amount of liquid in the industrialized yields for example pharmaceutical as well as petrochemicals complexes.

**Keywords:** ionic liquids, gas chromatography, static stages, Zwitterionic liquids, polymeric ionic liquids

## 1. Introduction

Ionic liquids are the utmost promising liquid green solvents with wide applications in separation science. The effect of the IL organic configuration as well as the stimulus of tributary factors, for example the IL temperature, pH, concentration, analysis time and voltage, are compatibly rationally talked concerning the accomplished parting enlargements. Gas chromatography is unique and extreme proficient, dependable, as well as stout methods for the study of unstable plus semivolatile composites. Effective along with rapid gas chromatographic investigation of objective analytes is mostly reliant on the enactment of the gas chromatography column. Though here have be present main active developments, there quiet a solid claim of extremely choosy, indolent, polar also thermally constant gas chromatography pillars intended for critically stimulating composites for example polychlorinated biphenyls, unrestricted fatty acids and unstable amines [1]. Furthermore, the physicochemical characteristics for instance surface tension, viscosity and melting point are too acute to yield extremely proficient gas chromatography columns. The viscosities of furthermost ionic liquids are frequently 1–3 remits of scale greater than outmoded biological diluents [2]. In demand for an ionic liquid to be measured as a immobile stage, the solid must have great viscosity that rests fixed above a comprehensive high temperature choice. Van der Waals as well as Hydrogen bonding kind interfaces amongst the anion plus cation of Ionic liquids rule the viscosityproperties. Furthermore, it is imperative to ruminate the surface tension

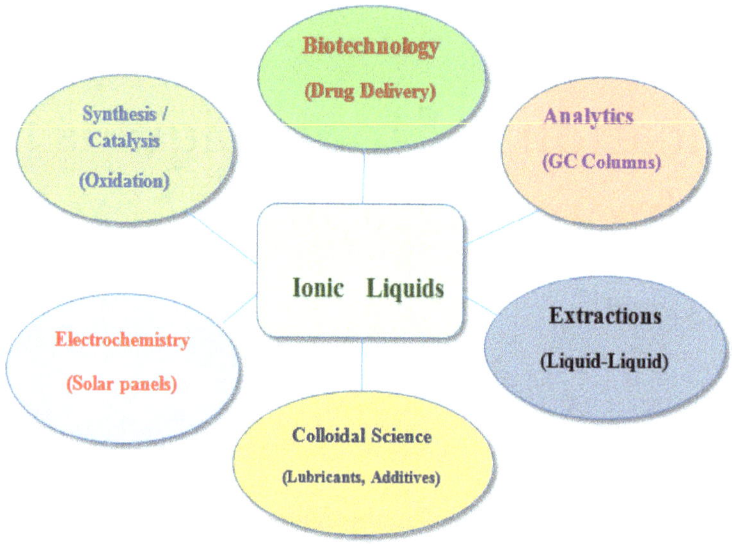


**Figure 1.**
*Applications of ionic liquids in various fields.*

of the Ionic liquids. Its values extending as of 30 to 50 dyne/cm usually display bigger wettability on the barrier of unprocessed tube pillars [2]. Ionic liquids establish an assembly of biological salts which are fluid lower than 100 °C, moreover, the ionic liquids that are fluid at room temperature are generally recognized as room temperature ionic liquids [3]. Ionic liquids are easy to manufacture, thermally steady, flameproof, chemically inactive, retain small vapor density, polar, and their discernment can be simply regulated by means of fluctuating the component anion or cation; and from now they have been extensively recycled as static stages in conservative gas chromatography (**Figure 1**) [4–10].

Ionic liquids can also be recycled as diluents for the suspension of various resources for example fiber [11], chitin [12], etc. The outstanding solubility of biological/inert composites in ionic liquids as well as a extensive variety of it as the fluid state brand them noble diluents for several responses. Furthermore, they displayed modest produces when related with conservative biological diluents [13, 14]. Thermodynamic factors of these ionic liquids were studied through chromatographic methods. ILs are centered on the numerous method, in company with the utmost extensively considered are N-alkylpyridinium, alkylammonium, N'N-dialkylimidazolium, and alkylphosphonium.

## 2. Stationary phase in gas chromatography using ILs

Owing to the ever-developing mandate for the great determination, high sympathy, as well as statistics amusing investigation of composite models for instance aromas, smells, petrochemicals, plus pharmacological uncooked supplies, continuous expansions of gas chromatography supports through exclusive discrimination, squat bleed, high dullness, and in addition varied high temperature operational series are desirable. Because of the high polarization then exceptional thermal constancy, IL-centered stationary stages have been employed to decide an extensive series of methodically stimulating complexes typically precise polar complexes with high steaming facts also fundamental resemblances comprising lengthy sequence oily acids, vital oils and polycyclic aromatic sulfur heterocycles [15, 16]. ILs are characteristically organized with a phosphorous- or nitrogen-comprising biological cation as well as an inorganic or else organic anion. Meanwhile the chief outline of IL-centered GC supports in 1999, ILs have been effectively engaged as stationary stages because of their trivial high polarity, tunable selectivity, high thermal stability and vapor pressure [17]. The ILs can be altered through diverse efficient collections to endure countless solvation interfaces in addition to display exclusive chromatographic discernment such as GC immobile phases. ILs have been exposed to have exclusive solvation competences plus discernment's on the distinctive solvent/solute interfaces [18]. Imidazolium-centered ILs be able to as per contrived to discrete equally nonpolar and polar analytes [17]. As, by means of varying the anionic lot of the imidazolium IL since chloride [Cl]- to hexafluorophosphate [$PF_6$]-, a substantial variance in discernment was perceived for polar analytes equaled to the non-polar complexes (**Figure 2**).

Additional examination of dissimilar modules of ILs, containing monocationic imidazolium, pyridinium, as well as pyrrolidinium exposed that the hydrogen contributor capability of the IL immobile phases was subjugated through IL cation. In contrast, the anionic lot was create to adopt the part of hydrogen acceptor anion as of proton giver analytes for example carboxylic acids plus alcohols [2]. Consequently, dicationic [19], tricationic [20], as well as phosphonium-centered cations [21] were oppressed to expand great thermal constancy plus fluid variety of ILs equated to customary monocationic static levels. Lately, in an effort to extend the applicability of IL static phases, task-specific ionic liquids (TSILs) were famil-iarized by functionalizing the IL cation with numerous agents [22]. For instance, the integration of aromatic segments in the IL cation improved the discernment for scented complexes, for example polycyclic aromatic hydrocarbons (PAHs). This is owing to enriched π-π sort communications amongst analytes as well as the aromatic clusters of the IL cation [4]. Overview of polar efficient clusters, for example hydroxyl segments, can effect in enlarged discernment for hydrogen compliant analytes [23]. Consequently, tweaking the IL-centered GC static phase configuration might augment choosiness essential for parting of precise compound model elements with comparable polarizations.

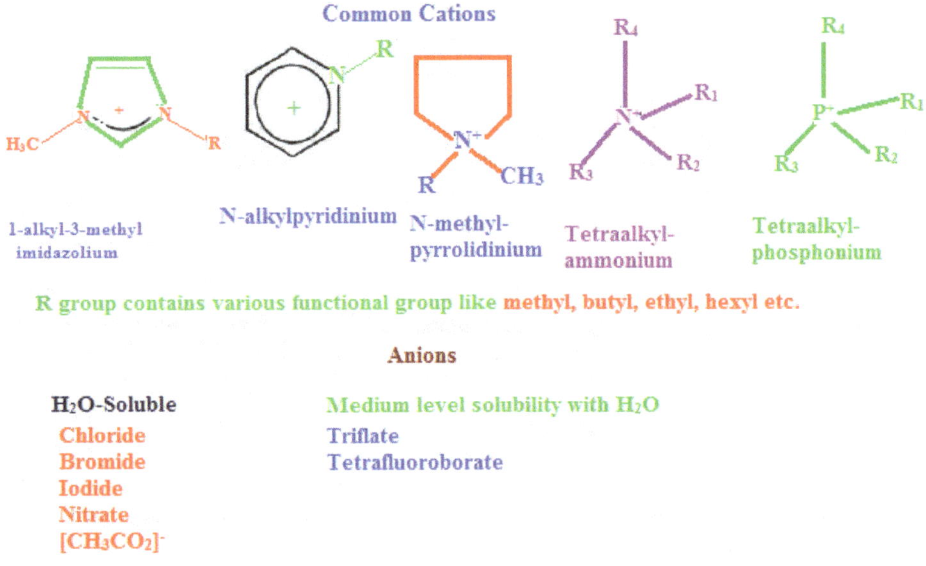


**Figure 2.**
*Common cations and anions.*

One of the important physical property of static stage is melting point as it basically commands the least effective high temperature of the ensued GC column. ILs with short melting points are extremely required as well as are usually acquired via integrating proportion-flouting sections also alkyl sideways chains with diverse dimensions [24–26]. Analytes naturally intermingle using IL-centered stationary segments over moreover partition or adsorption-kind contrivance [6, 27–28]. Better parting efficacies existed usually providing through the partition-kind retaining contrivance. Once the furnace temperature is lesser than the melting point of the IL-centered static stage, the molecular interface amongst the analytes as well as static stage is to be expected to be controlled by means of adsorption. Variance perusing calorimetry is usually operated to define the melting point of IL-centered stationary stages [25].

## 3. Incorporation of ionic liquids in multidimensional gas chromatography

Multidimensional gas chromatography is an influential method to accomplish progressive parting of impulsive as well as quasi-impulsive composites in compound environments [29–31]. By means of Multidimensional gas chromatography method, two or else extra gas chromatographic partings are engaged in a consecutive manner [29]. The paramount requirement to effectively enhance peak capacity in the composite system is to employ a combination of GC stationary phases possessing different selectivities. It was presented to the chromatographic state compromises advanced ultimate ability than conservative one-dimensional gas chromatography, permitting on behalf of the determination of model elements by means of comparable polarizations otherwise instabilities [32]. In this method, analytes are evaporated then exposed to a sequence of gas chromatography supports by means of chemically diverse static stages attached over an edge. In multidimensional gas chromatography analyte parting is preserved on both column, prominent to an upsurge in the parting control associated to the one [33]. Two types of multidimensional gas chromatography are usually engaged, specifically, core-wounding also inclusive. In core-wounding multidimensional gas chromatography, merely a choice rare portions of overflow after the first support are transported to the second support for extra parting [34].

Diverse support selectivity, which can be characterized using a liberated parting procedure, is the significant requisite to acquire advanced top capability in multidimensional gas chromatography methods [35]. Several customary non-ionic gas chromatography static segments are categorized as both non-polar and polar segments. These supports display a deficiency of variety in positions of solvation abilities, which can bound their capacity to decide composite models through gas chromatography× gas chromatography. Because of this hitch, ionic liquids centered supports have appeared as alternate gas chromatography × gas chromatography static stages. Utilizing ionic liquids centered stakes can permit exclusive solvation abilities in addition selectivities, moreover to advance thermal constancies comparative to customary segments. Ionic liquids have been applied as existing segments combined with customary non-ionic segments in numerous gas chromatography× gas chromatography partings [36]. Meanwhile maximum gas chromatography × gas chromatography partings can be controlled constructed on analyte instability in the first measurement monitored through involvement of dissimilar relations in the second measurement, it is collective to practice ionic liquids supports as the second measurement to estimate their enactment in relations of retaining contrivances. Presently, a huge amount of profitably accessible ionic liquids supports, for example the Supelco Low Bleed community, comprise numerous phosphonium and imidazolium grounded di cations which are typically combined with frequently tri fluoromethanesulfonate and bis[(trifluoromethyl)sulfonyl]imide anions [2, 20]. Economical ionic liquids static stages have been engaged in the parting of total of analytes for instance savor plus fragrance amalgams [36], aromatic hydrocarbons [37], alkyl halides [38], alkyl phosphonates [39], fatty acid methyl esters [40], as well as additional polar analytes (nitrogen, sulfur as well as oxygen-comprising composites). These analysis specify that ionic liquids supports establish considerable advanced selectivity plus retaining in the direction of commonly polar analytes associated to non-polar analytes owing to hydrogen-bonding interaction, electrostatic interactions and dipole–dipole relations, amongst ions [41].

## 4. Polymeric ionic liquids centered static stages in gas chromatography

Polymeric ionic liquids are stimulating family of composites that can be recycled as sorbent coverings in solid phase micro extraction. Polymeric ionic liquids are artificial polymers manufactured after ionic liquids monomers. Furthermore, Polymeric ionic liquids can be basically modified to display greater sensitivity and selectivity nearby diverse section of analytes. Polymeric ionic liquids are characteristically manufactured through functionalizing a polymerizable practical cluster on the cationic component of the ionic liquid by free radical polymerization in the attendance of a thermal originator. Polymeric ionic liquids reveal greater thermal constancy in addition to a confrontation to viscosity decrease at greater temperatures. These valuable structures can develop fiber lifespan, toughness plus eligibility of Polymeric ionic liq-uids while retentive the discrimination relics the intrinsic to ionic liquids. Polymeric ionic liquids have been displayed to extant extraordinary possessions as well as exhibitions [42–48], assisting novel plus stimulating parting procedures [49–51]. Ionic liquids have drawn abundant consideration in latest centuries as constituents for static stages in gas chromatography, because of stuffs similar their capability to create concurrent nonpolar as well as polar interfaces with the analytes, their extraordinary thermal constancy, before their insignificant air compressions then extensive fluid series [52–56]. Also, it is price revealing that these things can be effortlessly well-adjusted over minor fluctuations in the assembly of either the anion or cation, which, also, can intensely modify the choosiness or else the parting capability for the analyte of concern [36, 57–59]. The concern in ionic liquid-covered gas chromatography supports has enlarged afterward their marketable outline in 2008, also today, numerous ionic liquids glazed supports with dissimilar features are viably accessible. Though, one main task for the growth of static stages built on ionic liquids is the research of extremely standardized coverings, which would service decent ultimate regularities as well as extremely active complex partings, also concurrently, deliver extraordinary thermal constancies for the subsequent gas chromatography supports [60–62]. At extraordinary temperatures, identical ionic liquidglazed silica supports can practice flick commotion prominent to a diminution in the analyte retaining periods as well as efficacy. In this respect, polymerized ionic liquids can offer the compulsory replies, preserving the outstanding thermal constancy of the supports, in addition to uniting the chief structures of an ionic liquid as well as the distinctive polymer characteristics for example better automated constancy plus development capability [62–65].

## 5. Preference of Zwitterionic liquids in gas chromatography

Zwitterions, consequent after ionic liquids have inimitable characteristics for example reasonably small crystal conversion temperature, slight ion conductivity as well as exclusive stage conduct afterward partying with water. Moreover, the combination of convinced zwitterions plus negligible quantity of water can be observed as an exceptional liquid ideal of cell tissues. Zwitterionic liquids can be chemically precise like to predictable aprotic ionic liquids, excluding that the negative as well as positive charges exist in on the similar particle. Zwitterionic complexes are elements that have an entire clear charge of zero as well as are therefore electrically impartial. They transmit proper electrical charges of reverse symbol contained on diverse particles as well as formerly can be measured as internal salts. The utmost collective zwitterionic-kind ionic liquids are nitrogene heterocycles with sulfonate component. These Zwitter ionic Liquids have been competently considered as designable electrolyte constituents for fuel cells [66] as well as lithium batteries [67]. Not as much of discovered Zwitter ionic Liquids are configurations founded on imidazoliums with carboxylate occupations. They have been cast-off as forde sulfurization of fuels [68], Bronsted acidic catalytic agent [69], as liquid crystals [70] or for metallic oxides solubilization [71]. Therefore amino acids be existent typically as per zwitterions in a definite variety of pH then the pH at which the regular charge is zero is known as the particle's isoelectric fact (**Figure 3**).

Through zwitterionic composites, anions as well as cations are roped covalently. Imidazolium sulfonate is one of the example of zwitterionic liquids. The production of room temperature zwitterionic liquids in which both sulfonate anion as well as imidazolium cation elements attribute to the parental particle was conveyed in literature (**Figure 4**) [72, 73].

Motivated via means of the proclamation that these composites can be organized as fluids at room temperature, three zwitterionic liquids integrating alkyl side chain and oligoether substituents were intended then inspected as gas chromatography static stages. The fundamentally-regulated zwitterionic liquids - centered static stages offer distinctive choosiness, robust retaining, exceptional top regularity, also a reasonably widespread employed series appropriate for the study of volatile carboxylic acids. This comprising volatile fatty acids for instance lactic acid as well as butyric acid are significant for the construction of cosmetics, pharmaceuticals and fuels [74–76]. Gas chromatography is furthermost usually recycled for the quantification as well as parting of specific acids in acylated lipids. Derivatization of volatile carboxylic acids by means of numerous approaches for example alkylation plus acylation is characteristically accomplished to upsurge the explosive nature of these composites in addition to mark their investigation viable through gas chromatography (**Table 1**).

**Figure 3.**
*Zwitter ionic liquids.*

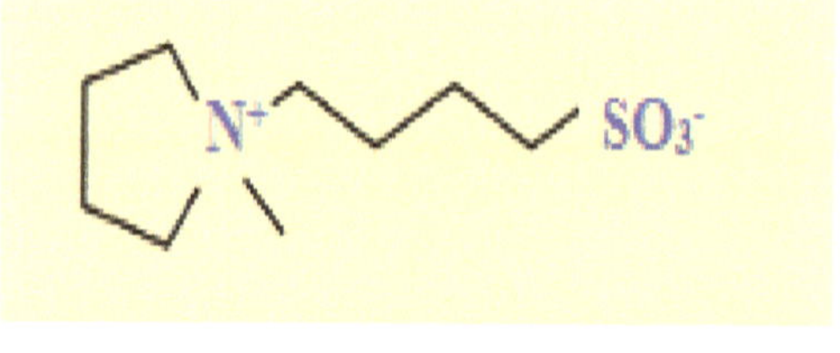

**Figure 4.**
*Imidazolium sulfonate Zwitterionic liquids.*

| S. No | Embattled Composites | Ionic liquids | Uses | Reference |
|---|---|---|---|---|
| 1. | Methyl esters / Fatty acids | 1,9-di(3-vinyl-imidazolium) nonane bis(trifluoromethyl) sulfonylimidate | Gas chromatography stage | [77] |
| 2. | Aromatic Composites / Alcohols | [BMIM][$NTf_2$], [BMIM]Cl | Gas chromatography stage | [78] |

**Table 1.**
*Use of ionic liquids as superficial-integrated static segments.*

## 6. Green aspects of ionic liquids

The adjustable physicochemical properties of ionic liquids have prolonged their usage addicted to a wide variety of diverse uses. Ionic liquids have an abundant prospective in biological amalgamation, electrochemistry, mass spectrometry, green chemistry as well as partings [79, 80]. In the field of analytical chemistry, ionic liquids have been recycled as static segments as well as diluents for headspace gas chromatography [80], movable segment extracts plus external-attached static segments in liquid chromatography [79, 81] for liquid–liquid abstractions as well as solid-phase micro abstraction [79, 80, 82]. Several "green-engrossed" manufacturing have originate that ionic liquids are outstanding applicants for their uses because of their precise small vapor compression [83, 84]. The overview of ionic liquids such as static segments has released up and around novel outlooks in this arena by means of their exclusive solvation features outcome in unusual discernment, which is entirely dissimilar to that of typical polyethylene glycol as well as poly dimethyl siloxane centered supports. Since of their atypical discernment plus extraordinary unresponsiveness, ionic liquids centered supports have previously establish numerous solicitations in the normal item for consumption ground in and multidimensional as well as mono gas chromatography in addition to preparative gas chromatography, prominent to the comprehensive investigation of composite sections (containing aqueous resolutions), plus the parting of stimulating sets of complexes. The speedily growing usage of Ionic liquids equally in educational plus manufacturing arenas have created an increasing apprehension approximately their effect on the environs. Meanwhile Ionic liquids are extremely solvable in liquid however are not continuously ecofriendly, a discharge of ionic liquids into the atmosphere might clue to substantial water contamination complications. Furthermore, Ionic liquids could develop insistent contaminants in discarded water seepage because of their great constancy in water. Intended for this purpose, research inspecting Ionic liquids biodegradability are of inordinate significance. The rising character of Ionic liquids in production as well as study plus the growing alarm nearby their green influence have advised a requisite for the progress of profligate, dependable as well as reproducible techniques for the classification plus investigation of Ionic liquids [85].

## 7. Conclusion

Ionic liquids have solicitations in various areas in chemistry. The applications of ionic liquids as extracts in chromatography displays abundant rewards equated to further extracts. Ionic liquids have been realistic in diverse extents of parting, for instance ionic liquid sustained tissues, as moveable segment extracts as well as external-joined static stages in chromatography partings also as the abstraction diluent in model provisions, since they can be collected from numerous anions in addition to cations that alter the things as well as stage conduct of fluids. The inflammable, non- explosive environment of ionic liquids marks them an outstanding optimal for the expansion of nontoxic methods. A substantial benefit of ionic liquids-centered stationary segments is their capability to have adequate to high updraft solidity though similarly unveiling a comprehensive host of solvation proficiencies, specific of their inimitable selectivities. In spite of their attainment, viable Ionic liquids-centered immobile segments dearth the solving authority for non-polar analytes, predominantly unsaturated as well as saturated hydrocarbons, cycloalkanes. This nonexistence of discernment has diminished fervor amongst certain parting experts who might modulate the feature of fundamental fine-tuning (in relations of anion/cation combining in addition to operational structures of every constituent) while emerging ionic liquids to display great discernment also robust solving influence. Ionic liquids have strained substantial consideration as gas chromatography immobile segments as of their tunable chemical plus physical properties. Conversely, profitable Ionic liquids-centered gas chromatography supports have not reconnoitered entirely of the solvation characteristics that can be obtainable through ionic liquids.

Moreover, their polarization, viscosity, hydrophobicity plus further physical and chemical properties can be designated by means of selecting the anionic and cationic component. Ionic liquids are considered as “exclusive diluents” as of this adjustable environment, which rises their prospective solicitations. The consumption of ionic liquids is maiden innovative prospects in diverse regions of parting discipline, with novel countless solicitations. Additional uses in partings are linked to the ecological, pharmacological, biomedical as well as various manufacturing trades. Ionic liquids have been discovered in partings for abstraction, reinforced fluid membranes, as extracts then as static segments in chromatography.

## Acknowledgements

The author received no financial support for writing this article. I want to acknowledge my mother **Mrs. Nargis Sultana** and husband **Mr. Safdar Hussain** for his encouragement and support for writing this article.

## Author details

Umaima Gazal
Department of Chemistry, Raja Bahadur Venkata Rama Reddy Women’s College, Affiliated to Osmania University, Hyderabad, India

*Address all correspondence to: dr.umaimagazal@gmail.com

## References

[1] Shen B, Semin D, Fang J, Guo G, Analysis of 4-bromo-3-fluorobenzaldehyde and separation of its region isomers by one-dimensional and two-dimensional gas chromatography, J. Chromatogr. A. 2016, 1462: 115-123.

[2] Yao C, Anderson J.L, Retention characteristics of organic compounds on molten salt and ionic liquid-based gas chromatography stationary phases. J. Chromatogr. A. 2009, 1216: 1658-1712.

[3] Hayes R, Warr G. G, Atkin R, Structure and nanostructure in ionic liquids, Chem. Rev. 2015,115: 6357-6426.

[4] Anderson J.L, Armstrong D.W, High-stability ionic liquids. A new class of stationary phases for gas chromatography, Anal. Chem. 2003, 75: 4851-4858.

[5] Anderson J.L, Armstrong D.W, Immobilized ionic liquids as high-selectivity/high temperature/high-stability gas chromatography stationary phases. Anal. Chem. 2005, 77: 6453-6462.

[6] Poole S. K, Shetty P. H, Poole C.F, Chromatographic and spectroscopic studies of the solvent properties of a new series of room-temperature liquid tetraalkylammonium sulfonates, Anal. Chim. Acta. 1989, 218: 241-264.

[7] Pomaville R. M, Poole C. F, Solute-solvent interactions in liquid tetrabutylammonium sulfonate salts studied by gas chromatography. Anal. Chem. 1988, 60: 1103-1108.

[8] Anderson J.L, Ding R, Ellern A, Armstrong D.W, Structure and properties of high stability geminal dicationic ionic liquids, J. Am. Chem. Soc. 2005, 127: 593-604.

[9] Tran C.D, Challa S, Fullerene-impregnated ionic liquid stationary phases for gas chromatography. Analyst. 2008, 133: 455-464.

[10] Breitbach Z.S, Armstrong D.W, Characterization of phosphonium ionic liquids through a linear solvation energy relationship and their use as GLC stationary phases. Anal. Bioanal. Chem. 2008, 390: 1605-1617.

[11] Zhu S, Wu Y, Chen Q, Yu Z, Wang C, Jin S, Ding Y, and Wu G, Utilization of Cellulose by combining two major green chemistry principles: using environmentally preferable solvents and bio-renewable feed-stocks. Green Chemistry, 2006, 8: 325-327.

[12] Xie H, Zhang S, and Li S, Chitin and chitosan dissolved in ionic liquids as reversible sorbents of $CO_2$. Green Chemistry. 2006, 8: 630-633.

[13] Morrison, D.W, Forbes, D.C, and Davis Jr, J.H, Base-promoted reactions in ionic liquid solvents. The Knoevenagel and Robinson annulation reactions, Tetrahedron Lett. 2001, 42: 6053-6055.

[14] Welton T, Room-Temperature Ionic Liquids. Solvents for Synthesis and Catalysis. Chemical Reviews. 1999, 99: 2071-2084.

[15] T.D. Ho, C. Zhang, L.W. Hantao, J.L. Anderson, Ionic Liquids in Analytical Chemistry: Fundamentals, Advances, and Perspectives, Anal. Chem., 86 (2014) 262-285.

[16] Trujillo-Rodríguez M. J, Nan H, Varona M, Emaus M. N, Souza I.D, Anderson J.L, Advances of Ionic Liquids in Analytical Chemistry. Anal. Chem. 2019, 91: 505-531.

[17] Armstrong D.W, He L.F, Liu Y.S, Examination of ionic liquids and their interaction with molecules, when used as stationary phases in gas chromatography. Anal. Chem. 1999, 71: 3873-3876.

[18] Anderson J. L, Ding J, Welton T, Armstrong D. W, Characterizing Ionic Liquids on the Basis of Multiple Solvation Interactions. J. Am. Chem. Soc. 2002, 124: 14247-14254.

[19] Anderson J. L, Ding R, Ellern A, Armstrong D. W, Structure and properties of high stability geminal dicationic ionic liquids. J. Am. Chem. Soc. 2004, 127: 593-604.

[20] Sharma P. S, Payagala T, Wanigasekara E, Wijeratne A. B, Huang J, Armstrong D. W, Trigonal Tricationic Ionic Liquids: Molecular Engineering of Trications to Control Physicochemical Properties. Chem. Mater. 2008, 20: 4182-4184.

[21] Breitbach Z. S, Armstrong D. W, Characterization of phosphonium ionic liquids through a linear solvation energy relationship and their use as GLC stationary phases. Anal. Bioanal. Chem. 2008, 390: 1605-1617.

[22] Davis J. H, Task-Specific Ionic Liquids. Chem. Letters. 2004, 33: 1072-1077.

[23] Huang K, Han X, Zhang X, Armstrong D, PEG-linked geminal dicationic ionic liquids as selective, high-stability gas chromatographic stationary phases. Anal. Bioanal. Chem. 2007, 389: 2265-2275.

[24] López-Martin I, Burello E, Davey P.N, Seddon K.R, Rothen - berg G, Anion and Cation Effects on Imidazolium Salt Melting Points: A Descriptor Modelling Study. ChemPhysChem, 2007, 8: 690-695.

[25] Murray S.M, O'Brien R.A, Mattson K.M, Ceccarelli C, Sykora R.E, West K.N, Davis J.H, The Fluid-Mosaic Model, Homeoviscous Adaptation, and Ionic Liquids: Dramatic Lowering of the Melting Point by Side- Chain Unsaturation, Angew Chem Int Edit. 2010, 49: 2755-2758.

[26] Mirjafari A, O'Brien R. A, West K.N, Davis J.H, Synthesis of New Lipid-Inspired Ionic Liquids by Thiol-ene Chemistry: Profound Solvent Effect on Reaction Pathway, Chemistry-a. European Journal. 2014, 20: 7576-7580.

[27] Dhanesar S.C, Coddens M.E, Poole C.F, Influence of phase loading on the performance of whisker-walled open tubular columns coated with organic molten salts. J. Chromatogr. A. 1985, 324: 415-421.

[28] Poole C. F, Furton K.G, Kersten B.R, Liquid Organic Salt Phases for Gas Chromatography. J. Chromatogr. Sci. 1986, 24: 400-409.

[29] Seeley J.V, Seeley S.K, Multi dimensional gas chromatography: fundamental advances and new applications. Anal. Chem. 2013, 85: 557-578.

[30] Marriott P.J, Chin S.T, Mai khunthod B, H.-G. Schmarr H.-G, Bieri S, Multidimensional gas chromatography Trac-Trends. Anal. Chem. 2012, 34: 1-21.

[31] Dimandja J. M.D, Peer Reviewed: Comprehensive 2-D GC provides high-performance separations in terms of selectivity, sensitivity, speed, and structure. Anal. Chem. 2004, 76: 167 A-174 A.

[32] Zakaria M, Gonnord M. F, Guiochon G, Applications of two-dimensional thin-layer chromatography. J. Chromatogr. A 1983, 271: 127-192.

[33] Marriott P. J, Chin S. T, Mai - khunthod B, Schmarr H. G, Bieri S, Multidimensional gas chromatography. Trends Anal. Chem. 2012, 34: 1-21.

[34] Simmons M. C, Snyder L. R, Two-stage Gas liquid chromatography. Anal. Chem. 1958, 30, 32-35.

[35] Seeley J. V, Bates C. T, McCurry J. D, Seeley S. K, Stationary phase selection

and comprehensive two-dimensional gas chromatographic analysis of Trace biodiesel in petroleum-based fuel. J. Chromatogr. A 2012, 1226: 103-109.

[36] Ragonese C, Sciarrone D, Tranchida P.Q , Dugo P, Mondello L, Use of ionic liquids as stationary phases in hyphenated gas chromatography techniques. J. Chromatogr. A 2012, 1255: 130-144.

[37] Krupčík J, Gorovenko R, Špánik I, Bočková I, Sandra P, Armstrong D. W, On the use of ionic liquid capillary columns for analysis of aromatic hydrocarbons in low-boiling petrochemical products by one-dimensional and comprehensive two-dimensional gas chromatography. J. Chromatogr. A. 2013, 1301: 225-236.

[38] Seeley J. V, Seeley S. K, Libby E. K, Breitbach Z. S, Armstrong D. W, Comprehensive two-dimensional gas chromatography using a high-temperature phosphonium ionic liquid column. Anal. Bioanal. Chem. 2008, 390: 323-332.

[39] Siegler W. C, Crank J. A, Armstrong D. W, Synovec R. E, Increasing selectivity in comprehensive three-dimensional gas chromatography via an ionic liquid stationary phase column in one dimension. J. Chromatogr. A 2010, 1217: 3144-3149.

[40] Nosheen A, Mitrevski B, Bano A, Marriott P. J, Fast comprehensive two-dimensional gas chromatography method for fatty acid methyl ester separation and quantification using dual ionic liquid columns. J. Chromatogr. A 2013, 1312: 118-123.

[41] Rodríguez-Sánchez S, Galindo-Iranzo P, Soria A. C, Sanz M. L, Quintanilla- López J. E, Lebrón-Aguilar R, Characterization by the solvation parameter model of the retention properties of commercial ionic liquid columns for gas chromatography J. Chromatogr. A 2014, 1326: 96-102.

[42] Yuan J.Y, Mecerreyes D, Antonietti M, Poly (ionic liquids): An update. Prog. Polym. Sci. 2013, 38: 1009-1036.

[43] Lin H, Zhang S, Sun J.-K, Antonietti M, Yuan J, Poly (ionic liquids) with engineered nanopores for energy and environmental applications. Polymer. 2020, 202: 122640.

[44] Zulfiqar S, Sarwar M.I, Mecerreyes D, Polymeric ionic liquids for CO2 capture and separation: Potential,progress and challenges. Polym. Chem. 2015, 36: 6435-6451.

[45] Morinaga T, Honma S, Ishizuka T, Kamijo T, Sato T, Tsujii Y, Synthesis of Monodisperse Silica Particles Grafted with Concentrated Ionic Liquid-Type Polymer Brushes by Surface-Initiated Atom Transfer Radical Polymerization for Use as a Solid State Polymer Electrolyte. Polymers 2016, 8: 146.

[46] Gupta N, Liang Y.N, Hu X, Thermally responsive ionic liquids and polymeric ionic liquids:emerging trends and possibilities. Curr. Opin. Chem. Eng. 2019, 25: 43-50.

[47] García-Verdugo E, Altava B, Burguete M.I, Lozano P, Luis S.V, Ionic liquids and continuous flow processes: A good marriage to design sustainable processes. Green Chem. 2015, 17: 2693-2713.

[48] Escorihuela J, Olvera-Mancilla J, Alexandrova L, Del Castillo L. F, Compan V, Recent Progress in the Development of Composite Membranes Based on Poly benzimidazole for High Temperature Proton Exchange Membrane (PEM) Fuel Cell Applications. Polymers. 2020, 12, 1861.

[49] Villa R, Alvarez E, Porcar R, Garcia-Verdugo E, Luis S.V, Lozano P, Ionic liquids as an enabling tool to integrate reaction and separation processes. Green Chem. 2019, 21: 6527-6544.

[50] Patinha D.J.S, Wang H, Yuan J, Rocha S.M, Silvestre A.J.D, Marrucho I.M, Thin Porous Poly(ionic liquid) Coatings for Enhanced Headspace Solid Phase Micro extraction. Polymers 2020, 12: 1909.

[51] Tian Y, Feng X.L, Zhang Y.P, Yu Q, Wang X.H, Tian M.K, Determination of VolatileWater Pollutants Using Cross-Linked Polymeric Ionic Liquid as Solid Phase Micro-Extraction Coatings. Polymers. 2020, 12: 292.

[52] Nan H, Anderson J.L, Ionic liquid stationary phases for multidimensional gas chromatography. Trends Anal. Chem. 2018, 105: 367-379.

[53] Trujillo-Rodríguez M.J, Nan H, Varona M, Emaus M.N, Souza I.D, Anderson J.L, Advances of ionic liquids in analytical chemistry. Anal Chem. 2019, 91: 505-531.

[54] Gónzalez-Álvarez J, Blanco-Gomis D, Arias-Abrodo P, Díaz-Llorente D, Busto E, Ríos-Lombardía N, Gotor-Fernández V, Gutierrez-Álvarez M.D, Polymeric imidazolium ionic liquids as valuable stationary phases in gas chromatography: Chemical synthesis and full characterization. Anal. Chim. Acta. 2012, 721: 173-181.

[55] Poole C.F, Poole S.W, Ionic liquid stationary phases for gas chromatography. J. Sep. Sci. 2011, 34: 880-900.

[56] Cagliero C, Mazzucotelli M, Rubiolo P, Marengo A, Galli S, Anderson J.L, Sgorbini B, Bicchi C, Can the selectivity of phosphonium based ionic liquids be exploited as stationary phase for routine gaschromatography? A case study: The use of trihexyl(tetradecyl) phosphonium chloride in the flavor, fragrance and natural product fields. J. Chromatogr. A 2020, 1619: 460969.

[57] Twu, P, Zhao Q , Pitner W.R, Acree W.E, Baker G.A, Anderson J.L, Evaluating the solvation properties of functionalized ionic liquids with varied cation/anion composition using the solvation parameter model. J. Chromatogr. A 2011, 1218: 5311-5338.

[58] Rodríguez-Sánchez S, Galindo-Iranzo P, Soria A.C, Sanz M.L, Quintanilla-López J.E, Lebrón-Aguilar R, Characterization by the solvation parameter model of the retention properties of commercial ionic liquid columns for gas chromatography. J. Chromatogr. A 2014, 1326: 96-102.

[59] Shashkov M.V, Sidelnikov V.N, Orthogonality and Quality of GC_GC Separations for Complex Samples with Ionic Liquid Stationary Phases in First Dimension. Chromatographia. 2019, 82: 615-624.

[60] Poole C.F, Lenca N, Gas chromatography on wall-coated open-tubular columns with ionic liquid stationary phases. J. Chromatogr. A. 2014, 1357: 87-109.

[61] Odugbesi G.A, Nana H, Soltani M, Davis J.H. Jr, Anderson J.L, Ultra-high thermal stability perarylated ionic liquids as gas chromatographic stationary phases for the selective separation of polyaromatic hydrocarbons and polychlorinated biphenyls. J. Chromatogr. A. 2019, 1604: 460-466.

[62] Roeleveld K, David F, Lynen F, Comparison between polymerized ionic liquids synthesized using chain-growth and step-growth mechanisms used as stationary phase in gas chromatography. J. Chromatogr. A. 2016, 1451: 135-144.

[63] Zhao Q , Anderson J.L, Highly selective GC stationary phases consisting of binary mixtures of polymeric ionic liquids. J. Sep. Sci. 2010, 33: 79-87.

[64] Ho W-Y, Hsieh Y.-N, Lin W.-C, Kao C.L, Huang P.C, Yeh C.F, Pane C.Y, Kuei C.H, High temperature imidazolium ionic polymer for gas chromatography. Anal. Methods. 2010, 2: 455-457.

[65] González-Álvarez J, Arias-Abrodo P, Puerto M, Viguri M.E, Pérez J, Gutiérrez-Álvarez M.D, Polymerized phosphonium-based ionic liquids as stationary phases in gas chromatography: Performance improvements by addition of graphene oxide. New J. Chem. 2015, 39: 8560-8568.

[66] Devanathan R, Recent developments in proton exchange membranes for fuel cells. Energy Environ.Sci.2008, 1:101-119.

[67] Quartarone E, Mustarelli P, Electrolytes for solid-state lithium rechargeable batteries: recent advances and perspectives. Chem.Soc.Rev.2011, 40: 2525-2540.

[68] Lissner E, De Souza W.F, Ferrera B, Dupont J, Oxidative Desulfurization of Fuels with Task-Specific Ionic Liquids. Chem.Sus.Chem. 2009, 2: 962-964.

[69] Fei Z.F, Zhao D.B, Geldbach T. J, Scopelliti R, Dyson P.J, Brønsted Acidic Ionic Liquids and Their Zwitterions: Synthesis, Characterization and pKa determination. Chem.Eur.J. 2004, 10: 4886-4893.

[70] Lin J.C.Y, Huang C.J, Lee Y.T, Lee K.M, Lin I.J.B, Carboxylic acid functionalized imidazolium salts: sequential formation of ionic, zwitterionic, acid-zwitterionic and lithium salt-zwitterionic liquid crystals. J. Mater. Chem. 2011, 21: 8110-8121.

[71] Nockemann P, Thijs B, Pittois S, Thoen J, Glorieux C, Van Hecke K, Van Meervelt L, Kirchner B, Binnemans K. Task-Specific Ionic Liquid for Solubilizing Metal Oxides. J. Phys. Chem. B 2006, 110: 20978-20992.

[72] Kuroda K, Satria H, Miyamura K, Tsuge Y, Ninomiya K, Takahashi K, Design of wall-destructive but membrane-compatible solvents. J. Am. Chem. Soc. 2017, 139: 16052-16055.

[73] Yoshizawa-Fujita M, Tamura T, Takeoka Y, Rikukawa M, Low-melting zwitterion: effect of oxyethylene units on thermal properties and conductivity. Chem. Commun., 2011, 47: 2345-2347.

[74] Straathof A. J. J, Transformation of biomass into commodity chemicals using enzymes or cells. Chem. Rev. 2014, 114: 1871-1908.

[75] Lange J.P, Price R, Ayoub P.M, Louis J, Petrus L, Clarke L, Gosselink H, Valeric biofuels: a platform of cellulosic transportation fuels. Angew. Chem. Int. Ed. 2010, 49: 4479-4483.

[76] Bond J.Q , Alonso D.M, Wang D, West R.M, Dumesic J.A, Integrated catalytic conversion of γ-valerolactone to liquid alkenes for transportation fuels. Science. 2010, 327: 1110-1114.

[77] Ragonese C, Tranchida P.Q , Sciarrone D, Mondello L, Conventional and fast gas Chromatography analysis of biodiesel blends using an ionic liquid stationary phase. J.Chromatogr. A 2009, 1216: 8992-8997.

[78] Baltazar Q .Q , Leininger S.K, Anderson J.L, Binary ionic liquid mixtures as gas chromatography stationary phases for improving the separation selectivity of alcohols and aromatic compounds. J. Chromatogr. A. 2008, 1182: 119-127.

[79] Joshi M. D, Anderson J. L, Recent advances of ionic liquids in separation science and mass spectrometry. RSC Adv. 2012, 2: 5470-5484.

[80] Anderson J. L, Armstron D. W, Wei G.T, Ionic liquids in Analytical Chemistry. Anal. Chem. 2006, 78: 2892-2902.

[81] Wang Y, Tian M, Bi W, Row K. H, Application of Ionic Liquids in High Performance Reversed-Phase Chromatography. I.J.M.S. 2009, 10: 2591-2610.

[82] Han X, Armstrong D. W, Ionic Liquids in Separations. Accounts Chem. Res. 2007, 40: 1079-1086.

[83] Tang S, Baker G. A, Zhao H, PEG-functionalized ionic liquids for cellulose dissolution and saccharification. Chem. Soc. Rev. 2012, 41: 4030-4066.

[84] Plechkova N. V, Seddon K. R, Applications of ionic liquids in the chemical industry. Chem. Soc. Rev. 2008, 37: 123-150.

[85] Cagliero C, Bicchi C, Ionic liquids as gas chromatographic stationary phases: how can theychange food and natural product analyses?, Analytical and Bioanalytical Chemistry 2020, 412: 17-25.

# 14

# Chromatographic Techniques for Estimation of Aflatoxins in Food Commodities

*Mateen Abbas*

**Abstract**

Aflatoxins, produced mainly by Aspergillus *flavus* Aspergillus *parasiticus*, have been documented as one of the major food contaminants throughout the world. Because of their toxic nature, these food contaminants have acknowledged considerable attention in recent years. Among the different types of Aflatoxins, the most prevalent and predominant Aflatoxins are AFB1, AFB2, AFG1, AFG2, AFM1, AFM2 which are considered the more lethal as compared to others. Several analytical and immunological methods are available for testing and estimating aflatoxins in different food commodities. However, chromatographic techniques have been considered superior regarding the estimation of aflatoxins both qualitatively and quantitatively. Chromatographic techniques have numerous applications for the separation and identification of chemical and biological compounds in food industry. It has grown to be the most popular and versatile of all analytical techniques in laboratories used for the analysis of multiple components in different matrices. For preliminary qualitative detection of Aflatoxins, Thin layer chromatography (TLC) is considered the best analytical technique which is being used broadly in food industry. However, liquid chromatographic techniques including High Performance Liquid Chromatography (HPLC) and Liquid chromatography-mass Spectrometry (LC–MS) are the best analytical techniques developed so far for the quantification of Aflatoxins in food commodities.

**Keywords:** Food, Aflatoxins; TLC, HPLC, LC–MS

## 1. Introduction

Aflatoxins are toxic substances formed by certain kind of fungi (molds) that have the potential to contaminate food, feed, crops and pose a serious health risk to humans and livestock. Aflatoxins are also assumed to be responsible for the annual loss of 25% or more of the world's food crops, which has significant economic implications. Various procedures for the detection and analysis of aflatoxins are available in feed and food, as they are highly specific, practical, and useful [1].

Aflatoxins are cancerous secondary metabolites produced primarily by Aspergillus *flavus* and Aspergillus *parasiticus* in foodstuff. Basically the chemical composition of aflatoxins contain the difurano-coumarin molecules which are synthesized following the polyketide pathway [2]. Eighteen different types of aflatoxins have been identified however six are well-known and recognized as B1, B2,

G1, G2, M1, and M2, respectively, [3]. Aflatoxins B1, B2, G1 & G2 found in different kinds of food & feed which are metabolized in animal body and then excreted mainly via milk as aflatoxins M1 and M2. All these aflatoxins have molecular differences; the aflatoxin B-group (B1 and B2) contains the cyclopentane ring which shows blue fluorescence under ultraviolet (UV) light whereas the aflatoxin G-group (G1 and G2) comprises the lactone ring and shows yellow-green fluorescence under UV light [4]. The different color fluorescence is important for identifying and differentiating between the aflatoxins B & G groups. Aflatoxin B1 is most commonly found in different kind of food matrixes [5] and widespread maximally [6, 7] in the world and accounts for 75% of all aflatoxins contamination in food commodities [8]. Aflatoxins M1 and M2 are hydroxylated products (metabolites) of aflatoxins B1 and B2, respectively, which are concomitant with animal milk upon ingestion of aflatoxins B1 and B2 contaminated feed. Furthermore, once converted from B1 and B2 forms, aflatoxins M1 and M2 remain stable during milk processing [9].

To protect consumer health, maximum levels (MLs) for mycotoxins in foodstuffs have been established worldwide. In particular, the European Union legislation (often considered as the most stringent one) has established MLs for aflatoxins [10].

International organizations for example AOAC (Association of Official Analytical Chemists), CEN (European Committee for Standardization) and ISO (International Organization for Standardization) have continued rendezvous experts over the years to develop internationally recognized analytical standards. The main objective is to evade the discrepancies in outcomes that may arise from the use of different analytical methodologies, with the risk to partial worldwide food trade. Currently, seventy-two official methods are offered from these organizations for scrutinizing the mycotoxins in food commodities.

A variety of methods to detect aflatoxins in food and feed are available for different needs and different techniques for their detection and analysis have been extensively researched to develop those that are highly specific, useful and practical.

Owing to its common incidence and toxic nature, numerous analytical and immunological methods were developed. However, there are minor modifications actually in most of these analytical methods from the officially adopted basic methods for certain food commodities. They differ only in the analytical techniques used for assessing the strength of fluorescence of the analyzed mycotoxins and in the extraction solvents used to extract the mycotoxins from different food matrixes. A plethora of methods are available for different needs, ranging from techniques/methods for regulatory control in Official laboratories starting from simple rapid test kits (AgraStrip®, CHARM EZ-M) to advanced methods [including immunochemical methods comprises radioimmunoassay (RIA), Enzyme-Linked Immunosorbent Assay (ELISA), Immunoaffinity column assay (ICA), Immunodipstick and immunosensors; Spectroscopic methods including Fluorometer, Spectrophotometer, Fourier-transform infrared spectroscopy (FTIR), Quartz Crystal Microbalances (QCMs), Surface Plasmon resonance Spectroscopy (SPRS); and some Chromatographic methods such as Thin Layer Chromatography (TLC) with densitometer, High-performance Thin Layer Chromatography (HPTLC), High-Performance Liquid Chromatography (HPLC), Gas Chromatography (GC), Liquid Chromatography-Mass Spectrometry (LC–MS)] for factories and grain silos (**Table 1**).

Quickness and effortlessness in analysis are the other features that have gain worldwide consideration recently. When a large number of samples have to be analyzed in a short time period then enzyme-linked immunosorbent assay (ELISA), mini-column quick methods, and radio-immunoassay (RIA) techniques may be used.

| Method | Sample preparation | LOD | Field Applicable | Reference |
|---|---|---|---|---|
| AgraStrip® | Simple extraction with Methanol | 4 ppb | Appropriate | [11] |
| Immunodipstick | Extraction only | 5 μg/Kg | Appropriate | [12] |
| CHARM EZ-M | Water based extraction | 1 ppb | Inappropriate | [13] |
| Fluorometer | IAC | 5–5000 μg/Kg | Inappropriate | [14] |
| TLC densitometer | Liquid Extraction, SPE | 1–20 ng/kg | Inappropriate | [15, 16] |
| HPTLC | Liquid Extraction | Pictogram | Inappropriate | [17] |
| RIA | Liquid Extraction | 1 μg/Kg | Inappropriate | [18] |
| FTIR | Liquid Extraction | <10 μg/Kg | Inappropriate | [19] |
| ELISA | Liquid Extraction | 1 μg/Kg | Inappropriate | [20] |
| HPLC | IAC or SPE | 0.5 μg/Kg | Inappropriate | [21] |
| LC-MS | Liquid Extraction | 0.1 μg/Kg | Inappropriate | [22] |
| QCMs | Liquid Extraction | 0.01–10 ng/mL | Inappropriate | [23, 24] |
| SPRS | Liquid Extraction | 3.0–98 ng/mL | Inappropriate | [25] |
| Electrochemical | Liquid Extraction | 0.1–2 μg/Kg | Inappropriate | [26, 27] |

**Table 1.**
*Evaluation of different methods for analysis of aflatoxins in food and feed.*

Potential innovative aflatoxins-detection techniques, based on the emerging techniques, include electronic noses, dip-stick kits, molecularly imprinted polymers, hyper-spectral imaging, and aptamer-based biosensors (small organic molecules that can bind specific target molecules). The latter techniques may have significance in remote areas because of their use, stability and ease of production. However, any method recommended for aflatoxin analysis should be economical and convenient to the handlers, taking into account their available laboratory facilities, as well as providing greater accuracy in the results.

All analytical methods for aflatoxins involve basically the same steps: sampling and sample extraction, clean-up, work-up, detection, and confirmation, as well as estimation of the toxin.

## 2. Sampling procedures are problematic

Adequate sampling techniques as well as appropriate sample preparation procedures are the most significant steps before performing the chemical analysis of aflatoxins. Aflatoxins are present in only a few grains and kernels obviously and have highly crooked distribution in food and feed commodities therefore, some variations in analytical results might be possible if the sample collected for analysis is not representative of the bulk [28–31].

As molds and aflatoxins aren't equally dispersed all through the bulk shipments and batches of stored grain, proper sampling is essential for obtaining a representative result. Proper protocols for sampling have been established, particularly in the perspective of regulatory control. For example, the Codex Alimentarius Commission has set the protocols to be used for various food commodities in setting maximum levels for aflatoxins. The United Nation's Food and Agriculture Organization (FAO) has established a mycotoxins sampling contrivance that is available on line. The use of recommended sampling methods is a problem, especially for subsistence farmers in rural areas who do not produce enough grain to allow for accurate testing. As a result, to improve surveillance and control in rural areas, low-cost, rapid and low-technology aflatoxins detection techniques are required. Food organizations trying their best to control aflatoxins in Africa and the World Food Programme's are also addressing these issues, for example, the World Food Programme has introduced the appropriate Purchase guidelines to ensure grain quality.

A precise and accurate sample can be selected by collecting a representative sample in large quantity and then dividing it into three equal parts. Differences in weight of selected samples may also be critical which depend on the regulations of a specific country. For example; the United Kingdom (UK) has proposed a sample weight of 10.5 kg, while the United States (US) has recommended the sample weight of 66 kg, greatly a larger amount. However, an average sample weight of 5–10 kg has been adopted by most of the countries. Precise grinding and sub-division of the sample would also be critical for accurate determination of aflatoxins. Spinning riffles, rotary sample divisors, and cascade samplers may also be used to select the representative sub-samples [31–33]. The size of the sub-samples may vary from 20 to 100 g. However, in most of the methods 50 g sample was used for analysis of aflatoxins, which looks to be the best in terms of economy in using costly extraction solvents.

## 3. Aflatoxins extraction and clean-up methods

The frequently used extraction and clean-up techniques for aflatoxins analysis are liquid–liquid extraction (LLE), solid-phase extraction (SPE) and "Quick, Easy, Cheap, Effective, Rugged, and Safe" (QuEChERS) methods. Furthermore, some other extraction methods are also offered in the literatures that are not commonly used in routine analysis at the moment.

### 3.1 Liquid–Liquid extraction (LLE)

Liquid–liquid extraction procedures are the simple, easy and cheap methods for the extraction of aflatoxins. It is based on the partition coefficient and different solubility properties of the mycotoxin in the organic or aqueous phase or in their combination mixtures. However, the shortcomings of these extraction techniques are that it does not provide appropriately clean analyte in all cases.

An efficient extraction method is required for the qualitative detection and quantification of aflatoxins in food and feed samples. Aflatoxins are commonly soluble in the polar-protic solvents like acetone, acetonitrile, chloroform and methanol. Hence, aflatoxins can be extracted by using either any of the mentioned pure solvents or in combination of these solvents as well as with small quantity of water [34, 35]. Several studies have been conducted on different food matrices to determine the extraction efficiency of various aqueous-organic solvents [36–38] and the different extraction recoveries have been reported. Since methanol has a minor negative effect on anti-bodies than other organic solvents like acetone and acetonitrile therefore aflatoxins extraction using a mixture of methanol with water (e.g.; 8 + 2 v/v) [37, 39] is required for determination of aflatoxins on immunoassay technique.

### 3.2 Liquid–Solid extraction (LSE)

Liquid–solid extraction technique is another simple and easy extraction method for the extraction of aflatoxins using solid matrices of different consistency.

Initially, the selection of an appropriate and the most effective extraction solvents is a crucial step to extract the component of interest. Most frequently used extraction solvents are mixtures of methanol/water or acetonitrile/water in different ratios [40]. For instance, the 80% methanol/water mixture proved to be the most optimal for extraction of aflatoxins in the case of nutmeg samples. The choice of methanol for further use is also preferable, because the antibodies better tolerate higher concentrations of methanol than acetonitrile. Methanol was also suitable for chromatographic separation, as aflatoxins were measurable without interference [41]. The extraction efficiency is significantly influenced by the composition of the extraction agents, the sample/solvent ratio, and the time of extraction. Sometimes, the use of only LSE method is inadequate to extract aflatoxins without interference and additionally some purification step(s) are required for proper extraction. The extraction process comprises the different steps including the weighing of homogenized sample which will be properly grind having appropriate particle size, addition of suitable extraction solvents and then dissolution or disintegrating the mixture applying, e.g., vortex, blender, shaker, or other approaches to extract the required components. After extraction, sample is filtered and cleaned prior to analysis.

### 3.3 Ultrasound extraction

Liquid–solid extraction efficiency can substantially be improved with the use of ultrasound. In the ultrasound extraction process the container (e.g., flask, centrifuge tube or vials) containing the sample to be extracted and the extraction solvent is most often immersed into an ultrasonic bath that contains water. After a few minutes, the acoustic cavitation under the influence of ultrasonic field in liquids significantly increases the transfer of the analytes and matrix components from the sample to the extraction solvent, thereby increasing the recovery and efficiency of extraction [42].

### 3.4 Supercritical fluid extraction (SFE)

Supercritical fluid extraction uses a supercritical fluid for the extraction of the required compound from the matrix. The SFE procedure is mainly used efficiently for the extraction of apolar organic molecules [43]. During the extraction of polar aflatoxins with SFE a number of problems have arisen, e.g., low recoveries and high concentrations of co-extracts. Furthermore, lipids may cause difficulties during further cleanup and chromatographic separation [44].

### 3.5 Solid phase extraction (SPE)

Solid phase extraction techniques are considered the most accurate and reliable approaches to clean-up the mixtures before qualitative and quantitative estimations. With the help of SPE, required analyte can be separated and unwanted components which may interfere during analysis can be removed accurately. Two types of SPE are used.

SPE is a multi-step process, starting from the conditioning then followed by the sample loading, washing and at the end elution of required analyte. In the SPE, the required analyte either bound to the matrix component(s) or removed from the sample [45]. Various extenders are used in the SPE columns. Aflatoxins are often analyzed by using C-18 (octadecylsilane) column. A specific application of SPE is the so-called immunoaffinity clean-up columns (IAC) and Multi-functional clean-up columns (MFC) including MultiSep®, MycoSep®, and Myco6in1 column [46]. The extraction of aflatoxins is usually followed by a cleanup step. The common cleanup technique used is immunoaffinity column (IAC) chromatography and Mycosep multi-functional cleanup (MFC) columns [47]. These purification techniques are considered the best choice for isolation of target analyte (like aflatoxins) and to clean-up or remove the unwanted components before their quantitative estmation using HPLC [48].

Immunoaffinity chromatographic technique proved to be the accurate and highly specific which reversibility of binding between an antigen and antibody to isolate, purify and separate the target molecule from matrices [49]. During the cleanup process, the extracted liquid sample is applied to the IAC which holding the specific antibodies to bind with aflatoxins that immobilized on a solid surface such as silica or agarose. As the extracted sample moves down the IAC column, the aflatoxins bind to the antibodies and are retained onto the column. To remove the unbound proteins and impurities washing step is generally required using appropriate ionic strength buffers or distilled water. Thereafter, the aflatoxins are recovered or removed from the IAC by using pure solvents like acetonitrile or methanol which breaks the bond between the antibody and the aflatoxins.

Mycosep multi-functional cleanup (MFC) columns are also recognized a best approach for purification of aflatoxins. It is simple, easy, handy to use and a rapid one-step purification technique. These columns are designed to retain certain groups of basic compounds that may create interferences in HPLC analysis. On the other hand, MFC purification columns allow the molecules of interest to pass through the columns. During the MFC cleanup procedure, after extraction of aflatoxins using suitable solvents a portion of the extract is passed through an MFC column designed particularly for aflatoxins analysis. Compounds that may create interferences are retained in MFC, whereas aflatoxins pass through the column. Ideal recovery (> 95%), precision and coefficient of variation (< 3%) of aflatoxins were observed by these columns [50].

## 4. Work-up

After the clean-up step, the extract must be worked up to make it suitable for the estimations. The purified pooled extract can be treated with sodium sulphate (anhydrous) to remove the moisture if present in the extract. To concentrate the extracted solvent evaporated it to dryness using nitrogen stream or in a rotary evaporator at 50°C. On the other hand, evaporation of solvents can be achieved with the help of steam bath under the nitrogen stream preferably. Finally, reconstituted the residues using pure organic solvents like acetonitrile or methanol and used for estimations.

## 5. Methods for detection and quantification of aflatoxins

The most commonly used chromatography techniques for analysis of aflatoxins are Thin-Layer Chromatography (TLC), High-Performance Liquid Chromatography (HPLC), Gas chromatography (GC) and Liquid ChromatographyMass Spectrometry (LC–MS). Although many of the chromatographic techniques are very sensitive, they require trained skilled technician, cumbersome pretreat-ment of sample, and expensive apparatus/ equipment [51].

### 5.1 Thin-layer chromatography (TLC)

Thin-layer chromatography is one of the most widely used separation techniques for detection of aflatoxins. TLC has been regarded by the Association of Official

Analytical Chemist (AOAC) as the method of choice since 1990. It consists of a solid immobilized stationary phase may contain either alumina or cellulose or silica on an inert material such as plastic or glass, called the matrix. The mobile phase is contained of acetonitrile: methanol: water mixture [52], which brings the sample along as it moves through the stationary phase. In TLC, aflatoxins are distributed between the mobile phase and stationary phases on the basis of partition coefficient or differences in solubility of the analytes in the two phases. Different types of aflatoxins (B1, B2, G1 & G2), according to their interaction with the stationary and mobile phases as well as due to the different molecular structures, either adhere to the solid surface of stationary phase more or remain in the mobile phase, thus allowing for effective and quick separation. TLC technique has been commonly used in food industry for the determination of aflatoxins [53–55] and detection limit of 1–20 ppb of different types of aflatoxins has been reported. The major advantage the TLC is that it can detect different types of mycotoxins with good resolution and excellent sensitivities [56]. It also requires pre-treatment of sample, skilled and trained technician, and expensive equipment as well [57]. In addition, there are also some drawbacks of TLC which may probably be occurring during spotting, TLC plate development, and interpretation.

Quantification of aflatoxins on TLC plates using fluoro-densitometer is considered to be a more precise and accurate method than visual estimates [58] with the minimum limit of detection (LOD) is 1 μg/kg. Although fluoro-densitometers are commercially available, but not commonly used due to its high cost and visual fluorescence identification method is still to be continue for identification of aflatoxins [59].

Attempts to improve TLC have led to the development of automated form of TLC, called the high-performance thin-layer chromatography (HPTLC).

### 5.2 High performance thin layer chromatography (HPTLC)

The conventional TLC method has improved through the automation of sample spotting, plate development and interpretation in HPTLC. Currently, HPTLC is one of the best analytical methods for estimation of aflatoxins [60, 61].

Automated sample applicator, digital scanner, and a computing integrator, lead to improve the sensitivity and precision in the quantification of aflatoxins. The other benefit of HPTLC method is the use of minimum amount (only 1 μl) for sample spotting, instead of 10-20 μl used for the conventional TLC method. With the use of HPTLC minimum concentrations of aflatoxins (5 pg) can be possibly detected [59].

However, the costly equipment, extensive sample treatment procedure and the requirement for skilled researcher, limit the HPTLC technique to the laboratory and thus it is inapplicable in field situations.

### 5.3 High-performance liquid chromatography (HPLC)

The most commonly used chromatographic technique for separation and determination of organic compounds is High Performance Liquid Chromatography (HPLC). Worldwide, approximately 80% of all organic compounds are estimated using HPLC [62].

The HPLC technique for estimation of aflatoxins has high automation, high sensitivity and high precision. There are two types of phase systems comprising normal phase (wherein mobile phase: non-polar & stationary phase: polar) and reverse phase (wherein mobile phase: polar & stationary phase: non-polar)) in combination with UV/VIS absorption and fluorescence detection. Reverse phase HPLC is broadly used for estimation of aflatoxins [59].

In HPLC, the stationary phase is confined to either a plastic or glass tube and the mobile phase containing the organic/aqueous solvents that pass through the solid adsorbent. The sample to be examined is introduced on top of the column which passes through and distributes between both the stationary and mobile and phases.

The components present in the sample move through the column with the mobile phase at different speeds because of their different relative affinities and interactions. Separate fractions containing individual components in the sample elute from the HPLC column at different rates. The HPLC technique involves the use of a stationary phase (polar or non-polar columns), a pump that moves the mobile phase(s) through the column or other parts of HPLC at constant flow rate, a degasser to remove the trapped gases or air bubbles in the mobile phase, a detector to quantify the analytes and read out device to display the retention times of individual components.

Reversed phase chromatographic mode is most commonly used in HPLC for the identification and quantification of aflatoxins. Chemical derivatization of aflatoxins B1 and G1 typically required to improve the sensitivity because the natural fluorescence of aflatoxins B1 and G1 may be inadequate to meet the necessary detection limit [63]. **Figure 1** depicts derivatization reactions of aflatoxin B1 with the acid and halogens. In the first reaction, Trifluoro Acetic Acid (TFA) hydrolyzes the second furan ring of aflatoxin B1 to produce highly fluorescent aflatoxin B2a, while bromine and iodine are used as chemical reagents in the second and third derivatization reactions, respectively. When these halogens react with aflatoxin B1, they produced highly fluorescent aflatoxin B1 derivatives.

HPLC provides quick, accurate and reliable aflatoxins results within a short time. FLD has been presented an excellent sensitivity of 0.1 ng/kg [65]. However, the shortcoming of using HPLC to analyze the aflatoxins is the requirement of laborious purification columns to clean-up the sample. Furthermore, HPLC involves the tedious pre-column or post-column derivatization processes to improve the sensitivity of aflatoxins [62]. To overcome the challenges of derivatization processes

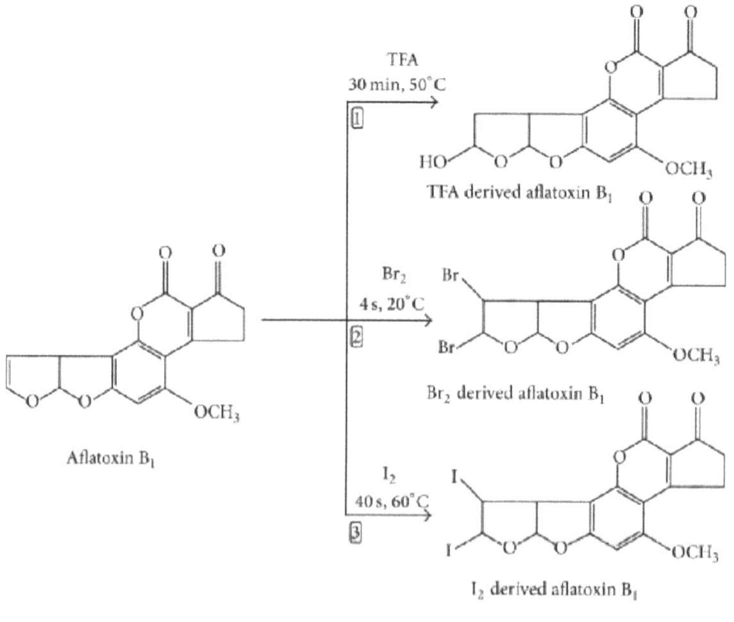


**Figure 1.**
*Derivatization of aflatoxin B1 with trifluoroacetic acid, bromine ($Br_2$) and iodine ($I_2$) [64].*

in aflatoxins testing, a modification of the HPLC protocol in which the HPLC is coupled to mass spectrometry has been developed and is currently used in aflatoxin determination [66].

Since the mass spectrometry does not requires the use of UV/VIS fluorescence or absorbance of analyte, thus chemical derivatization of compounds is no longer required. HPLC–MS/MS produces structural information using small amount of sample and has low detection limits developed up to now [21]. On the other hand, HPLC–MS/MS is costly equipment that can only be handled by trained, qualified and professional person. Furthermore, this also restricts its use to only well-equipped laboratory environment and not field conditions.

### 5.4 Liquid chromatography-mass spectrometry (LC: MS)

Although different HPLC methods are available for quantitative determination of aflatoxins with selective sample clean-up techniques, still the methods are required to confirm the identity of the substances. A method other than the commonly used UV/VIS and fluorescent methods, for the confirmation is mass spectrometry method that coupled with HPLC.

LC–MS technique has become the fastest growing technique available for analysis of mycotoxins. The potential benefits of LC–MS technique for mycotoxin analysis have long been recognized and exploited. Simultaneous determination of multi-mycotoxins can be possible with LC–MS according to the mass to charge ratio (m/z) of analytes, an intrinsic property that provides more specific identification based on molecular weight of the target analyte. The impact of modern LC–MS technique has been signified by the unmatchable sensitivity in quantitation, specificity in identification and number of mycotoxins that could be analyzed in one analysis [67].

A modern LC–MS instrument, particularly LC–MS-triple quadrupole (LC–MS-QQQ ), has been developed and introduced with increasing sensitivity for quantitative analysis of mycotoxins. Despite high capital costs of LC–MS instruments, many efforts have been exerted to quantitate aflatoxins using this technique [68].

### 5.5 Gas chromatography (GC)

In gas chromatography, an inert gas is used as the mobile phase instead of liquid and the stationary phase may be a liquid coated onto inert solid particles or solid. GC analysis, like other chromatographic approaches, is based on the differential partitioning of analytes between the two phases. The stationary process is made up of inert particles covered with a liquid layer that confined in a long stainless steel or glass tube known as a column fixed in oven to maintain the specific temperature. The sample to be tested is vaporized into a gaseous form and transported by a carrier gas into the stationary phase.

The different chemical components within the sample will distribute themselves between the stationary phase and mobile phase. Components of the sample mixture with a higher affinity for the stationary phase travel through the column more slowly, while those with a lower affinity move through the column faster. Each portion of the analyte should, in reality, have its own partition coefficient, which will dictate how quickly it passes through the column [69]. After the separation of volatile compounds, these are detected using a universal GC detector known as Flame Ionization Detector (FID) or an Electron Capture Detector (ECD) and the most recent and advanced mass spectrometer (MS) detector [70].

Since aflatoxins are non-volatile, thus derivatization will be required to be detected [71]. However, GC is not commonly used in commercial analysis of aflatoxins because some other cheaper and simple chromatographic techniques are existed [72]. Furthermore, gas chromatography is limited to the analysis of a few mycotoxins, such as A-trichothecenes and B-trichothecenes, due to the requirement of preliminary cleanup step prior to analysis. GC technique has some other disadvantages including drifting responses, non-linearity of calibration curves, memory effects from previous samples, and high variation in repeatability and reproducibility [73].

Earlier, gas chromatography mass spectrometry with negative ion chemical ionization has been used for confirmation of aflatoxin B1 [74], injection was applied using an on-column injector, which is necessary because of the thermoslability of the aflatoxins. Gas chromatography mass spectrometry have also been used with electron impact for aflatoxins B1, B2, G1 and G2 [75].

## 6. Conclusions

Several qualitative as well as quantitative methods have been explored for analysis of aflatoxins in food commodities, crops and feeds. Among all the different developed methods, chromatographic techniques are considered the most appropriate methods in aflatoxins analysis. Analytical methods based on immunochemistry and spectroscopy have also been added to the chromatographic methods, some of which emerged as better alternatives for routine and on-site estimation of aflatoxins. Although a large number of analytical techniques are constantly being optimized, the LC/MS/MS technique is considered the most valuable confirmation technique for analyzing multiple mycotoxins as it is high specific, sensitive, accurate and reliable.

## Acknowledgements

I would like to thanks the Director, Quality Operations Laboratory (QOL), Institute of Microbiology, University of Veterinary and Animal Sciences (UVAS) and all colleagues for helpful suggestions and support for the preparation of manuscript.

## Author details

Mateen Abbas
Quality Operations Laboratory (QOL), Institute of Microbiology, University of Veterinary and Animal Sciences, Lahore, Pakistan

*Address all correspondence to: mateen.abbas@uvas.edu.pk

## References

[1] Evaluation of certain contaminants in food (Eighty-third report of the Joint FAO/WHO Expert Committee on Food Additives). WHO Technical Report Series WHO Technical Report Series, No.1002, 2017.

[2] Bennett JW, Klich M. "Mycotoxins," Clinical Microbiology Reviews. 2003; 16(3): 497-516.

[3] Dors GC, Caldas S, Feddern V. "Aflatoxins: contamination, analysis and control," in Aflatoxins-Biochemistry and Molecular Biology. InTech, Shanghai, China, 2011. 415-438 p.

[4] Gourama H, Bullerman LB. "Detection of molds in foods and feeds: potential rapid and selective methods," Journal of Food Protection. 1995; 58(12): 1389-1394.

[5] Hussein HS, Brasel JM. "Toxicity, metabolism, and impact of mycotoxins on humans and animals," Toxicology. 2001; 167(2): 101-134.

[6] Cullen JM, Newberne PM, Eaton D, Groopman J. "Acute hepatotoxicity of aflatoxins," in The Toxicology of Aflatoxins: Human Health, Veterinary and Agricultural Significance. 1993; 3-26.

[7] Kok WT. "Derivatization reactions for the determination of aflatoxins by liquid chromatography with fluorescence detection," Journal of Chromatography B: Biomedical Sciences and Applications. 1994; 659(1-2): 127-137.

[8] Ayub M, Sachan D. "Dietary factors affecting aflatoxin bi carcinogenicity," Malaysian Journal of Nutrition. 1997; 3: 161-197.

[9] Stroka J, Anklam E. "New strategies for the screening and determination of aflatoxins and the detection of aflatoxin producing moulds in food and feed," TrAC—Trends in Analytical Chemistry. 2002; 21(2): 90-95.

[10] European Union. Commission Regulation (EC) No 1881/2006 of 19 December 2006 setting maximum levels for certain contaminants in foodstuffs. Off. J. Eur. Union 2006, L 364, 5-24.

[11] Romer Labs Division Holding GmbH Erber Campus 13131 Getzersdorf, Austria, https://www.romerlabs. com/en/ analytes/ mycotoxins/aflatoxin-testing/

[12] Luppa PB, Sokoll LJ, Chan DW. "Immunosensor-principles and applications to clinical chemistry," Clinica Chimica Acta. 2001; 314 (1-2): 1-26.

[13] Charm Sciences, Inc. (Headquarters) 659 Andover Street Lawrence, MA 01843-1032 USA, https://www.charm.com/products/test-and-kits/mycotoxin-tests/

[14] Malone BR, Humphrey CW, Romer TR, Richard JL. "Determination of aflatoxins in grains and raw peanuts by a rapid procedure with fluorometric analysis," Journal of AOAC International. 2000; 83(1): 95-98.

[15] Balzer I, Bogdani´c C, Pepeljnjak S. "Rapid thin layer chromatographic method for determining aflatoxin B1, ochratoxin A, and zearalenone in corn," Journal of the Association of Official Analytical Chemists. 1978; 61(3): 584-585.

[16] Park DL, Trucksess MW, Nesheim S, Stack M, Newell RF. "Solvent-efficient thin-layer chromatographic method for the determination of aflatoxins B1, B2, G1, and G2 in corn and peanut products: collaborative study," Journal of AOAC International. 1994; 77(3): 637-646.

[17] Liang Y, Baker ME, Yeager BT, Denton MB. "Quantitative analysis of aflatoxins by high-performance thin layer chromatography utilizing a scientifically operated charge coupled device detector," Analytical Chemistry. 1996; 68(22): 3885-3891.

[18] Sun P, Chu Y. "A simple solid-phase radioimmunoassay for aflatoxin b1," Journal of Food Safety. 1979; 1: 67-75.

[19] Pearson T, Wicklow D, Maghirang E, Xie F, Dowell F. "Detecting aflatoxin in single corn kernels by transmittance and reflectance spectroscopy," Transactions of the American Society of Agricultural Engineers. 2001; 44(5): 1247-1254.

[20] Omar SS, Moawiya AH, Salvatore P. "Validation of HPLC and Enzyme-Linked Immunosorbent Assay (ELISA) Techniques for Detection and Quantification of Aflatoxins in Different Food Samples", Foods. 2020; 9(661). doi:10.3390/foods9050661

[21] Rahmani A, Jinap S, Soleimany F. "Qualitative and quantitative analysis of mycotoxins," Comprehensive Reviews in Food Science and Food Safety. 2009; 8(3): 202-251.

[22] Kokkonen M, Jestoi M, Rizzo A. "Determination of selected mycotoxins in mould cheeses with liquid chromatography coupled tandem with mass spectrometry," Food Additives and Contaminants. 2005; 22(5): 449-456.

[23] Gaag BV, Spath S, Dietrich H. "Biosensors and multiple mycotoxin analysis," Food Control. 2003; 14(4): 251-254.

[24] Liron Z, Tender LM, Golden JP, Ligler FS. "Voltage induced inhibition of antigen-antibody binding at conducting optical waveguides," Biosensors and Bioelectronics. 2002; 17(6-7): 489-494.

[25] Yu H, Eggleston CM, Chen J, Wang W, Dai Q , Tang J. "Optical waveguide lightmode spectroscopy (OWLS) as a sensor for thin film and quantum dot corrosion," Sensors. 2012; 12(12): 17330-17342.

[26] Masoomi L, Sadeghi O, Banitaba MH, Shahrjerdi A, Davarani SSH. "A non-enzymatic nanomagnetic electroimmunosensor for determination of Aflatoxin B1 as a model antigen," Sensors and Actuators B: Chemical. 2013; 177: 1122-1127.

[27] Linting Z, Ruiyi L, Zaijun L, Qianfang X, Yinjun F, Junkang L. "An immunosensor for ultrasensitive detection of aflatoxin B1 with an enhanced electrochemical performance based on graphene/conducting polymer/gold nanoparticles/the ionic liquid composite film on modified gold electrode with electrodeposition," Sensors and Actuators B: Chemical. 2012; 174: 359-365.

[28] Whitaker TB, Dickens JW, Monroe RJ. Variability of aflatoxin test results. J Am Oil Chem Soc. 1974; 51: 214-218.

[29] Horwitz W, Albert R. The reliability of aflatoxin assay. Assoc Food Drug Officials Q Bull. 1982; 46: 14-24.

[30] Coker RD. Control of aflatoxin in groundnut products with emphasis on sampling, analysis and detoxification. In: McDonald D, Mehan VK, eds. Aflatoxin contamination of groundnut: Proceedings of the International Workshop. Patancheru, India: International Crops Research Institute for the Semi-Arid Tropics, 1989:124-132.

[31] Coker RD. Design of sampling plans for determination of mycotoxins in foods and feeds. In: Sinha KK, Bhatnagar D, eds. Mycotoxins in agricultural and food safety. New York: Marcel Dekker, 1998:109-133.

[32] Coker RD, Jones BD, Nagler MJ, Gilman GA, Wallbridge AJ, Panigrahi S. Mycotoxin training manual. London: Tropical Development and Research Institute, 1984.

[33] Coker RD. High performance liquid chromatography and other chemical quantification methods used in analysis of mycotoxins in foods. In: Gilbert J, ed. Analysis of food contaminants. London: Elsevier, 1984: 207-263 p.

[34] Bertuzzi T, Rastelli S, Mulazzi A, Pietri A. "Evaluation and improvement of extraction methods for the analysis of aflatoxins B1, B2, G1 and G2 from naturally contaminated maize," Food Analytical Methods. 2012; 5(3): 512-519.

[35] Taylor SL, King JW, Richard JL, Greer JI. "Analyticalscale supercritical fluid extraction of aflatoxin b1 from fieldinoculated corn," Journal of Agricultural and Food Chemistry. 1993; 41(6): 910-913.

[36] Stroka J, Petz M, Joerissen U, Anklam E. "Investigation of various extractants for the analysis of aflatoxin B1 in different food and feed matrices," Food Additives and Contaminants.1999; 16(8): 331-338.

[37] Arranz I, Sizoo E, Van Egmond H. "Determination of aflatoxin B1 in medical herbs: interlaboratory study," Journal of AOAC International. 2006; 89(3): 595-605.

[38] Gallo A, Masoero F, Bertuzzi T, Piva G, Pietri A. "Effect of the inclusion of adsorbents on aflatoxin B-1 quantification in animal feedstuffs," Food Additives and Contaminants—Part A Chemistry, Analysis, Control, Exposure and Risk Assessment. 2010; 27(1): 54-63.

[39] Lee NA, Wang S, Allan RD, Kennedy IR. "A rapid aflatoxin B1 Elisa: development and validation with reduced matrix effects for peanuts, corn, pistachio, and soybeans," Journal of Agricultural and Food Chemistry. 2004; 52(10): 2746-2755.

[40] Sheibani A, Ghaziaskar HS. Pressurized fluid extraction for quantitative recovery of aflatoxins B1 and B2 from pistachio. Food Contr. 2009; 20: 124-128. doi: 10.1016/j.foodcont.2008.03.001

[41] Kong WJ, Liu SY, Qiu F, Xiao XH, Yang MH. Simultaneous multi-mycotoxin determination in nutmeg by ultrasound-assisted solid liquid extraction and immunoaffinity column clean-up coupled with liquid chromatography and on-line post-column photochemical derivatization fluorescence detection. Analyst. 2013; 7: 2729-2739. doi: 10.1039/c3an00059a

[42] Xie L, Chen M, Ying Y. Development of methods for determination of aflatoxins. Food Sci. Nutr. 2016; 56: 2642-2664. doi: 10.1080/10408398.2014. 907234

[43] Anklam E, Berg H, Matthiasson L, Sharman M, Ulberth F. Supercritical fluid extraction (SFE) in food analysis: a review. Food Addit. Contam. 1998; 15: 729-750. doi: 10.1080/ 0265203 9809374703

[44] Shephard GS. Aflatoxin analysis at the beginning of the twenty-first century. Anal. Bioanal. Chem. 2009; 395: 1215-1224. doi: 10.1007/ s00216-009-2857-y

[45] Yao H, Hruska Z, Diana D, Mavungu J. Developments in detection and determination of aflatoxins. World Mycotoxin J. 2015; 8: 181-191.

[46] Tang YY, Lin YH, Chen CY, Su TW, Wang CS, Chiueh CL. Development of a quantitative multi-mycotoxin method in rice, maize, wheat and peanut using UPLC-MS/MS. Food Anal. Meth. 2013; 6: 727-736. doi: 10.1007/s12161-012-9473-8

[47] Ma F, Chen R, Li P, Zhang Q , Zhang W, Hu W. "Preparation of an immunoaffinity column with amino-silica gel microparticles and its application in sample cleanup for aflatoxin detection in agri-products," Molecules. 2013; 18(2): 2222-2235.

[48] Scott PM, Trucksess MW. "Application of immunoaffinity columns tomycotoxin analysis," Journal of AOAC International. 1997; 80(5): 941-949.

[49] Shelver WL, Larsen GL, Huwe JK. "Use of an immunoaffinity column for tetrachlorodibenzo-p-dioxin serum sample cleanup," Journal of Chromatography B: Biomedical Applications. 1998; 705(2): 261-268.

[50] Wilson TJ, Romer TR. Use of the Mycosep multifunctional cleanup column for liquid chromatographic determination of aflatoxins in agricultural products. J Assoc Off Anal Chem. 1991; 74(6): 951-956.

[51] Sapsford KE, Taitt CR, Fertig S. "Indirect competitive immunoassay for detection of aflatoxin B1 in corn and nut products using the array biosensor," Biosensors and Bioelectronics. 200; 21(12): 2298-2305.

[52] Betina V, "Thin-layer chromatography of mycotoxins," Journal of Chromatography. 1985; 334(3); 211-276.

[53] Gulyas H. "Determination of aflatoxins b1, b2, g1, g2 and m1 by high pressure thin layer chromatography," Journal of Chromatography A. 1985; 319: 105-111.

[54] Abdel-Gawad KM, Zohri AA. "Fungal flora and mycotoxins of six kinds of nut seeds for human consumption in Saudi Arabia," Mycopathologia,. 1993; 124(1): 55-64.

[55] Younis YMH, Malik KM. "Tlc and hplc analysis of aflatoxin contamination in sudanese peanut and peanut products," Kuwait Journal of Science and Engineering. 2003;30(1): 2003.

[56] Trucksess M, Brumley W, Nesheim S. "Rapid quantitation and confirmation of aflatoxins in corn and peanut butter, using a disposable silica gel column, thin layer chromatography, and gas chromatography/mass spectromety," Journal of the Association of Official Analytical Chemists. 1984; 67(5): 973-975.

[57] Papp E, H-Otta K, Z′aray G, Mincsovics E. "Liquid chromatographic determination of aflatoxins," Microchemical Journal. 2002; 73 (1-2): 39-46.

[58] Pons WA Jr, Goldblatt LA. Instrumental evaluation of aflatoxin resolution on TLC plates. J Assoc Off Anal Chem. 1968; 51: 1194-1197.

[59] Mehan VK. Analytical and immuno-chemical methods for the analysis of aflatoxins in groundnuts and groundnut products. In: Mehan VK, Gowda CLL, eds. Aflatoxin contamination problems in groundnut in Asia. Proceedings of the First Asia Working Group Meeting. Patancheru, India: International Crops Research Institute for the Semi-Arid Tropics. 1997; 49-65.

[60] Ramesh J, Sarathchandra G, Sureshkumar V. "Analysis of feed samples for aflatoxin b1 contamination by hptlc-a validated method," International Journal of Current Microbiology and Applied Sciences. 2013; 2: 373-377.

[61] Nawaz S, Coker RD, Haswell SJ. "Development and evaluation of analytical methodology for the determination of aflatoxins in palm kernels," Analyst. 1992; 117(1): 67-74.

[62] Li P, Zhang Q, Zhang D. "Aflatoxin measurement and analysis," in Aflatoxins—Detection, Measurement

and Control, I. Torres-Pacheco, Ed., Intech, Shanghai, China. 2011. 183-208 p.

[63] Kok WT, Van Neer TCH, Traag WA, Tuinstra LGT. "Determination of aflatoxins in cattle feed by liquid chromatography and post-column derivatization with electrochemically generated bromine," Journal of Chromatography A. 1986; 367: 231-236.

[64] Kok WT. "Derivatization reactions for the determination of aflatoxins by liquid chromatography with fluorescence detection," Journal of Chromatography B: Biomedical Sciences and Applications. 194; 659(1-2): 127-137.

[65] Herzallah SM. "Determination of aflatoxins in eggs, milk, meat and meat products using HPLC fluorescent and UV detectors," Food Chemistry. 2009; 114(3): 1141-1146.

[66] Takino M and Tanaka T. Determination of Aflatoxins in Food by LC-MS/MS, Agilent Technologies, 2008.

[67] Di Stefano V, Avellone G, Bongiorno D, Cunsolo V, Muccilli V, Sforza S, Dossena A, Drahos L, Vékey K. Applications of liquid chromatography—Mass spectrometry for food analysis. J. Chromatogr. A 2012; 1259: 74-85.

[68] Sforza, S, Dall'Asta C, Marchelli R. Recent advances in mycotoxin determination in food and feed by hyphenated chromatographic techniques/mass spectrometry. Mass Spectrom. Rev. 2005; 25: 54-76.

[69] Boyer RF. Modern Experimental Biochemistry, Benjamin/Cummings Publishing Company, 1993.

[70] Pascale MN. "Detection methods for mycotoxins in cereal grains and cereal products," Matica Srpska Proceedings for Natural Sciences, 2009. vol. 117, 15-25 p.

[71] Scott P. "Mycotoxin methodology," Food Additives & Contaminants, 1995. vol. 12, no. 3, 395-403 p.

[72] Liang Y, Zhang C, and Liu L. "Chromatographic analysis of mycotoxins," ZhongguoWeisheng Jianyan Zazhi; 2005. vol. 15, 1273 p.

[73] Petterson H and Langseth W. "Intercomparison of trichothecenes analysis and feasibility to produce certified calibrants," EU Reports EUR 20285, European Commission BCR Information Project, 2002.

[74] Trucksess MW, Brumley WC and Nesheim S. Rapid quantitation and confirmation of aflatoxins in com and peanut butter, using a disposable silica gel column, thin layer chromatography, and gas chromatography/mass spectrometry. J. Assoc. Off Anal. Chem., 1984; 67: 973-975.

[75] Goto T, Matsui M and Kitsuwa T. Analysis of Aspergillus mycotoxins by gas chromatography using fused silica capillary column. Maikotokishin. 1990; 31: 43-47.

# PERMISSIONS

All chapters in this book were first published by InTech Open; hereby published with permission under the Creative Commons Attribution License or equivalent. Every chapter published in this book has been scrutinized by our experts. Their significance has been extensively debated. The topics covered herein carry significant findings which will fuel the growth of the discipline. They may even be implemented as practical applications or may be referred to as a beginning point for another development.

The contributors of this book come from diverse backgrounds, making this book a truly international effort. This book will bring forth new frontiers with its revolutionizing research information and detailed analysis of the nascent developments around the world.

We would like to thank all the contributing authors for lending their expertise to make the book truly unique. They have played a crucial role in the development of this book. Without their invaluable contributions this book wouldn't have been possible. They have made vital efforts to compile up to date information on the varied aspects of this subject to make this book a valuable addition to the collection of many professionals and students.

This book was conceptualized with the vision of imparting up-to-date information and advanced data in this field. To ensure the same, a matchless editorial board was set up. Every individual on the board went through rigorous rounds of assessment to prove their worth. After which they invested a large part of their time researching and compiling the most relevant data for our readers.

The editorial board has been involved in producing this book since its inception. They have spent rigorous hours researching and exploring the diverse topics which have resulted in the successful publishing of this book. They have passed on their knowledge of decades through this book. To expedite this challenging task, the publisher supported the team at every step. A small team of assistant editors was also appointed to further simplify the editing procedure and attain best results for the readers.

Apart from the editorial board, the designing team has also invested a significant amount of their time in understanding the subject and creating the most relevant covers. They scrutinized every image to scout for the most suitable representation of the subject and create an appropriate cover for the book.

The publishing team has been an ardent support to the editorial, designing and production team. Their endless efforts to recruit the best for this project, has resulted in the accomplishment of this book. They are a veteran in the field of academics and their pool of knowledge is as vast as their experience in printing. Their expertise and guidance has proved useful at every step. Their uncompromising quality standards have made this book an exceptional effort. Their encouragement from time to time has been an inspiration for everyone.

The publisher and the editorial board hope that this book will prove to be a valuable piece of knowledge for researchers, students, practitioners and scholars across the globe.

# LIST OF CONTRIBUTORS

**Irma Pilar Herrera-Camacho, Lourdes Millán-Pérez-Peña and Nora Hilda Rosas Murrieta**
Biochemistry and Molecular Biology Laboratory, Chemistry Center, Science Institute, Autonomous University of Puebla, Puebla, México

**Francisca Sosa-Jurado**
Virology and Molecular Biology Laboratory, Eastern Biomedical Research Center, Mexican Social Security Institute, Puebla, México

**Nancy Martínez-Montiel and Rebeca Débora Martínez-Contreras**
Microbial Molecular Ecology Laboratory, Center for Research in Microbiological Sciences (CICM) - Science Institute, Autonomous University of Puebla, Puebla, Mexico

**Alice Abdel Aleem**
Weill Cornell Medicine Qatar, Brain and Mind Research Institute, Education City, Doha, Qatar

**Deepak Jose and Mahadevan Harikrishnan**
School of Industrial Fisheries, Cochin University of Science and Technology, Kerala, India

**Małgorzata Natonek-Wiśniewska and Piotr Krzyścin**
Department of Animal Molecular Biology, National Research Institute of Animal Production, Balice, Poland

**Benito Ramos González and Miranda Córdova Mercado**
Scientific Research Department, Criminalistics and Forensic Services Institute, Attorney General's Office of the State of Nuevo Leon, Monterrey, Mexico

**Orlando Salas Salas**
Forensic Genetic Laboratory, Criminalistics and Forensic Services Institute, Attorney General's Office of the State of Nuevo Leon, Monterrey, Mexico

**Juan Carlos Hernández Reyes and Martín Guardiola Ramos**
Evidence Analysis Laboratory, Criminalistics and Forensic Services Institute, Attorney General's Office of the State of Nuevo Leon, Monterrey, Mexico

**Elton Solis Esquivel**
Forensic Chemistry Laboratory and Quality Assurance, Criminalistics and Forensic Services Institute, Attorney General's Office of the State of Nuevo Leon, Monterrey, Mexico

**Gerardo Castellanos Aguilar and Porfirio Diaz Torres**
Crime Investigation Group, Criminalistics and Forensic Services Institute, Attorney General's Office of the State of Nuevo Leon, Monterrey, Mexico

**Jina Heikrujam**
Plant Biotechnology Laboratory, Department of Biotechnology, Assam University, Silchar, Assam, India
Kwaklei and Khonggunmelei Orchids Pvt. Ltd., Imphal, Manipur, India

**Rajkumar Kishor**
Kwaklei and Khonggunmelei Orchids Pvt. Ltd., Imphal, Manipur, India

**Pranab Behari Mazumder**
Plant Biotechnology Laboratory, Department of Biotechnology, Assam University, Silchar, Assam, India

**Oana Maria Boldura**
Faculty of Veterinary Medicine, Department of Chemistry, Biochemistry and Molecular Biology, Banat University of Agricultural Sciences and Veterinary Medicine "King Mihai I of Romania", Timisoara, Romania

**Cristina Petrine and Alin Mihu**
Emergency County Clinical Hospital, Arad, Romania

**Cornel Balta**
Emergency County Clinical Hospital, Arad, Romania
Institute of Life Sciences, Vasile Goldis Western University of Arad, Romania

**Armaity Nasarabadi Fouts, Alejandro Romero, James Nelson, Mike Hogan and Shanavaz Nasarabadi**
GenTegra, LLC, Pleasanton, California, United States

**Hiroyuki Kataoka, Atsushi Ishizaki and Keita Saito**
School of Pharmacy, Shujitsu University, Nishigawara, Okayama, Japan

**Nainar Kohila, Kasi Sathiyaseelan and Mariyathanislas Sagaya Lourdhu Sumithra**
Department of Chemistry, Aditanar College of Arts and Science, M.S. University, Tirunelveli, Tamil Nadu, India

**Narasimha S. Lakka**
Department of Science and Humanities, VIGNAN'S Foundation for Science, Technology and Research (VFSTR), Guntur, Andhra Pradesh, India
Department of Analytical Research and Development, Jodas Expoim Private Limited, Hyderabad, India

**Chandrasekar Kuppan**
Department of Science and Humanities, VIGNAN'S Foundation for Science, Technology and Research (VFSTR), Guntur, Andhra Pradesh, India

**Pasquale Avino, Ivan Notardonato and Mario Vincenzo Russo**
Department of Agricultural, Environmental and Food Sciences, University of Campobasso, Campobasso, Italy

**Umaima Gazal**
Department of Chemistry, Raja Bahadur Venkata Rama Reddy Women's College, Affiliated to Osmania University, Hyderabad, India

**Mateen Abbas**
Quality Operations Laboratory (QOL), Institute of Microbiology, University of Veterinary and Animal Sciences, Lahore, Pakistan

# Index

www.ingramcontent.com/pod-product-compliance
Lightning Source LLC
LaVergne TN
LVHW080907210926
848485LV00004B/609